Induction Motors Handbook

Induction Motors Handbook

Edited by **Maurice Willis**

Clanrye International

New Jersey

Published by Clanrye International,
55 Van Reypen Street,
Jersey City, NJ 07306, USA
www.clanryeinternational.com

Induction Motors Handbook
Edited by Maurice Willis

International Standard Book Number: 978-1-63240-305-6 (Hardback)

Printed in the United States of America.

Contents

Preface

Induction motors are capable of providing high manufacturing outcome with energy efficiency in varied industrial uses and are the core of advanced automation. Prompted by the necessity of power-efficient progresses, procedure optimization, soft-start capacity and more ecological profits, there is a growing need to use induction motors for multiple functions at steady adaptable paces. This book disseminates knowledge about growing concerns in this area, like noise and control schemes applied for high-performance functions and diagnostics. The collaboration of contributions by renowned scientists make this book not just a mechanical account of information but an international exploration on the technology of induction motors for readers. It provides them with a mixture of theory, implementation problems and practical examples.

This book is a result of research of several months to collate the most relevant data in the field.

When I was approached with the idea of this book and the proposal to edit it, I was overwhelmed. It gave me an opportunity to reach out to all those who share a common interest with me in this field. I had 3 main parameters for editing this text:

1. Accuracy - The data and information provided in this book should be up-to-date and valuable to the readers.
2. Structure - The data must be presented in a structured format for easy understanding and better grasping of the readers.
3. Universal Approach - This book not only targets students but also experts and innovators in the field, thus my aim was to present topics which are of use to all.

Thus, it took me a couple of months to finish the editing of this book.

I would like to make a special mention of my publisher who considered me worthy of this opportunity and also supported me throughout the editing process. I would also like to thank the editing team at the back-end who extended their help whenever required.

Editor

Modelling

Chapter 1

Mathematical Model of the Three-Phase Induction Machine for the Study of Steady-State and Transient Duty Under Balanced and Unbalanced States

Alecsandru Simion, Leonard Livadaru and Adrian Munteanu

Additional information is available at the end of the chapter

1. Introduction

A proper study of the induction machine operation, especially when it comes to transients and unbalanced duties, requires effective mathematical models above all. The mathematical model of an electric machine represents all the equations that describe the relationships between electromagnetic torque and the main electrical and mechanical quantities.

The theory of electrical machines, and particularly of induction machine, has mathematical models with *distributed* parameters and with *concentrated* parameters respectively. The first mentioned models start with the cognition of the magnetic field of the machine components. Their most important advantages consist in the high generality degree and accuracy. However, two major disadvantages have to be mentioned. On one hand, the computing time is rather high, which somehow discountenance their use for the real-time control. On the other hand, the distributed parameters models do not take into consideration the influence of the temperature variation or mechanical processing upon the material properties, which can vary up to 25% in comparison to the initial state. Moreover, particular constructive details (for example slots or air-gap dimensions), which essentially affects the parameters evaluation, cannot be always realized from technological point of view.

The mathematical models with concentrated parameters are the most popular and consequently employed both in scientific literature and practice. The equations stand on resistances and inductances, which can be used further for defining magnetic fluxes, electromagnetic torque, and et.al. These models offer results, which are globally acceptable but cannot detect important information concerning local effects (Ahmad, 2010; Chiasson, 2005; Krause et al., 2002; Ong, 1998; Sul, 2011).

The family of mathematical models with concentrated parameters comprises different approaches but two of them are more popular: *the phase coordinate* model and the *orthogonal (dq)* model (Ahmad, 2010; Bose, 2006; Chiasson, 2005; De Doncker et al., 2011; Krause et al., 2002; Marino et al., 2010; Ong, 1998; Sul, 2011; Wach, 2011).

The first category works with the real machine. The equations include, among other parameters, the mutual stator-rotor inductances with variable values according to the rotor position. As consequence, the model becomes non-linear and complicates the study of dynamic processes (Bose, 2006; Marino et al., 2010; Wach, 2011).

The orthogonal (dq) model has begun with Park's theory nine decades ago. These models use parameters that are often independent to rotor position. The result is a significant simplification of the calculus, which became more convenient with the defining of the *space phasor* concept (Boldea & Tutelea, 2010; Marino et al., 2010; Sul, 2011).

Starting with the "classic" theory we deduce in this contribution a mathematical model that exclude the presence of the currents and angular velocity in voltage equations and uses total fluxes alone. Based on this approach, we take into discussion two control strategies of induction motor by principle of constant total flux of the stator and rotor, respectively.

The most consistent part of this work is dedicated to the study of unbalanced duties generated by supply asymmetries. It is presented a comparative analysis, which confronts a balanced duty with two unbalanced duties of different unbalance degrees. The study uses as working tool the Matlab-Simulink environment and provides variation characteristics of the electric, magnetic and mechanical quantities under transient operation.

2. The equations of the three-phase induction machine in phase coordinates

The structure of the analyzed induction machine contains: 3 identical phase windings placed on the stator in an 120 electric degrees angle of phase difference configuration; 3 identical phase windings placed on the rotor with a similar difference of phase; a constant air-gap (close slots in an ideal approach); an unsaturated (linear) magnetic circuit that allow to each winding to be characterized by a main and a leakage inductance. Each phase winding has W_s turns on stator and W_R turns on rotor and a harmonic distribution. All inductances are considered constant. The schematic view of the machine is presented in Fig. 1a.

The voltage equations that describe the 3+3 circuits are:

$$u_{as} = R_s i_{as} + \frac{d\psi_{as}}{dt}, \quad u_{bs} = R_s i_{bs} + \frac{d\psi_{bs}}{dt}, \quad u_{cs} = R_s i_{cs} + \frac{d\psi_{cs}}{dt} \tag{1}$$

$$u_{AR} = R_R i_{AR} + \frac{d\psi_{AR}}{dt}, \quad u_{BR} = R_R i_{BR} + \frac{d\psi_{BR}}{dt}, \quad u_{CR} = R_R i_{CR} + \frac{d\psi_{CR}}{dt} \tag{2}$$

In a matrix form, the equations become:

$$\left[u_{abcs}\right]=\left[R_s\right]\left[i_{abcs}\right]+\frac{d\left[\psi_{abcs}\right]}{dt} \tag{3}$$

$$\left[u_{ABCR}\right]=\left[R_R\right]\left[i_{ABCR}\right]+\frac{d\left[\psi_{ABCR}\right]}{dt} \tag{4}$$

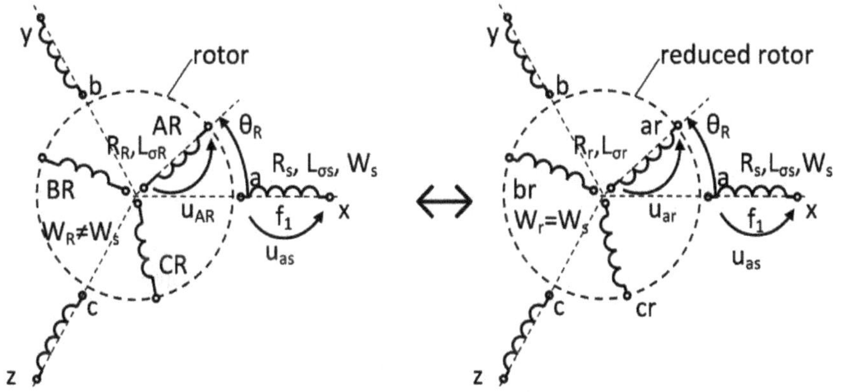

(a) (b)

Figure 1. Schematic model of three-phase induction machine: a. real; b. reduced rotor

The quantities in brackets represent the matrices of voltages, currents, resistances and total flux linkages for the stator and rotor. Obviously, the total fluxes include both main and mutual components. Further, we define the self-phase inductances, which have a leakage and a main component: $L_{jj}=L_{\sigma s}+L_{hs}$ for stator and $L_{JJ}=L_{\Sigma R}+L_{HR}$ for rotor. The mutual inductances of two phases placed on the same part (stator or rotor) have negative values, which are equal to half of the maximum mutual inductances and with the main self-phase component: $M_{jk}=L_{jk}=L_{hj}=L_{hk}$. The expressions in matrix form are:

$$\left[L_{ss}\right]=\begin{bmatrix} L_{\sigma s}+L_{hs} & -(1/2)L_{hs} & -(1/2)L_{hs} \\ -(1/2)L_{hs} & L_{\sigma s}+L_{hs} & -(1/2)L_{hs} \\ -(1/2)L_{hs} & -(1/2)L_{hs} & L_{\sigma s}+L_{hs} \end{bmatrix} \tag{5-1}$$

$$\left[L_{RR}\right]=\begin{bmatrix} L_{\Sigma R}+L_{HR} & -(1/2)L_{HR} & -(1/2)L_{HR} \\ -(1/2)L_{HR} & L_{\Sigma R}+L_{HR} & -(1/2)L_{HR} \\ -(1/2)L_{HR} & -(1/2)L_{HR} & L_{\Sigma R}+L_{HR} \end{bmatrix} \tag{5-2}$$

$$\left[L_{sR}\right]=\left[L_{Rs}\right]_t=L_{sR}\cdot\begin{bmatrix} \cos\theta_R & \cos\left(\theta_R+u\right) & \cos\left(\theta_R+2u\right) \\ \cos\left(\theta_R+2u\right) & \cos\theta_R & \cos\left(\theta_R+u\right) \\ \cos\left(\theta_R+u\right) & \cos\left(\theta_R+2u\right) & \cos\theta_R \end{bmatrix} \tag{5-3}$$

where u denotes the angle of 120^0 (or $2\pi/3$ rad).

The analysis of the induction machine usually reduces the rotor circuit to the stator one. This operation requires the alteration of the rotor quantities with the coefficient $k=W_s/W_R$ by complying with the conservation rules. The new values are:

$$u_{abcr} = k \cdot u_{ABCR};\ \psi_{abcr} = k \cdot \psi_{ABCR};\ i_{abcr} = (1/k) \cdot i_{ABCR};$$

$$R_r = k^2 \cdot R_R;\ L_{hr} = k^2 \cdot L_{HR} = \left(\frac{W_s}{W_R}\right)^2 \cdot \frac{W_R^2}{\Re_h} = \frac{W_s^2}{\Re_h} = L_{hs}; \tag{6}$$

$$L_{\sigma r} = k^2 L_{\Sigma R} = \left(\frac{W_s}{W_R}\right)^2 \frac{W_R^2}{\Re_{\sigma R}} = \frac{W_s^2}{\Re_{\sigma r}} \approx L_{\sigma s};\ L_{sr} = kL_{sR} = \left(\frac{W_s}{W_R}\right) \frac{W_s W_R}{\Re_h} = L_{hs}$$

where the reluctances of the flux paths have been used. The new matrices, with rotor quantities denoted with lowercase letters are:

$$\left[L_{rr}\right] = k^2\left[L_{RR}\right] = \begin{bmatrix} L_{\sigma r} + L_{hs} & -(1/2)L_{hs} & -(1/2)L_{hs} \\ -(1/2)L_{hs} & L_{\sigma r} + L_{hs} & -(1/2)L_{hs} \\ -(1/2)L_{hs} & -(1/2)L_{hs} & L_{\sigma r} + L_{hs} \end{bmatrix} \tag{7-1}$$

$$\left[L_{sr}\right] = k\left[L_{sR}\right] = \left[L_{rs}\right]_t = L_{hs} \cdot \begin{bmatrix} \cos\theta_R & \cos(\theta_R + u) & \cos(\theta_R + 2u) \\ \cos(\theta_R + 2u) & \cos\theta_R & \cos(\theta_R + u) \\ \cos(\theta_R + u) & \cos(\theta_R + 2u) & \cos\theta_R \end{bmatrix} \tag{7-2}$$

By virtue of these transformations, the voltage equations become:

$$\begin{cases} \left[u_{abcs}\right] = \left[R_s\right]\left[i_{abcs}\right] + \dfrac{d\left[\psi_{abcs}\right]}{dt} = \left[R_s\right]\left[i_{abcs}\right] + \left[L_{ss}\right]\dfrac{d\left[i_{abcs}\right]}{dt} + \dfrac{d\{\left[L_{sr}\right]\left[i_{abcr}\right]\}}{dt} \\ \left[u_{abcr}\right] = \left[R_r\right]\left[i_{abcr}\right] + \dfrac{d\left[\psi_{abcr}\right]}{dt} = \left[R_r\right]\left[i_{abcr}\right] + \left[L_{rr}\right]\dfrac{d\left[i_{abcr}\right]}{dt} + \dfrac{d\{\left[L_{sr}\right]_t\left[i_{abcs}\right]\}}{dt} \end{cases} \tag{8}$$

By using the notations:

$$\begin{aligned} (\Sigma L_\Pi) &= L_{\sigma r}(3L_{hs} + L_{\sigma s}) + L_{\sigma s}(3L_{hs} + L_{\sigma r}) \\ (\Pi L_s) &= L_{\sigma r}(L_{hs} + L_{\sigma s}) + L_{\sigma s}(3L_{hs} + L_{\sigma r}) \\ (\Pi L_r) &= L_{\sigma s}(L_{hs} + L_{\sigma r}) + L_{\sigma r}(3L_{hs} + L_{\sigma s}) \end{aligned} \tag{9}$$

and after the separation of the currents derivatives, (8) can be written under operational form as follows:

$$\left(\bar{s}+\frac{R_s(\Pi L_s)}{L_{\sigma s}(\Sigma L_\Pi)}\right)\bar{i}_{as}=-\frac{R_s L_{hs} L_{\sigma r}}{L_{\sigma s}(\Sigma L_\Pi)}\left(\bar{i}_{bs}+\bar{i}_{cs}\right)+\frac{2R_r L_{hs}}{(\Sigma L_\Pi)}\left[\bar{i}_{ar}\cos\theta_R+\bar{i}_{br}\cos(\theta_R+u)+\bar{i}_{cr}\cos(\theta_R+2u)\right]+$$
$$+\dot{\theta}_R\frac{L_{hs}(3L_{hs}+2L_{\sigma r})}{(\Sigma L_\Pi)}\left[\bar{i}_{ar}\sin\theta_R+\bar{i}_{br}\sin(\theta_R+u)+\bar{i}_{cr}\sin(\theta_R+2u)\right]+$$
$$+2{,}6\dot{\theta}_R\frac{L_{hs}^2}{(\Sigma L_\Pi)}\left(\bar{i}_{bs}-\bar{i}_{cs}\right)+\frac{L_{\sigma r}L_{hs}}{L_{\sigma s}(\Sigma L_\Pi)}\left(\bar{u}_{as}+\bar{u}_{bs}+\bar{u}_{cs}\right)+\frac{3L_{hs}+2L_{\sigma r}}{(\Sigma L_\Pi)}\bar{u}_{as}-$$
$$-\frac{2L_{hs}}{(\Sigma L_\Pi)}\left[\bar{u}_{ar}\cos\theta_R+\bar{u}_{br}\cos(\theta_R+u)+\bar{u}_{cr}\cos(\theta_R+2u)\right],$$

$$\left(\bar{s}+\frac{R_s(\Pi L_s)}{L_{\sigma s}(\Sigma L_\Pi)}\right)\bar{i}_{bs}=-\frac{R_s L_{hs} L_{\sigma r}}{L_{\sigma s}(\Sigma L_\Pi)}\left(\bar{i}_{cs}+\bar{i}_{as}\right)+\frac{2R_r L_{hs}}{(\Sigma L_\Pi)}\left[\bar{i}_{br}\cos\theta_R+\bar{i}_{cr}\cos(\theta_R+u)+\bar{i}_{ar}\cos(\theta_R+2u)\right]+$$
$$+\dot{\theta}_R\frac{L_{hs}(3L_{hs}+2L_{\sigma r})}{(\Sigma L_\Pi)}\left[\bar{i}_{br}\sin\theta_R+\bar{i}_{cr}\sin(\theta_R+u)+\bar{i}_{ar}\sin(\theta_R+2u)\right]+$$
$$+2{,}6\dot{\theta}_R\frac{L_{hs}^2}{(\Sigma L_\Pi)}\left(\bar{i}_{cs}-\bar{i}_{as}\right)+\frac{L_{\sigma r}L_{hs}}{L_{\sigma s}(\Sigma L_\Pi)}\left(\bar{u}_{as}+\bar{u}_{bs}+\bar{u}_{cs}\right)+\frac{3L_{hs}+2L_{\sigma r}}{(\Sigma L_\Pi)}\bar{u}_{bs}-$$
$$-\frac{2L_{hs}}{(\Sigma L_\Pi)}\left[\bar{u}_{br}\cos\theta_R+\bar{u}_{cr}\cos(\theta_R+u)+\bar{u}_{ar}\cos(\theta_R+2u)\right],$$

$$\left(\bar{s}+\frac{R_s(\Pi L_s)}{L_{\sigma s}(\Sigma L_\Pi)}\right)\bar{i}_{cs}=-\frac{R_s L_{hs} L_{\sigma r}}{L_{\sigma s}(\Sigma L_\Pi)}\left(\bar{i}_{as}+\bar{i}_{bs}\right)+\frac{2R_r L_{hs}}{(\Sigma L_\Pi)}\left[\bar{i}_{cr}\cos\theta_R+\bar{i}_{ar}\cos(\theta_R+u)+\bar{i}_{br}\cos(\theta_R+2u)\right]+$$
$$+\dot{\theta}_R\frac{L_{hs}(3L_{hs}+2L_{\sigma r})}{(\Sigma L_\Pi)}\left[\bar{i}_{cr}\sin\theta_R+\bar{i}_{ar}\sin(\theta_R+u)+\bar{i}_{br}\sin(\theta_R+2u)\right]+$$
$$+2{,}6\dot{\theta}_R\frac{L_{hs}^2}{(\Sigma L_\Pi)}\left(\bar{i}_{as}-\bar{i}_{bs}\right)+\frac{L_{\sigma r}L_{hs}}{L_{\sigma s}(\Sigma L_\Pi)}\left(\bar{u}_{as}+\bar{u}_{bs}+\bar{u}_{cs}\right)+\frac{3L_{hs}+2L_{\sigma r}}{(\Sigma L_\Pi)}\bar{u}_{cs}-$$
$$-\frac{2L_{hs}}{(\Sigma L_\Pi)}\left[\bar{u}_{cr}\cos\theta_R+\bar{u}_{ar}\cos(\theta_R+u)+\bar{u}_{br}\cos(\theta_R+2u)\right],$$

$$\left(\bar{s}+\frac{R_r(\Pi L_r)}{L_{\sigma r}(\Sigma L_\Pi)}\right)\bar{i}_{ar}=\frac{2L_{hs}R_s}{(\Sigma L_\Pi)}\left[\bar{i}_{as}\cos\theta_R+\bar{i}_{bs}\cos(\theta_R+2u)+\bar{i}_{cs}\cos(\theta_R+u)\right]+2{,}6\dot{\theta}_R\frac{L_{hs}^2}{(\Sigma L_\Pi)}\left(-\bar{i}_{br}+\bar{i}_{cr}\right)-$$
$$-\frac{2L_{hs}}{(\Sigma L_\Pi)}\left[\bar{u}_{as}\cos\theta_R+\bar{u}_{bs}\cos(\theta_R+2u)+\bar{u}_{cs}\cos(\theta_R+u)\right]+\frac{L_{\sigma s}L_{hs}}{L_{\sigma r}(\Sigma L_\Pi)}\left(\bar{u}_{ar}+\bar{u}_{br}+\bar{u}_{cr}\right)+$$
$$+\frac{2L_{\sigma s}+3L_{hs}}{(\Sigma L_\Pi)}\bar{u}_{ar}+\dot{\theta}_R L_{hs}\frac{2L_{\sigma s}+3L_{hs}}{(\Sigma L_\Pi)}\left[\bar{i}_{as}\sin\theta_R+\bar{i}_{bs}\sin(\theta_R+2u)+\bar{i}_{cs}\sin(\theta_R+u)\right]-$$
$$-\frac{L_{\sigma s}L_{hs}R_r}{L_{\sigma r}(\Sigma L_\Pi)}\left(\bar{i}_{br}+\bar{i}_{cr}\right),$$

$$\left(\bar{s}+\frac{R_r(\Pi L_r)}{L_{\sigma r}(\Sigma L_\Pi)}\right)\bar{i}_{br}=\frac{2L_{hs}R_s}{(\Sigma L_\Pi)}\left[\bar{i}_{bs}\cos\theta_R+\bar{i}_{cs}\cos(\theta_R+2u)+\bar{i}_{as}\cos(\theta_R+u)\right]+2{,}6\dot{\theta}_R\frac{L_{hs}^2}{(\Sigma L_\Pi)}\left(-\bar{i}_{cr}+\bar{i}_{ar}\right)-$$
$$-\frac{2L_{hs}}{(\Sigma L_\Pi)}\left[\bar{u}_{bs}\cos\theta_R+\bar{u}_{cs}\cos(\theta_R+2u)+\bar{u}_{as}\cos(\theta_R+u)\right]+\frac{L_{\sigma s}L_{hs}}{L_{\sigma r}(\Sigma L_\Pi)}\left(\bar{u}_{ar}+\bar{u}_{br}+\bar{u}_{cr}\right)+$$
$$+\frac{2L_{\sigma s}+3L_{hs}}{(\Sigma L_\Pi)}\bar{u}_{br}+\dot{\theta}_R L_{hs}\frac{2L_{\sigma s}+3L_{hs}}{(\Sigma L_\Pi)}\left[\bar{i}_{bs}\sin\theta_R+\bar{i}_{cs}\sin(\theta_R+2u)+\bar{i}_{as}\sin(\theta_R+u)\right]-$$
$$-\frac{L_{\sigma s}L_{hs}R_r}{L_{\sigma r}(\Sigma L_\Pi)}\left(\bar{i}_{cr}+\bar{i}_{ar}\right),$$

$$\left(\bar{s}+\frac{R_r(\Pi L_r)}{L_{\sigma r}(\Sigma L_{\Pi})}\right)\bar{i}_{cr}=\frac{2L_{hs}R_s}{(\Sigma L_{\Pi})}\left[\bar{i}_{cs}\cos\theta_R+\bar{i}_{as}\cos(\theta_R+2u)+\bar{i}_{bs}\cos(\theta_R+u)\right]+2{,}6\dot{\theta}_R\frac{L_{hs}^2}{(\Sigma L_{\Pi})}\left(-\bar{i}_{ar}+\bar{i}_{br}\right)-$$
$$-\frac{2L_{hs}}{(\Sigma L_{\Pi})}\left[\bar{u}_{cs}\cos\theta_R+\bar{u}_{as}\cos(\theta_R+2u)+\bar{u}_{bs}\cos(\theta_R+u)\right]+\frac{L_{\sigma s}L_{hs}}{L_{\sigma r}(\Sigma L_{\Pi})}\left(\bar{u}_{ar}+\bar{u}_{br}+\bar{u}_{cr}\right)+ \quad (10)$$
$$+\frac{2L_{\sigma s}+3L_{hs}}{(\Sigma L_{\Pi})}\bar{u}_{cr}+\dot{\theta}_R L_{hs}\frac{2L_{\sigma s}+3L_{hs}}{(\Sigma L_{\Pi})}\left[\bar{i}_{cs}\sin\theta_R+\bar{i}_{as}\sin(\theta_R+2u)+\bar{i}_{bs}\sin(\theta_R+u)\right]-$$
$$-\frac{L_{\sigma s}L_{hs}R_r}{L_{\sigma r}(\Sigma L_{\Pi})}\left(\bar{i}_{ar}+\bar{i}_{br}\right),$$

Besides (10), the equations concerning mechanical quantities must be added. To this end, the electromagnetic torque has to be calculated. To this effect, we start from the coenergy expression, W'_m, of the 6 circuits (3 are placed on stator and the other 3 on rotor) and we take into consideration that the leakage fluxes, which are independent of rotation angle of the rotor, do not generate electromagnetic torque, that is:

$$W'_m=\frac{1}{2}\left[i_{abcs}\right]_t\left(\left[L_{ss}\right]-L_{\sigma s}[1]\right)\left[i_{abcs}\right]+\frac{1}{2}\left[i_{abcr}\right]_t\left(\left[L_{rr}\right]-L_{\sigma r}[1]\right)\left[i_{abcr}\right]+\left[i_{abcs}\right]_t\left[L_{sr}(\theta_R)\right]\left[i_{abcr}\right] \quad (11)$$

The magnetic energy of the stator and the rotor does not depend on the rotation angle and consequently, for the electromagnetic torque calculus nothing but the last term of (11) is used. One obtains:

$$T_e=\frac{1}{2}p\left[i_{abcs}\right]_t\frac{d\left[L_{sr}(\theta_R)\right]}{d\theta_R}\left[i_{abcr}\right]=$$
$$=\frac{1}{2}pL_{hs}\sin\theta_R\left[i_{as}\left(-2i_{ar}+i_{br}+i_{cr}\right)+i_{bs}\left(+i_{ar}-2i_{br}+i_{cr}\right)+i_{cs}\left(+i_{ar}+i_{br}-2i_{cr}\right)\right]+ \quad (12)$$
$$+\frac{\sqrt{3}}{2}pL_{hs}\cos\theta_R\left[i_{as}\left(i_{cr}-i_{br}\right)+i_{bs}\left(i_{ar}-i_{cr}\right)+i_{cs}\left(i_{br}-i_{ar}\right)\right]$$

The equation of torque equilibrium can now be written under operational form as:

$$\bar{\omega}_R\left(\frac{J\bar{s}+k_z}{p}\right)=\frac{1}{2}pL_{hs}\left\{\sin\theta_R\cdot\left[i_{as}\left(-2i_{ar}+i_{br}+i_{cr}\right)+i_{bs}\left(i_{ar}-2i_{br}+i_{cr}\right)+\right.\right.$$
$$\left.\left.+i_{cs}\left(i_{ar}+i_{br}-2i_{cr}\right)\right]+\sqrt{3}\cos\theta_R\cdot\left[i_{as}\left(i_{cr}-i_{br}\right)+i_{bs}\left(i_{ar}-i_{cr}\right)+i_{cs}\left(i_{br}-i_{ar}\right)\right]\right\}-T_{st}$$

$$\bar{s}\theta_R=\omega_R=\dot{\theta}_R \quad (13)$$

where ω_R represents the *rotational pulsatance* (or *rotational pulsation*).

The simulation of the induction machine operation in Matlab-Simulink environment on the basis of the above equations system is rather complicated. Moreover, since all equations depend on the angular speed than the precision of the results could be questionable mainly for the study of rapid transients. Consequently, the use of other variables is understandable. Further, we shall use the *total fluxes* of the windings (3 motionless windings on stator and other rotating 3 windings on rotor).

It is well known that the total fluxes have a self-component and a mutual one. Taking into consideration the rules of reducing the rotor circuit to the stator one, the matrix of inductances can be written as follows:

$$\left[L_{abcabc}\right]=L_{hs}\cdot\begin{bmatrix} 1+l_{\sigma s} & -(1/2) & -(1/2) & \cos\theta_R & \cos(\theta_R+u) & \cos(\theta_R+2u) \\ -(1/2) & 1+l_{\sigma s} & -(1/2) & \cos(\theta_R+2u) & \cos\theta_R & \cos(\theta_R+u) \\ -(1/2) & -(1/2) & 1+l_{\sigma s} & \cos(\theta_R+u) & \cos(\theta_R+2u) & \cos\theta_R \\ \cos\theta_R & \cos(\theta_R+2u) & \cos(\theta_R+u) & 1+l_{\sigma r} & -(1/2) & -(1/2) \\ \cos(\theta_R+u) & L_{hs}\cos\theta_R & \cos(\theta_R+2u) & -(1/2) & 1+l_{\sigma r} & -(1/2) \\ \cos(\theta_R+2u) & \cos(\theta_R+u) & \cos\theta_R & -(1/2) & -(1/2) & 1+l_{\sigma r} \end{bmatrix} \quad (14)$$

Now, the equation system (8) can be written shortly as:

$$\left[u_{abcabc}\right]=\left[R_{s,r}\right]\left[i_{abcabc}\right]+\frac{d\left[\psi_{abcabc}\right]}{dt}, \quad where: \ \left[\psi_{abcabc}\right]=\left[L_{abcabc}\right]\left[i_{abcabc}\right] \quad (15)$$

By using the multiplication with the reciprocal matrix:

$$\left[L_{abcabc}\right]^{-1}\left[\psi_{abcabc}\right]=\left[L_{abcabc}\right]^{-1}\left[L_{abcabc}\right]\left[i_{abcabc}\right], \ or \ \left[i_{abcabc}\right]=\left[L_{abcabc}\right]^{-1}\left[\psi_{abcabc}\right] \quad (16)$$

than (15) becomes:

$$\left[u_{abcabc}\right]=\left[R_{s,r}\right]\left[L_{abcabc}\right]^{-1}\left[\psi_{abcabc}\right]+\frac{d\left[\psi_{abcabc}\right]}{dt} \quad (17)$$

This is an expression that connects the voltages to the total fluxes with no currents involvement. Now, practically the reciprocal matrix must be found. To this effect, we suppose that the reciprocal matrix has a similar form with the direct matrix. If we use the condition: $\left[L_{abcabc}\right]^{-1}\left[L_{abcabc}\right]=\left[1\right]$, than through term by term identification is obtained:

$$\left[L_{abcabc}\right]^{-1}=\frac{1}{(\Pi LD)}\cdot \begin{bmatrix} \Pi L_{s\sigma} & L_{hs}L_{\sigma r}^2 & L_{hs}L_{\sigma r}^2 & \Gamma\cos\theta_R & \Gamma\cos(\theta_R+u) & \Gamma\cos(\theta_R+2u) \\ L_{hs}L_{\sigma r}^2 & \Pi L_{s\sigma} & L_{hs}L_{\sigma r}^2 & \Gamma\cos(\theta_R+2u) & \Gamma\cos\theta_R & \Gamma\cos(\theta_R+u) \\ L_{hs}L_{\sigma r}^2 & L_{hs}L_{\sigma r}^2 & \Pi L_{s\sigma} & \Gamma\cos(\theta_R+u) & \Gamma\cos(\theta_R+2u) & \Gamma\cos\theta_R \\ \Gamma\cos\theta_R & \Gamma\cos(\theta_R+2u) & \Gamma\cos(\theta_R+u) & \Pi L_{r\sigma} & L_{hs}L_{\sigma s}^2 & L_{hs}L_{\sigma s}^2 \\ \Gamma\cos(\theta_R+u) & \Gamma\cos\theta_R & \Gamma\cos(\theta_R+2u) & L_{hs}L_{\sigma s}^2 & \Pi L_{r\sigma} & L_{hs}L_{\sigma s}^2 \\ \Gamma\cos(\theta_R+2u) & \Gamma\cos(\theta_R+u) & \Gamma\cos\theta_R & L_{hs}L_{\sigma s}^2 & L_{hs}L_{\sigma s}^2 & \Pi L_{r\sigma} \end{bmatrix} \quad (18)$$

where the following notations have been used:

$$(\Pi LD)=\left(3L_{hs}L_{\sigma s}+3L_{hs}L_{\sigma r}+2L_{\sigma r}L_{\sigma s}\right)L_{\sigma r}L_{\sigma s}; \qquad \Gamma=-2L_{hs}L_{\sigma s}L_{\sigma r};$$
$$\Pi L_{s\sigma}=\left(L_{hs}L_{\sigma r}+3L_{hs}L_{\sigma s}+2L_{\sigma r}L_{\sigma s}\right)L_{\sigma r}; \quad \Pi L_{r\sigma}=\left(L_{hs}L_{\sigma s}+3L_{hs}L_{\sigma r}+2L_{\sigma r}L_{\sigma s}\right)L_{\sigma s} \quad (19)$$

Further, the matrix product is calculated: $\left[R_{s,r}\right]\left[L_{abcabc}\right]^{-1}\left[\psi_{abcabc}\right]$, which is used in (17). After a convenient grouping, the system becomes:

$$\frac{d\psi_{as}}{dt}+\frac{\Pi L_{s\sigma}R_s}{(\Pi LD)}\psi_{as}=u_{as}-\frac{L_{hs}L_{\sigma r}^2R_s}{(\Pi LD)}(\psi_{bs}+\psi_{cs})+\frac{L_{hs}L_{\sigma s}L_{\sigma r}R_s}{(\Pi LD)}\times$$
$$\times\left[(2\psi_{ar}-\psi_{br}-\psi_{cr})\cos\theta_R+\sqrt{3}(\psi_{cr}-\psi_{br})\sin\theta_R\right] \tag{20-1}$$

$$\frac{d\psi_{bs}}{dt}+\frac{\Pi L_{s\sigma}R_s}{(\Pi LD)}\psi_{bs}=u_{bs}-\frac{L_{hs}L_{\sigma r}^2R_s}{(\Pi LD)}(\psi_{cs}+\psi_{as})+\frac{L_{hs}L_{\sigma s}L_{\sigma r}R_s}{(\Pi LD)}\times$$
$$\times\left[(-\psi_{ar}+2\psi_{br}-\psi_{cr})\cos\theta_R+\sqrt{3}(\psi_{ar}-\psi_{cr})\sin\theta_R\right] \tag{20-2}$$

$$\frac{d\psi_{cs}}{dt}+\frac{\Pi L_{s\sigma}R_s}{(\Pi LD)}\psi_{cs}=u_{cs}-\frac{L_{hs}L_{\sigma r}^2R_s}{(\Pi LD)}(\psi_{as}+\psi_{bs})+\frac{L_{hs}L_{\sigma s}L_{\sigma r}R_s}{(\Pi LD)}\times$$
$$\times\left[(-\psi_{ar}-\psi_{br}+2\psi_{cr})\cos\theta_R+\sqrt{3}(\psi_{br}-\psi_{ar})\sin\theta_R\right] \tag{20-3}$$

$$\frac{d\psi_{ar}}{dt}+\frac{\Pi L_{r\sigma}R_r}{(\Pi LD)}\psi_{ar}=u_{ar}-\frac{L_{hs}L_{\sigma s}^2R_r}{(\Pi LD)}(\psi_{br}+\psi_{cr})+\frac{L_{hs}L_{\sigma s}L_{\sigma r}R_r}{(\Pi LD)}\times$$
$$\times\left[(2\psi_{as}-\psi_{bs}-\psi_{cs})\cos\theta_R+\sqrt{3}(\psi_{bs}-\psi_{cs})\sin\theta_R\right] \tag{20-4}$$

$$\frac{d\psi_{br}}{dt}+\frac{\Pi L_{r\sigma}R_r}{(\Pi LD)}\psi_{br}=u_{br}-\frac{L_{hs}L_{\sigma s}^2R_r}{(\Pi LD)}(\psi_{cr}+\psi_{ar})+\frac{L_{hs}L_{\sigma s}L_{\sigma r}R_r}{(\Pi LD)}\times$$
$$\times\left[(-\psi_{as}+2\psi_{bs}-\psi_{cs})\cos\theta_R+\sqrt{3}(\psi_{cs}-\psi_{as})\sin\theta_R\right] \tag{20-5}$$

$$\frac{d\psi_{cr}}{dt}+\frac{\Pi L_{r\sigma}R_r}{(\Pi LD)}\psi_{cr}=u_{cr}-\frac{L_{hs}L_{\sigma s}^2R_r}{(\Pi LD)}(\psi_{ar}+\psi_{br})+\frac{L_{hs}L_{\sigma s}L_{\sigma r}R_r}{(\Pi LD)}\times$$
$$\times\left[(-\psi_{as}-\psi_{bs}+2\psi_{cs})\cos\theta_R+\sqrt{3}(\psi_{as}-\psi_{bs})\sin\theta_R\right] \tag{20-6}$$

For the calculation of the *electromagnetic torque* we can use the principle of energy conservation or the expression of stored magnetic energy. The expression of the electromagnetic torque corresponding to a multipolar machine (p is the number of pole pairs) can be written in a matrix form as follows:

$$T_e=-\frac{p}{2}\cdot\left\{\left[\psi_{abcabc}\right]_t\cdot\frac{d\left[L_{abcabc}\right]^{-1}}{d\theta_R}\cdot\left[\psi_{abcabc}\right]\right\} \tag{21}$$

To demonstrate the validity of (21), one uses the expression of the matrix $\left[L_{abcabc}\right]^{-1}$, (18), in order to calculate its derivative:

$$\frac{d}{d\theta_R}\left[L_{abcabc}\right]^{-1} = \Lambda_3 \cdot$$
$$\cdot\begin{bmatrix} 0 & 0 & 0 & \sin\theta_R & \sin(\theta_R+u) & \sin(\theta_R+2u) \\ 0 & 0 & 0 & \sin(\theta_R+2u) & \sin\theta_R & \sin(\theta_R+u) \\ 0 & 0 & 0 & \sin(\theta_R+u) & \sin(\theta_R+2u) & \sin\theta_R \\ \sin\theta_R & \sin(\theta_R+2u) & \sin(\theta_R+u) & 0 & 0 & 0 \\ \sin(\theta_R+u) & \sin\theta_R & \sin(\theta_R+2u) & 0 & 0 & 0 \\ \sin(\theta_R+2u) & \sin(\theta_R+u) & \sin\theta_R & 0 & 0 & 0 \end{bmatrix} \quad (22)$$

where the following notation has been used:

$$\Lambda_3 = \frac{1}{(3/2)(L_{\sigma s}+L_{\sigma r})+L_{\sigma r}L_{\sigma s}/L_{hs}} \quad (23)$$

This expression defines the permeance of a three-phase machine for the mathematical model in total fluxes.

Observation: One can use the general expression of the electromagnetic torque where the direct and reciprocal matrices of the inductances (which link the currents with the fluxes) should be replaced, that is:

$$T_e = \frac{1}{2}p\left[i_{abcabc}\right]_t \frac{d\left[L_{abcabc}\right]}{d\theta_R}\left[i_{abcabc}\right] = \frac{1}{2}p\cdot\left[\psi_{abcabc}\right]_t\left[L\right]_t^{-1}\cdot\frac{d\left[L_{abcabc}\right]}{d\theta_R}\cdot\left[L\right]^{-1}\left[\psi_{abcabc}\right]$$
$$T_e = -\frac{1}{2}p\left[\psi_{abcabc}\right]_t \frac{d\left[L_{abcabc}\right]^{-1}}{d\theta_R}\left[\psi_{abcabc}\right] \quad (24)$$

A more convenient expression that depends on sinθR and cosθR, leads to the electromagnetic torque equation *in fluxes* alone:

$$T_e = -(1/2)p\Lambda_3\left\{\left[\psi_{as}(2\psi_{ar}-\psi_{br}-\psi_{cr})+\psi_{bs}(2\psi_{br}-\psi_{cr}-\psi_{ar})+\psi_{cs}(2\psi_{cr}-\psi_{ar}-\psi_{br})\right]\sin\theta_R+\right.$$
$$\left.+\sqrt{3}\left[\psi_{as}(\psi_{br}-\psi_{cr})+\psi_{bs}(\psi_{cr}-\psi_{ar})+\psi_{cs}(\psi_{ar}-\psi_{br})\right]\cos\theta_R\right\} \quad (25)$$

Ultimately, by getting together the equations of the 6 electric circuits and the movement equations we obtain an 8 equation system, which can be written under operational form:

$$\overline{\psi}_{as}\left(\overline{s}+\frac{\Pi L_{s\sigma}R_s}{(\Pi LD)}\right) = \overline{u}_{as} - \frac{L_{hs}L_{\sigma r}^2 R_s}{(\Pi LD)}\left(\overline{\psi}_{bs}+\overline{\psi}_{cs}\right)+\frac{L_{hs}L_{\sigma s}L_{\sigma r}R_s}{(\Pi LD)}\times$$
$$\times\left[\left(2\overline{\psi}_{ar}-\overline{\psi}_{br}-\overline{\psi}_{cr}\right)\cos\theta_R+\sqrt{3}\left(\overline{\psi}_{cr}-\overline{\psi}_{br}\right)\sin\theta_R\right] \quad (26\text{-}1)$$

$$\overline{\psi}_{bs}\left(\overline{s}+\frac{\Pi L_{s\sigma}R_s}{(\Pi LD)}\right) = \overline{u}_{bs} - \frac{L_{hs}L_{\sigma r}^2 R_s}{(\Pi LD)}\left(\overline{\psi}_{cs}+\overline{\psi}_{as}\right)+\frac{L_{hs}L_{\sigma s}L_{\sigma r}R_s}{(\Pi LD)}\times$$
$$\times\left[\left(-\overline{\psi}_{ar}+2\overline{\psi}_{br}-\overline{\psi}_{cr}\right)\cos\theta_R+\sqrt{3}\left(\overline{\psi}_{ar}-\overline{\psi}_{cr}\right)\sin\theta_R\right] \quad (26\text{-}2)$$

$$\overline{\psi}_{cs}\left(\bar{s}+\frac{\Pi L_{s\sigma}R_s}{(\Pi LD)}\right)=\bar{u}_{cs}-\frac{L_{hs}L_{\sigma r}^2R_s}{(\Pi LD)}\left(\overline{\psi}_{as}+\overline{\psi}_{bs}\right)+\frac{L_{hs}L_{\sigma s}L_{\sigma r}R_s}{(\Pi LD)}\times$$
$$\times\left[\left(-\overline{\psi}_{ar}-\overline{\psi}_{br}+2\overline{\psi}_{cr}\right)\cos\theta_R+\sqrt{3}\left(\overline{\psi}_{br}-\overline{\psi}_{ar}\right)\sin\theta_R\right] \tag{26-3}$$

$$\overline{\psi}_{ar}\left(\bar{s}+\frac{\Pi L_{r\sigma}R_r}{(\Pi LD)}\right)=\bar{u}_{ar}-\frac{L_{hs}L_{\sigma s}^2R_r}{(\Pi LD)}\left(\overline{\psi}_{br}+\overline{\psi}_{cr}\right)+\frac{L_{hs}L_{\sigma s}L_{\sigma r}R_r}{(\Pi LD)}\times$$
$$\times\left[\left(2\overline{\psi}_{as}-\overline{\psi}_{bs}-\overline{\psi}_{cs}\right)\cos\theta_R+\sqrt{3}\left(\overline{\psi}_{bs}-\overline{\psi}_{cs}\right)\sin\theta_R\right] \tag{26-4}$$

$$\overline{\psi}_{br}\left(\bar{s}+\frac{\Pi L_{r\sigma}R_r}{(\Pi LD)}\right)=\bar{u}_{br}-\frac{L_{hs}L_{\sigma s}^2R_r}{(\Pi LD)}\left(\overline{\psi}_{cr}+\overline{\psi}_{ar}\right)+\frac{L_{hs}L_{\sigma s}L_{\sigma r}R_r}{(\Pi LD)}\times$$
$$\times\left[\left(-\overline{\psi}_{as}+2\overline{\psi}_{bs}-\overline{\psi}_{cs}\right)\cos\theta_R+\sqrt{3}\left(\overline{\psi}_{cs}-\overline{\psi}_{as}\right)\sin\theta_R\right] \tag{26-5}$$

$$\overline{\psi}_{cr}\left(\bar{s}+\frac{\Pi L_{r\sigma}R_r}{(\Pi LD)}\right)=\bar{u}_{cr}-\frac{L_{hs}L_{\sigma s}^2R_r}{(\Pi LD)}\left(\overline{\psi}_{ar}+\overline{\psi}_{br}\right)+\frac{L_{hs}L_{\sigma s}L_{\sigma r}R_r}{(\Pi LD)}\times$$
$$\times\left[\left(-\overline{\psi}_{as}-\overline{\psi}_{bs}+2\overline{\psi}_{cs}\right)\cos\theta_R+\sqrt{3}\left(\overline{\psi}_{as}-\overline{\psi}_{bs}\right)\sin\theta_R\right] \tag{26-6}$$

$$\dot{\theta}_R\left(\bar{s}+k_z/J\right)=(p/J)\langle-(1/2)p\Lambda_3\left\{\sin\theta_R\left[\overline{\psi}_{as}\left(2\overline{\psi}_{ar}-\overline{\psi}_{br}-\overline{\psi}_{cr}\right)+\right.\right.$$
$$\left.+\overline{\psi}_{bs}\left(2\overline{\psi}_{br}-\overline{\psi}_{cr}-\overline{\psi}_{ar}\right)+\overline{\psi}_{cs}\left(2\overline{\psi}_{cr}-\overline{\psi}_{ar}-\overline{\psi}_{br}\right)\right]+\sqrt{3}\cos\theta_R$$
$$\left.\cdot\left[\overline{\psi}_{as}\left(\overline{\psi}_{br}-\overline{\psi}_{cr}\right)+\overline{\psi}_{bs}\left(\overline{\psi}_{cr}-\overline{\psi}_{ar}\right)+\overline{\psi}_{cs}\left(\overline{\psi}_{ar}-\overline{\psi}_{br}\right)\right]\right\}-T_{st}\rangle \tag{26-7}$$

$$\frac{d\theta_R}{dt}=\dot{\theta}_R=\omega_R \tag{26-8}$$

This equation system, (26-1)-(26-8) allows the study of any operation duty of the three-phase induction machine: steady state or transients under balanced or unbalanced condition, with simple or double feeding.

3. Mathematical models used for the study of steady-state under balanced and unbalanced conditions

Generally, the symmetrical three-phase squirrel cage induction machine has the stator windings connected to a supply system, which provides variable voltages according to certain laws but have the same pulsation. Practically, this is the case with 4 wires connection, 3 phases and the neutral. The sum of the phase currents gives the current along neutral and the homopolar component can be immediately defined. The analysis of such a machine can use the symmetric components theory. This is the case of the machine with *two unbalances* as concerns the supply. The study can be done either using the equation system (26-1...8) or on the basis of symmetric components theory with three distinct mathematical models for each component (positive sequence, negative sequence and homopolar).

The vast majority of electric drives uses however the 3 wires connection (no neutral). Consequently, there is no homopolar current component, the homopolar fluxes are zero as well and the sum of the 3 phase total fluxes is null. This is an asymmetric condition with *single unbalance,* which can be studied by using the direct and inverse sequence components when the transformation from 3 to 2 axes is mandatory. This approach practically replaces the three-phase machine with unbalanced supply with two symmetric three-phase machines. One of them produces the positive torque and the other provides the negative torque. The resultant torque comes out through superposition of the effects.

3.1. The abc-αβ0 model in total fluxes

The operation of the machine with 2 unbalances can be analyzed by considering certain expressions for the instantaneous values of the stator and rotor quantities (voltages, total fluxes and currents eventually, which can be transformed from (a, b, c) to (α, β, 0) reference frames in accordance with the following procedure :

$$\begin{bmatrix} \psi_{\alpha s} \\ \psi_{\beta s} \\ \psi_{0s} \end{bmatrix} = \sqrt{\frac{2}{3}} \cdot \begin{bmatrix} 1 & -1/2 & -1/2 \\ 0 & \sqrt{3}/2 & -\sqrt{3}/2 \\ \sqrt{2}/2 & \sqrt{2}/2 & \sqrt{2}/2 \end{bmatrix} \cdot \begin{bmatrix} \psi_{as} \\ \psi_{bs} \\ \psi_{cs} \end{bmatrix} \tag{27}$$

We define the following notations:

$$\begin{aligned} &\frac{\Pi L_{s\sigma} R_s}{(\Pi LD)} = \frac{\left(L_{hs}L_{\sigma r} + 3L_{hs}L_{\sigma s} + 2L_{\sigma r}L_{\sigma s}\right)}{\left(3L_{hs}L_{\sigma r} + 3L_{hs}L_{\sigma s} + 2L_{\sigma r}L_{\sigma s}\right)}\left(\frac{R_s}{L_{\sigma s}}\right) \cong \frac{2}{3}\left(\frac{R_s}{L_{\sigma s}}\right) = v_{st}; \\ &\frac{L_{hs}L_{\sigma r}^2 R_s}{(\Pi LD)} = \frac{L_{hs}L_{\sigma r}R_s}{\left(3L_{hs}L_{\sigma s} + 3L_{hs}L_{\sigma r} + 2L_{\sigma r}L_{\sigma s}\right)L_{\sigma s}} \cong \frac{1}{6}\left(\frac{R_s}{L_{\sigma s}}\right) = v_{sr}; \\ &v_{st} - v_{sr} = \frac{3 + 2\left(L_{\sigma r}/L_{hs}\right)}{3\left(L_{\sigma r}/L_{\sigma s}\right) + 3 + 2\left(L_{\sigma r}/L_{hs}\right)}\left(\frac{R_s}{L_{\sigma s}}\right) \cong \frac{1}{2}\left(\frac{R_s}{L_{\sigma s}}\right) = v_s \end{aligned} \tag{28-1}$$

$$\begin{aligned} &\frac{\Pi L_{r\sigma} R_r}{(\Pi LD)} = \frac{\left(L_{hs}L_{\sigma s} + 3L_{hs}L_{\sigma r} + 2L_{\sigma r}L_{\sigma s}\right)}{\left(3L_{hs}L_{\sigma r} + 3L_{hs}L_{\sigma s} + 2L_{\sigma r}L_{\sigma s}\right)}\left(\frac{R_r}{L_{\sigma r}}\right) \cong \frac{2}{3}\left(\frac{R_r}{L_{\sigma r}}\right) = v_{rt}; \\ &\frac{L_{hs}L_{\sigma s}^2 R_r}{(\Pi LD)} = \frac{L_{hs}L_{\sigma s}R_r}{\left(3L_{hs}L_{\sigma s} + 3L_{hs}L_{\sigma r} + 2L_{\sigma r}L_{\sigma s}\right)L_{\sigma r}} \cong \frac{1}{6}\left(\frac{R_r}{L_{\sigma r}}\right) = v_{rs}; \\ &v_{rt} - v_{rs} = \frac{3 + 2\left(L_{\sigma s}/L_{hs}\right)}{3\left(L_{\sigma s}/L_{\sigma r}\right) + 3 + 2\left(L_{\sigma s}/L_{hs}\right)}\left(\frac{R_r}{L_{\sigma r}}\right) \cong \frac{1}{2}\left(\frac{R_r}{L_{\sigma r}}\right) = v_r \end{aligned} \tag{28-2}$$

$$\frac{3L_{hs}L_{\sigma s}L_{\sigma r}R_s}{(\Pi LD)}=\frac{3L_{\sigma s}}{\left(3L_{\sigma s}+3L_{\sigma r}+2L_{\sigma r}L_{\sigma s}/L_{hs}\right)}\left(\frac{R_s}{L_{\sigma s}}\right)\cong\frac{1}{2}\left(\frac{R_s}{L_{\sigma s}}\right)=\nu_{\sigma s}; \quad (28\text{-}3)$$

$$\frac{3L_{hs}L_{\sigma s}L_{\sigma r}R_r}{(\Pi LD)}=\frac{3L_{\sigma r}}{\left(3L_{\sigma s}+3L_{\sigma r}+2L_{\sigma r}L_{\sigma s}/L_{hs}\right)}\left(\frac{R_r}{L_{\sigma r}}\right)\cong\frac{1}{2}\left(\frac{R_r}{L_{\sigma r}}\right)=\nu_{\sigma r}; \quad (28\text{-}4)$$

By using these notations in (17) and after convenient groupings we obtain:

$$\frac{d\psi_{as}}{dt}+\nu_{st}\psi_{as}=u_{as}-\nu_{sr}\left(\psi_{bs}+\psi_{cs}\right)+\frac{1}{3}\nu_{\sigma s}\times$$
$$\times\left[\left(2\psi_{ar}-\psi_{br}-\psi_{cr}\right)\cos\theta_R+\sqrt{3}\left(\psi_{cr}-\psi_{br}\right)\sin\theta_R\right] \quad (29\text{-}1)$$

$$\frac{d\psi_{bs}}{dt}+\nu_{st}\psi_{bs}=u_{bs}-\nu_{sr}\left(\psi_{cs}+\psi_{as}\right)+\frac{1}{3}\nu_{\sigma s}\times$$
$$\times\left[\left(-\psi_{ar}+2\psi_{br}-\psi_{cr}\right)\cos\theta_R+\sqrt{3}\left(\psi_{ar}-\psi_{cr}\right)\sin\theta_R\right] \quad (29\text{-}2)$$

$$\frac{d\psi_{cs}}{dt}+\nu_{st}\psi_{cs}=u_{cs}-\nu_{sr}\left(\psi_{as}+\psi_{bs}\right)+\frac{1}{3}\nu_{\sigma s}\times$$
$$\times\left[\left(-\psi_{ar}-\psi_{br}+2\psi_{cr}\right)\cos\theta_R+\sqrt{3}\left(\psi_{br}-\psi_{ar}\right)\sin\theta_R\right] \quad (29\text{-}3)$$

$$\frac{d\psi_{ar}}{dt}+\nu_{rt}\psi_{ar}=u_{ar}-\nu_{rs}\left(\psi_{br}+\psi_{cr}\right)+\frac{1}{3}\nu_{\sigma r}\times$$
$$\times\left[\left(2\psi_{as}-\psi_{bs}-\psi_{cs}\right)\cos\theta_R+\sqrt{3}\left(\psi_{bs}-\psi_{cs}\right)\sin\theta_R\right] \quad (29\text{-}4)$$

$$\frac{d\psi_{br}}{dt}+\nu_{rt}\psi_{br}=u_{br}-\nu_{rs}\left(\psi_{cr}+\psi_{ar}\right)+\frac{1}{3}\nu_{\sigma r}\times$$
$$\times\left[\left(-\psi_{as}+2\psi_{bs}-\psi_{cs}\right)\cos\theta_R+\sqrt{3}\left(\psi_{cs}-\psi_{as}\right)\sin\theta_R\right] \quad (29\text{-}5)$$

$$\frac{d\psi_{cr}}{dt}+\nu_{rt}\psi_{cr}=u_{cr}-\nu_{rs}\left(\psi_{ar}+\psi_{br}\right)+\frac{1}{3}\nu_{\sigma r}\times$$
$$\times\left[\left(-\psi_{bs}+2\psi_{cs}-\psi_{as}\right)\cos\theta_R+\sqrt{3}\left(\psi_{as}-\psi_{bs}\right)\sin\theta_R\right] \quad (29\text{-}6)$$

Typical for the cage machine or even for the wound rotor after the starting rheostat is short-circuited is the fact that the *rotor voltages become zero*. The equations of the six circuits get different as a result of certain convenient math operations. (29-2) and (29-3) are multiplied by (-1/2) and afterwards added to (29-1); (29-3) is subtracted from (29-2); (29-1), (29-2) and (29-3) are added together. We obtain three equations that describe the stator. Similarly, (29-4), (29-5) and (29-6) are used for the rotor equations. The new equation system is:

$$\begin{cases} \dfrac{d\psi_{\alpha s}}{dt} + v_s\psi_{\alpha s} = u_{\alpha s} + v_{\sigma s}\left(\psi_{\alpha r}\cos\theta_R - \psi_{\beta r}\sin\theta_R\right) \\ \dfrac{d\psi_{\beta s}}{dt} + v_s\psi_{\beta s} = u_{\beta s} + v_{\sigma s}\left(\psi_{\alpha r}\sin\theta_R + \psi_{\beta r}\cos\theta_R\right) \\ \dfrac{d\psi_{0s}}{dt} + \left(v_{st} + 2v_{sr}\right)\psi_{0s} = u_{0s} \end{cases} \quad (30\text{-}1, 2, 3)$$

$$\begin{cases} \dfrac{d\psi_{\alpha r}}{dt} + v_r\psi_{\alpha r} = u_{\alpha r} + v_{\sigma r}\left(\psi_{\alpha s}\cos\theta_R + \psi_{\beta s}\sin\theta_R\right) \\ \dfrac{d\psi_{\beta r}}{dt} + v_r\psi_{\beta r} = u_{\beta r} + v_{\sigma r}\left(-\psi_{\alpha s}\sin\theta_R + \psi_{\beta s}\cos\theta_R\right) \\ \dfrac{d\psi_{0r}}{dt} + \left(v_{rt} + 2v_{rs}\right)\psi_{0r} = u_{0r} \end{cases} \quad (30\text{-}4, 5, 6)$$

Further, the movement equation has to be attached. It is necessary to establish the detailed expression of the electromagnetic torque in *fluxes* alone starting with (25) and using convenient transformations:

$$T_e = -(3/2)p\Lambda_3\left[\left(\psi_{\alpha s}\psi_{\alpha r} + \psi_{\beta s}\psi_{\beta r}\right)\sin\theta_R + \left(\psi_{\alpha s}\psi_{\beta r} - \psi_{\beta s}\psi_{\alpha r}\right)\cos\theta_R\right] \quad (31)$$

Ultimately, the 8 equation system under operational form is:

$$\overline{\psi}_{\alpha s}\left(\overline{s} + v_s\right) = \overline{u}_{\alpha s} + v_{\sigma s}\left(\overline{\psi}_{\alpha r}\cos\theta_R - \overline{\psi}_{\beta r}\sin\theta_R\right) \quad (32\text{-}1)$$

$$\overline{\psi}_{\beta s}\left(\overline{s} + v_s\right) = \overline{u}_{\beta s} + v_{\sigma s}\left(\overline{\psi}_{\alpha r}\sin\theta_R + \overline{\psi}_{\beta r}\cos\theta_R\right) \quad (32\text{-}2)$$

$$\overline{\psi}_{0s}\left(\overline{s} + v_{st} + 2v_{sr}\right) = \overline{u}_{0s} \quad (32\text{-}3)$$

$$\overline{\psi}_{\alpha r}\left(\overline{s} + v_r\right) = \overline{u}_{\alpha r} + v_{\sigma r}\left(\overline{\psi}_{\alpha s}\cos\theta_R + \overline{\psi}_{\beta s}\sin\theta_R\right) \quad (32\text{-}4)$$

$$\overline{\psi}_{\beta r}\left(\overline{s} + v_r\right) = \overline{u}_{\beta r} + v_{\sigma r}\left(-\overline{\psi}_{\alpha s}\sin\theta_R + \overline{\psi}_{\beta s}\cos\theta_R\right) \quad (32\text{-}5)$$

$$\overline{\psi}_{0r}\left(\overline{s} + v_{rt} + 2v_{rs}\right) = \overline{u}_{0r} \quad (32\text{-}6)$$

$$\dot{\theta}_R\left(\overline{s} + k_z / J\right) = (p/J)\cdot\left\{-(3/2)p\Lambda_3\left[\left(\overline{\psi}_{\alpha s}\overline{\psi}_{\alpha r} + \overline{\psi}_{\beta s}\overline{\psi}_{\beta r}\right)\sin\theta_R + \left(\overline{\psi}_{\alpha s}\overline{\psi}_{\beta r} - \overline{\psi}_{\beta s}\overline{\psi}_{\alpha r}\right)\cos\theta_R\right] - T_{st}\right\} \quad (32\text{-}7)$$

$$\frac{d\theta_R}{dt} = \dot{\theta}_R = \omega_R \quad (32\text{-}8)$$

These equations allow the study of three-phase induction machine for any duty. It has to be mentioned that the electromagnetic torque expression has no homopolar components of the total fluxes.

3.2. The abc-dq model in total fluxes

For the study of the single unbalance condition is necessary to consider expressions of the instantaneous values of the stator and rotor quantities (voltages, total fluxes and eventually currents in *a,b,c* reference frame) whose sum is null. The real quantities can be transformed to (d,q) reference frame (Simion et al., 2011). By using the notations (28-1), (28-2), (28-3) and (28-4) then after convenient grouping we obtain (Simion, 2010):

$$\frac{d\psi_{\alpha s}}{dt} + \nu_s \psi_{\alpha s} = u_{\alpha s} + \nu_{\sigma s}\left(\psi_{\alpha r}\cos\theta_R - \psi_{\beta r}\sin\theta_R\right)$$
$$\frac{d\psi_{\beta s}}{dt} + \nu_s \psi_{\beta s} = u_{\beta s} + \nu_{\sigma s}\left(\psi_{\alpha r}\sin\theta_R + \psi_{\beta r}\cos\theta_R\right) \quad (33\text{-}1, 2)$$

$$\frac{d\psi_{\alpha r}}{dt} + \nu_r \psi_{\alpha r} = \nu_{\sigma r}\left(\psi_{\alpha s}\cos\theta_R + \psi_{\beta s}\sin\theta_R\right)$$
$$\frac{d\psi_{\beta r}}{dt} + \nu_r \psi_{\beta r} = \nu_{\sigma r}\left(-\psi_{\alpha s}\sin\theta_R + \psi_{\beta s}\cos\theta_R\right) \quad (33\text{-}3, 4)$$

Further, the movement equation (31) must be attached. The operational form of the equation system (4 electric circuits and 2 movement equations) is:

$$\overline{\psi}_{\alpha s}\left(\overline{s} + \nu_s\right) = \overline{u}_{\alpha s} + \nu_{\sigma s}\left(\overline{\psi}_{\alpha r}\cos\theta_R - \overline{\psi}_{\beta r}\sin\theta_R\right) \quad (34\text{-}1)$$

$$\overline{\psi}_{\beta s}\left(\overline{s} + \nu_s\right) = \overline{u}_{\beta s} + \nu_{\sigma s}\left(\overline{\psi}_{\alpha r}\sin\theta_R + \overline{\psi}_{\beta r}\cos\theta_R\right) \quad (34\text{-}2)$$

$$\overline{\psi}_{\alpha r}\left(\overline{s} + \nu_r\right) = \nu_{\sigma r}\left(\overline{\psi}_{\alpha s}\cos\theta_R + \overline{\psi}_{\beta s}\sin\theta_R\right) \quad (34\text{-}3)$$

$$\overline{\psi}_{\beta r}\left(\overline{s} + \nu_r\right) = \nu_{\sigma r}\left(-\overline{\psi}_{\alpha s}\sin\theta_R + \overline{\psi}_{\beta s}\cos\theta_R\right) \quad (34\text{-}4)$$

$$\dot{\theta}_R\left(\overline{s} + k_z / J\right) = (p/J)\cdot\left\{-(3/2)p\Lambda_3\left[\left(\overline{\psi}_{\alpha s}\overline{\psi}_{\alpha r} + \overline{\psi}_{\beta s}\overline{\psi}_{\beta r}\right)\sin\theta_R + \left(\overline{\psi}_{\alpha s}\overline{\psi}_{\beta r} - \overline{\psi}_{\beta s}\overline{\psi}_{\alpha r}\right)\cos\theta_R\right] - T_{st}\right\}$$

(34-5)

$$\frac{d\theta_R}{dt} = \dot{\theta}_R = \omega_R \quad (34\text{-}6)$$

The equation sets (33-1...4) and (34-1...6) prove that a three-phase induction machine connected to the supply system by 3 wires can be studied similarly to a two-phase machine (*two-phase* mathematical model). Its parameters can be deduced by linear transformations of the original parameters including the supply voltages (Fig. 2a).

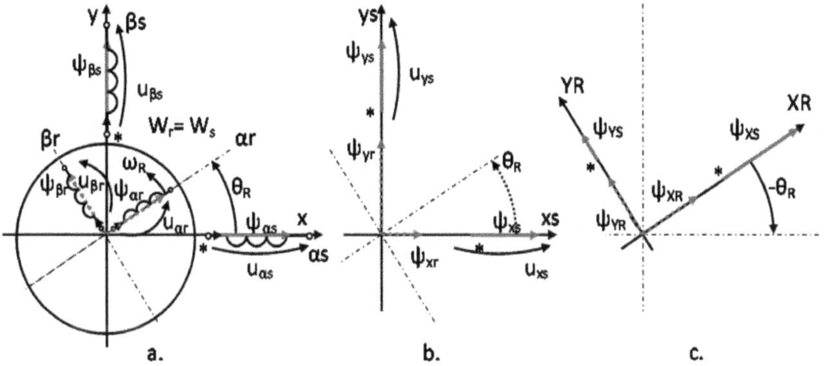

Figure 2. Induction machine schematic view: a.Two-phase model; b. Simplified view of the total fluxes in stator reference frame; c. Idem, but in rotor reference frame

The windings of two-phase model are denoted with (*αs, βs*) and (*αr, βr*) in order to trace a correspondence with the real two-phase machine, whose subscripts are (*as, bs*) and (*ar, br*) respectively. We shall use the subscripts *xs* and *ys* for the quantities that corresponds to the three-phase machine but transformed in its two-phase model. This is a rightful assumption since (*αs, βs*) axes are collinear with (*x, y*) axes, which are commonly used in analytic geometry. Further, new notations (35) for the flux linkages of the right member of the equations (33-1...4) will be defined by following the next rules:

- projection sums corresponding to rotor flux linkages from (*αr, βr*) axes along the two stator axes (denoted with *x* and *y* that is ψ_{xr}, ψ_{yr}) when they refer to the flux linkages from the right member of the first two equations, Fig. 2b.
- projection sums corresponding to stator flux linkages from (*αs, βs*) axes along the two rotor axes (denoted with X and Y that is ψ_{XS}, ψ_{YS}) when they refer to the flux linkages from the last two equations, Fig. 2c.

$$\begin{cases} \psi_{xr} = \psi_{\alpha r}\cos\theta_R - \psi_{\beta r}\sin\theta_R, \quad \psi_{yr} = \psi_{\alpha r}\sin\theta_R + \psi_{\beta r}\cos\theta_R \\ \psi_{XS} = \psi_{\alpha s}\cos\theta_R + \psi_{\beta s}\sin\theta_R, \quad \psi_{YS} = -\psi_{\alpha s}\sin\theta_R + \psi_{\beta s}\cos\theta_R \end{cases} \tag{35}$$

Some aspects have to be pointed out. When the machine operates under motoring duty, the pulsation of the stator flux linkages from (*αs, βs*) axes is equal to ω_s. Since the rotational pulsation is ω_R then the pulsation of the rotor quantities from (*αr, βr*) axes is equal to $\omega_r = s\omega_s = \omega_s - \omega_R$. The pulsation of the rotor quantities projected along the stator axes with the subscripts *xr* and *yr* is equal to ω_s. The pulsation of the stator quantities projected along the rotor axes with the subscripts *XS* and *YS* is equal to ω_r. The equations (33-1...4) become:

$$\overline{\psi}_{\alpha s}(\overline{s}+v_s)=\overline{u}_{\alpha s}+v_{\sigma s}\overline{\psi}_{xr} \tag{36-1}$$

$$\overline{\psi}_{\beta s}(\overline{s}+v_s)=\overline{u}_{\beta s}+v_{\sigma s}\overline{\psi}_{yr} \tag{36-2}$$

$$\overline{\psi}_{\alpha r}(\overline{s}+v_r)=v_{\sigma r}\overline{\psi}_{XS} \tag{36-3}$$

$$\overline{\psi}_{\beta r}(\overline{s}+v_r)=v_{\sigma r}\overline{\psi}_{YS} \tag{36-4}$$

The first two equations join the quantities with the pulsation ω_s and the other two, the quantities with the pulsation $\omega_r = s\omega_s$. The expression of the magnetic torque, in *total fluxes* and *rotor position angle* becomes:

$$T_e=-(3/2)p\Lambda_3\left(\overline{\psi}_{\alpha s}\overline{\psi}_{yr}-\overline{\psi}_{\beta s}\overline{\psi}_{xr}\right) \tag{37}$$

or a second equivalent expression:

$$T_e=(3/2)p\Lambda_3\left(\overline{\psi}_{\alpha r}\overline{\psi}_{YS}-\overline{\psi}_{\beta r}\overline{\psi}_{XS}\right) \tag{38}$$

which shows the "total symmetry" of the two-phase model of the three-phase machine regarding both stator and rotor. The equations of the four circuits together with the movement equation (37) under operational form give:

$$\overline{\psi}_{\alpha s}\left(\overline{s}+v_s\right)=\overline{u}_{\alpha s}+v_{\sigma s}\overline{\psi}_{xr} \tag{39-1}$$

$$\overline{\psi}_{\beta s}\left(\overline{s}+v_s\right)=\overline{u}_{\beta s}+v_{\sigma s}\overline{\psi}_{yr} \tag{39-2}$$

$$\overline{\psi}_{\alpha r}\left(\overline{s}+v_r\right)=v_{\sigma r}\overline{\psi}_{XS} \tag{39-3}$$

$$\overline{\psi}_{\beta r}\left(\overline{s}+v_r\right)=v_{\sigma r}\overline{\psi}_{YS} \tag{39-4}$$

$$\dot{\theta}_R\left(\overline{s}+k_z/J\right)=(p/J)\cdot\left\{\ (3/2)p\Lambda_3\left(\overline{\psi}_{\beta s}\overline{\psi}_{xr}-\overline{\psi}_{\alpha s}\overline{\psi}_{yr}\right)-T_{st}\right\} \tag{39-5}$$

$$\frac{d\theta_R}{dt}=\dot{\theta}_R=\omega_R \tag{39-6}$$

This last equation system allows the study of transients under single unbalance condition. It is similar with the frequently used equations (Park) but contains as variables only total fluxes and the rotation angle. There are no currents or angular speed in the voltage equations.

4. Expressions of electromagnetic torque

For the steady state analysis of the symmetric three-phase induction machine, one can define the simplified space phasor of the stator flux, which is collinear to the total flux of the

(αs) axis and has a $\sqrt{3}$ times higher modulus. In a similar way can be obtained the space phasors of the stator voltages and rotor fluxes and the system equation (39-1...6) that describe the steady state becomes:

$$\begin{aligned} \underline{U}_{sR3} &= (\nu_s + j\omega_s)\underline{\Psi}_{sR3} - \nu_{\sigma s}\underline{\Psi}_{rR3} = (\omega_s - j\nu_s)\Psi_{sR3}e^{j\alpha_s} + j\nu_{\sigma s}\Psi_{rR3}e^{j\alpha_r} \\ 0 &= \nu_{\sigma r}\underline{\Psi}_{sR3} - (\nu_r + js\omega_s)\underline{\Psi}_{rR3} = -j\nu_{\sigma r}\Psi_{sR3}e^{j\alpha_s} + (j\nu_r - s\omega_s)\Psi_{rR3}e^{j\alpha_r} \end{aligned} \tag{40}$$

$$T_e = (3/2)p\Lambda_3\Psi_{sR3}\Psi_{rR3}\sin(\alpha_s - \alpha_r)$$

When the speed regulation of the cage induction machine is employed by means of voltage and/or frequency variation then the simultaneous control of the two total flux space vectors is difficult. As consequence, new strategies more convenient can be chosen. To this effect, we shall deduce expressions of the electromagnetic torque that include only one of the total flux space vectors either from stator or rotor.

4.1. Variation of the torque with the stator total flux space vector

One of the methods used for the control of induction machine consists in the operation with *constant stator total flux space vector*. From (40), the rotor total flux space vector is:

$$\begin{aligned} \underline{\Psi}_{rR3} &= \frac{\nu_{\sigma r}}{\nu_r + js\omega_s}\underline{\Psi}_{sR3} = \frac{\nu_{\sigma r}\underline{\Psi}_{sR3}}{\sqrt{\omega_s^2 s^2 + \nu_r^2}}\left(\frac{\nu_r}{\sqrt{\omega_s^2 s^2 + \nu_r^2}} - j\frac{s\omega_s}{\sqrt{\omega_s^2 s^2 + \nu_r^2}}\right) = \\ &= \frac{\nu_{\sigma r}\underline{\Psi}_{sR3}}{\sqrt{\omega_s^2 s^2 + \nu_r^2}}e^{-j\theta}; (\theta = \alpha_s - \alpha_r); \sin\theta = \frac{s\omega_s}{\sqrt{\omega_s^2 s^2 + \nu_r^2}}; \cos\theta = \frac{\nu_r}{\sqrt{\omega_s^2 s^2 + \nu_r^2}} \end{aligned} \tag{41}$$

where θ is the angle between stator and rotor total flux space vectors. This angle has the meaning of an *internal angle of the machine*.

The expression of the magnetic torque that depends with the stator total flux space vector becomes:

$$\begin{aligned} T_e &= -\left(\frac{3}{2}\right)p\Lambda_3 \operatorname{Re}\left(j\underline{\Psi}_{sR3}\underline{\Psi}^*_{rR3}\right) = -\left(\frac{3}{2}\right)p\Lambda_3 \operatorname{Re}\left\{j\underline{\Psi}_{sR3}\frac{\nu_{\sigma r}}{\sqrt{\omega_s^2 s^2 + \nu_r^2}} \cdot \underline{\Psi}^*_{sR3}(\cos\theta + j\sin\theta)\right\} = \\ &= \frac{3}{2}\frac{\nu_{\sigma r}}{\nu_r}p\Lambda_3\Psi_{sR3}^2\frac{s\omega_s\nu_r}{\omega_s^2 s^2 + \nu_r^2} = \frac{3}{4}\frac{\nu_{\sigma r}}{\nu_r}p\Lambda_3\Psi_{sR3}^2\sin 2\theta. \end{aligned} \tag{42}$$

Assuming the ideal hypothesis of maintaining constant the stator flux, for example equal to the no-load value, then the pull-out torque, T_{emax}, corresponds to $\sin 2\theta = 1$ that is:

$$\begin{aligned} &2\sin\theta\cos\theta = 1 \leftrightarrow s_{cr}\omega_s = \nu_r, and\ T_{e\max} = \frac{3}{4}\frac{\nu_{\sigma r}}{\nu_r}p\Lambda_3\Psi_{sR3}^2; \\ &or\ T_{e\max} = \frac{3}{2}\nu_{\sigma r}p\Lambda_3\left(\frac{U_{sR3}}{\omega_s}\right)^2\frac{\nu_r}{\nu_s^2(\nu_r/\omega_s)^2 + 2\nu_{\sigma s}\nu_{\sigma r}\nu_r/\omega_s + \nu_{tt}^2 + 2\nu_r^2} \end{aligned} \tag{43}$$

Now an observation can be formulated. Let us suppose an ideal static converter that operates with a *U/f=constant=k₁* strategy. For low supply frequencies, the pull-out torque decreases in value since the denominator increases with the pulsatance decrease, ω_s (Fig. 3). Within certain limits at low frequencies, an increase of the supply voltage is necessary in order to maintain the pull-out torque value. In other words, $U/f = k_2$, and $k_2 > k_1$.

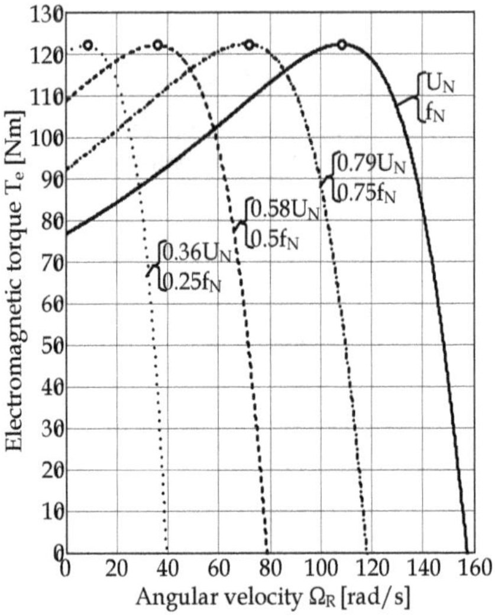

Figure 3. Mechanical characteristics, $M_e = f(\Omega_R)$ at $\Psi_{sR3} = const.$

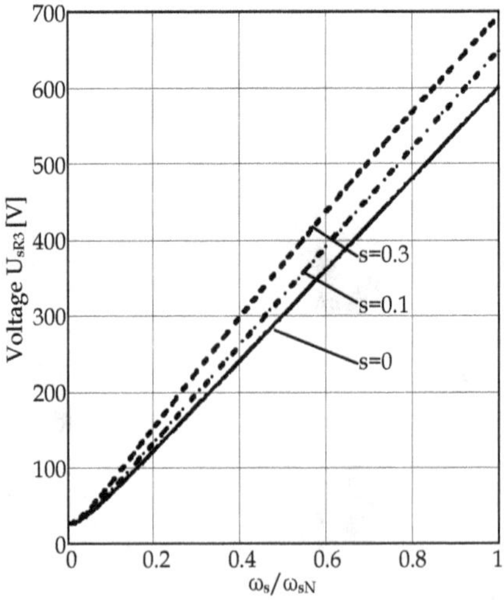

Figure 4. Resultant stator voltage vs. pulsatance $U_{sR3} = f(\omega_s)$ *at* $\Psi_{sR3} = const.$ *(1,91Wb)*

A proper control of the induction machine requires a strategy based on *U/f = variable.* More precisely, for low frequency values it is necessary to increase the supply voltage with respect to the values that result from *U/f = const.* strategy. At a pinch, when the frequency becomes zero, the supply voltage must have a value capable to compensate the voltage drops upon the equivalent resistance of the windings. Lately, the modern static converters can be parameterized on the basis of the catalog parameters of the induction machine or on the basis of some laboratory tests results.

From (40) we can deduce:

$$\underline{\Psi}_{sR3} = -j\frac{\underline{U}_{sR3}}{\omega_s}\frac{(s\omega_s - j\nu_r)}{(s\omega_s - \nu_{tt}) - j(\nu_r + s\nu_s)} \leftrightarrow U_{sR3}^2 = \frac{\Psi_{sR3}^2\omega_s^2\left(As^2 + 2Bs + C\right)}{\omega_s^2 s^2 + \nu_r^2} \qquad (44)$$

and further:

$$\frac{U_{sR3}^2}{\omega_s^2} = \Psi_{sR3}^2\left[1 + \frac{\nu_s^2 s^2 + 2\nu_{\sigma s}\nu_{\sigma r}s + \nu_{tt}^2}{\omega_s^2 s^2 + \nu_r^2}\right] \leftrightarrow \Psi_{sR3} = \frac{U_{sR3}}{\omega_s}\sqrt{\frac{F(s)}{F(s) + sG(s)}} \qquad (45)$$

$$where: \quad F(s) = \omega_s^2 s^2 + \nu_r^2; \quad G(s) \approx \nu_s^2 s + 2\nu_{\sigma s}\nu_{\sigma r}$$

if the term ν_{tt} was neglected. By inspecting the square root term, which is variable with the slip (and load as well), we can point out the following observations.

- Constant maintaining of the stator flux for *low pulsations* (that is low angular velocity values including start-up) can be obtained with a significant increase of the supply voltage. The "additional" increasing of the voltage depends proportionally on the load value. Analytically, this fact is caused by the predominance of the term G against F, (45). From the viewpoint of physical phenomena, a higher voltage in case of severe start-up or low frequency operation is necessary for the compensation of the leakage fluxes after which the stator flux must keep its prescribed value.
- Constant maintaining of the stator flux for *high pulsations* (that is angular speeds close or even over the rated value) requires an insignificant rise of the supply voltage. The *U/f* ratio is close to its rated value (rated values of *U* and *f*) especially for low load torque values. However, a certain increase of the voltage is required proportionally with the load degree. Analytically, this fact is now caused by the predominance of the term F against G, (45).
- In conclusion, the resultant stator flux remain constant for *U/f =constant=*k_1 strategy if the load torque is small. For high loads (especially if the operation is close to the pull-out point), the maintaining of the stator flux requires an increase of the *U/f* ratio, which means a significant rise of the voltage and current.

If the machine parameters are established, then a variation rule of the supply voltage can be settled in order to have a constant stator flux (equal, for example, to its no-load value) both for frequency and load variation.

Fig. 4 presents (for a machine with predetermined parameters: supply voltage with the amplitude of 490 V (U_{as}=346.5V); R_s=R_r=2; L_{hs}=0,09; $L_{\sigma s}$= $L_{\sigma r}$=0,01; J=0,05; p=2; k_z=0,02;

ω_1=314,1 (SI units)) the variation of the resultant stator voltage with the pulsatance (in per unit description) for three constant slip values. The variation is a straight line for reduced loads and has a certain inflection for low frequency values (a few Hz). For under-load operation, a significant increase of the voltage with the frequency is necessary. This fact is more visible at high slip values, close to pull-out value (in our example the pull-out slip is of 0,33).

The variation rule based on $U_{sR}=f(\omega_s)$ strategy (applied to the upper curve from Fig. 4) provide an operation of the motor within a large range of angular speeds (from start-up to rated point) under a developed torque, whose value is close to the pull-out one. Obviously, the input current is rather high (4-5 I_{1N}) and has to be reduced. Practically, the operation points must be placed within the upper and the lower curves, Fig. 4. It is also easy to notice that the operation with higher frequency values than the rated one does not generally require an increase of the supply voltage but the developed torque is lower and lower. In this case, the output power keeps the rated value.

4.2. Variation of the torque with the rotor total flux space vector

Usually, the electric drives that demand high value starting torque use *constant rotor total flux space vector* strategy. The stator total flux space vector can be written from (41) as:

$$\underline{\Psi}_{sR3} = \frac{\nu_r + js\omega_s}{\nu_{\sigma r}}\underline{\Psi}_{rR3} \Leftrightarrow \Psi_{sR3} = \Psi_{rR3}\frac{\sqrt{\omega_s^2 s^2 + \nu_r^2}}{\nu_{\sigma r}}; \underline{\Psi}_{sR3} = \frac{\sqrt{\omega_s^2 s^2 + \nu_r^2}}{\nu_{\sigma r}}\underline{\Psi}_{rR3}e^{j\theta};$$
$$(\theta = \alpha_s - \alpha_r); \sin\theta = \frac{s\omega_s}{\sqrt{\omega_s^2 s^2 + \nu_r^2}}; \cos\theta = \frac{\nu_r}{\sqrt{\omega_s^2 s^2 + \nu_r^2}} \tag{46}$$

and the expression of the electromagnetic torque on the basis of rotor flux alone becomes:

$$T_e = -\left(\frac{3}{2}\right)p\Lambda_3 \operatorname{Re}\left(j\underline{\Psi}_{sR3}\underline{\Psi}^*_{rR3}\right) = \frac{3}{2}\frac{p\Lambda_3}{\nu_{\sigma r}}\Psi_{rR3}^2 s\omega_s \tag{47}$$

Assuming the ideal hypothesis of maintaining constant the rotor flux, for example equal to the no-load value, then the electromagnetic torque expression is:

$$T_e = \frac{3}{2}\frac{p\Lambda_3}{\nu_{\sigma r}}\Psi_{rR30}^2 s\omega_s \approx \frac{3}{2}\frac{p\Lambda_3}{\nu_{\sigma r}}\left(\frac{\nu_{\sigma r}U_{sR3}}{\nu_r\omega_s}\right)^2 s\omega_s = \frac{3}{2}\frac{p\Lambda_3\nu_{\sigma r}}{\nu_r^2}\left(\frac{U_{sR3}^2}{\omega_s^2}\right)(\omega_s - p\Omega_R) \tag{48}$$

where the voltage and pulsation is supposed to have rated values. Taking into discussion a machine with predetermined parameters (supply voltage with the amplitude of 490 V (U_{as}=346.5V); R_s=R_r=2; L_{hs}=0,09; $L_{\sigma s}$= $L_{\sigma r}$=0,01; J=0,05; p=2; k_z=0,02; ω_1=314,1 (SI units)) then the expression of the mechanical characteristic is:

$$T_e = \frac{3}{2}\frac{2\cdot 32{,}14\cdot 96{,}43}{103{,}57^2}\left(\frac{U_{sR3N}^2}{\omega_{sN}^2}\right)(\omega_s - 2\Omega_R) = 3{,}17(\omega_s - 2\Omega_R) \tag{49}$$

which is a straight line, A1 in Fig. 5. The two intersection points with the axes correspond to synchronism (T_e=0, $\Omega_R=\omega_s/2$=157) and start-up (T_e=995 Nm, Ω_R=0) respectively.

The pull-out torque is extremely high and acts at start-up. This behavior is caused by the hypothesis of maintaing constant the rotor flux at a value that corresponds to no-load operation (when the rotor reaction is null) no matter the load is. The compensation of the magnetic reaction of the rotor under load is *hypothetical* possible through an unreasonable increase of the supply voltage. Practically, the pull-out torque is much lower.

Another unreasonable possibility is the maintaining of the rotor flux to a value that corresponds to start-up (*s* = *1*) and the supply voltage has its rated value. In this case the expression of the mechanical characteristic is (50) and the intersection points with the axes (line A2, Fig. 5) correspond to synchronism (T_e=0, $\Omega_R=\omega_s/2$=157) and start-up (T_e=78 Nm, Ω_R=0) respectively.

$$T_e = \frac{3}{2}\frac{2\cdot 32,14\cdot}{96,43}\Psi_{rRk}^2\left(\omega_s - 2\Omega_R\right) = 0,25\left(\omega_s - 2\Omega_R\right) \tag{50}$$

The supply of the stator winding with constant voltage and rated pulsation determines a variation of the resultant rotor flux within the short-circuit value (Ψ_{rRk}=0,5Wb) and the synchronism value (Ψ_{rR0}=1,78Wb). The operation points lie between the two lines, A1 and A2, on a position that depends on the load torque. When the supply pulsation is two times smaller (and the voltage itself is two times smaller as well) and the resultant rotor flux is maintained constant to the value Ψ_{rR0}=1,78Wb, then the mechanical characteristic is described by the straight line B1, which is parallel to the line A1. Similarly, for Ψ_{rRk}=0,5Wb, the mechanical characteristic become the line B2, which is parallel to A2.

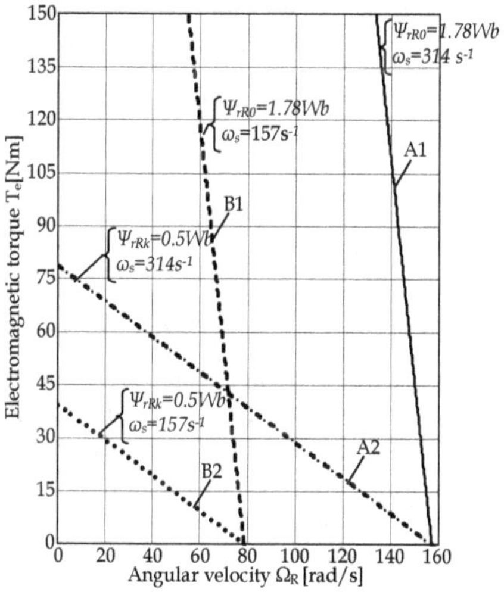

Figure 5. Mechanical characteristics $T_e=f(\Omega_R)$, $\Psi_{rR}=const.$

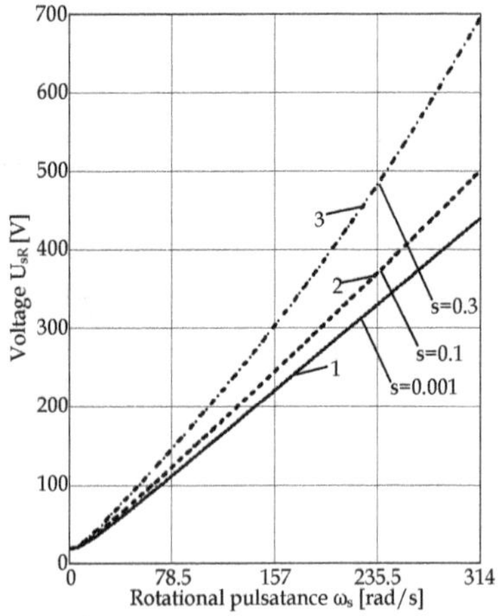

Figure 6. Resultant stator voltage vs. pulsatance, $U_{sR}=f(\omega_s)$ at $\Psi_{rR}=const.$ *(1.3Wb)*

When the applied voltage and pulsation are two times smaller regarding the rated values then the operation points lie between B1 and B2 since the rotor flux varies within Ψ_{rRk}=0,5Wb (short-circuit) and Ψ_{rR0}=1,78Wb (synchronism).

The control based on constant rotor flux strategy ensures parallel mechanical characteristics. This is an important advantage since the induction machine behaves like shunt D.C. motor. A second aspect is also favorable in the behavior under this strategy. The mechanical characteristic has no sector of unstable operation as the usual induction machine has.

The modification of the flux value (generally with decrease) leads to a different slope of the characteristics, which means a significant decrease of the torque for a certain angular speed.

The question is "what variation rule of U_{sR}/ω_s must be used in order to have constant rotor flux"? The expression of the modulus of the resultant rotor flux can be written as:

$$\frac{U_{sR3}^2}{\omega_s^2} = \Psi_{rR3}^2 \frac{As^2 + 2Bs + C}{v_{\sigma r}^2} \leftrightarrow \frac{U_{sR3}}{\omega_s} = \frac{\Psi_{rR3}}{v_{\sigma r}} \sqrt{As^2 + 2Bs + C} \tag{51}$$

$$with: \quad A = \omega_s^2 + v_s^2; B = v_{\sigma s} v_{\sigma r}; C = v_r^2 + v_{tt}^2; v_{tt}^2 = (v_s v_r - v_{\sigma s} v_{\sigma r})^2 / \omega_s^2 \, .$$

Fig. 6 presents the variation of the stator voltage with pulsatance at constant resultant rotor flux (1,3 Wb), which are called the *control characteristics* of the static converter connected to the induction machine. The presented characteristics correspond to three constant slip values, s=0,001 (no-load)-curve 1, s=0,1 (rated duty)-curve 2 and s=0,3 (close to pull-out point)-curve 3. It can be seen that the operation with high slip values (high loads) require an increased stator voltage for a certain pulsation. As a matter of fact, the ratio U_{sR3}/ω_s must be increased with the load when the pulsatance (pulsation) and the angular speed rise as well. Such a strategy is indicated for fans, pumps or load machines with speed-dependent torque.

When the pulsation of the stator voltage is low (small angular velocities) then the torque that has to be overcame is small too, but it will rise with the speed and the frequency along a parabolic variation. Since the upper limit of the torque is given by the limited power of the machine (thermal considerations) then this strategy requires additional precautions as concern the safety devices that protect both the static converter and the supply source itself.

The analysis of the square root term from (51) generates similar remarks as in the above discussed control strategy.

Finally is important to say that a control characteristic must be prescribed for the static converter. This characteristic should be simplified and generally reduced to a straight line placed between the curves 1 and 2 from Fig. 6.

5. Study of the unbalanced duties

The unbalanced duties (generated by supply asymmetries) are generally analyzed by using the theory of symmetric components, according to which any asymmetric three-phase system with *single unbalance* (the sum of the applied instantaneous voltages is always zero) can be equated with two symmetric systems of opposite sequences: positive (+) (or direct) and negative (-) (or inverse) respectively. There are two possible ways for the analysis of this problem.

a. When the amplitudes of the phase voltages are different and/or the angles of phase difference are not equal to 2π/3 then the *unbalanced three-phase* system can be replaced with an equivalent *unbalanced two-phase* system, which further is taken apart in two systems, one of *direct sequence* with higher two-phase amplitude voltages and the other of *inverse sequence* with lower two-phase amplitude voltages. Usually, this equivalence process is obtained by using an orthogonal transformation. Not only voltages but also the total fluxes and eventually the currents must be established for the two resulted systems. The quantities of the unbalanced two-phase system can be written as follows:

$$\begin{bmatrix} \underline{U}_{\alpha s} \\ \underline{U}_{\beta s} \\ \underline{U}_{0s} \end{bmatrix} = \sqrt{\frac{2}{3}} \begin{bmatrix} 1 & -1/2 & -1/2 \\ 0 & \sqrt{3}/2 & -\sqrt{3}/2 \\ 1/\sqrt{2} & 1/\sqrt{2} & 1/\sqrt{2} \end{bmatrix} \begin{bmatrix} \underline{U}_{as} \\ \underline{U}_{bs} \\ \underline{U}_{cs} \end{bmatrix} \leftrightarrow \begin{cases} \underline{U}_{\alpha s} = \sqrt{\frac{3}{2}} \underline{U}_{as}; \underline{U}_{\beta s} = \sqrt{\frac{3}{2}} \frac{\underline{U}_{bs} - \underline{U}_{cs}}{\sqrt{3}} \\ \underline{U}_{0s} = 0; \quad \underline{U}_{as} + \underline{U}_{bs} + \underline{U}_{cs} = 0 \end{cases} \tag{52}$$

Further, the unbalanced quantities are transformed to balanced quantities and we obtain:

$$\begin{bmatrix} \underline{U}_{\alpha s(+)} \\ \underline{U}_{\alpha s(-)} \end{bmatrix} = \frac{1}{2} \begin{bmatrix} 1 & j \\ 1 & -j \end{bmatrix} \begin{bmatrix} \underline{U}_{\alpha s} \\ \underline{U}_{\beta s} \end{bmatrix}, \text{ or: } \begin{cases} \underline{U}_{\alpha s(+)} = \left(\underline{U}_{as} e^{j\pi/6} + j\underline{U}_{bs} \right) / \sqrt{2}; \\ \underline{U}_{\alpha s(-)} = \left(\underline{U}_{as} e^{-j\pi/6} - j\underline{U}_{bs} \right) / \sqrt{2} \end{cases} \tag{53}$$

The quantities of the three-phase system with *single unbalance* can be written as follows:

$$u_{as} = U\sqrt{2} \cos \omega t \Leftrightarrow \underline{U}_{as} = Ue^{j0}; \underline{U}_{bs} = kUe^{-j\beta}; \underline{U}_{cs} = -U(1 + ke^{-j\beta}) \tag{54}$$

and further:

$$\underline{U}_{\alpha s(+)} = U(e^{j\pi/6} + ke^{j(\pi/2-\beta)}) / \sqrt{2}; \underline{U}_{\alpha s(-)} = U(e^{-j\pi/6} - ke^{j(\pi/2-\beta)}) / \sqrt{2} \tag{55}$$

Modulus of these components can be determined at once with:

$$U_{\alpha s(+)} = U\sqrt{1+k^2+2k\sin(\beta+\pi/6)} / \sqrt{2}; U_{\alpha s(-)} = U\sqrt{1+k^2-2k\sin(\beta-\pi/6)} / \sqrt{2} \tag{56}$$

For the transformation of the unbalanced two-phase quantities in balanced two-phase components (53) must be used:

$$\begin{cases} \underline{U}_{\alpha s} = \underline{U}_{\alpha s(+)} + \underline{U}_{\alpha s(-)} \\ \underline{U}_{\beta s} = -j\underline{U}_{\alpha s(+)} + j\underline{U}_{\alpha s(-)} \end{cases} \tag{57}$$

The matrix equation of the two-phase model is written in a convenient way hereinafter:

$$\begin{bmatrix} \underline{U}_{\alpha s} \\ \underline{U}_{\beta s} \\ 0 \\ 0 \end{bmatrix} = \begin{bmatrix} \nu_s + j\omega_s & 0 & 0 & -\nu_{\sigma s} \\ 0 & \nu_s + j\omega_s & -\nu_{\sigma s} & 0 \\ 0 & \nu_{\sigma r} & -(\nu_r + j\omega_s) & \omega_R \\ \nu_{\sigma r} & 0 & -\omega_R & -(\nu_r + j\omega_s) \end{bmatrix} \times \begin{bmatrix} \underline{\Psi}_{\alpha s} \\ \underline{\Psi}_{\beta s} \\ \underline{\Psi}_{yr} \\ \underline{\Psi}_{xr} \end{bmatrix} \tag{58}$$

Using elementary math (multiplications with constants, addition and subtraction of different equations) we can obtain the equations of the two-phase *direct* (M2D) and *inverse* (M2I) models:

$$\text{(M2D)} \quad \begin{bmatrix} \underline{U}_{\alpha s(+)} \\ 0 \end{bmatrix} = \begin{bmatrix} \nu_s + j\omega_s & -\nu_{\sigma s} \\ \nu_{\sigma r} & -(\nu_r + js_d\omega_s) \end{bmatrix} \times \begin{bmatrix} \underline{\Psi}_{\alpha s(+)} \\ \underline{\Psi}_{xr(+)} \end{bmatrix} \tag{59}$$

$$\text{(M2I)} \quad \begin{bmatrix} \underline{U}_{\alpha s(-)} \\ 0 \end{bmatrix} = \begin{bmatrix} \nu_s + j\omega_s & -\nu_{\sigma s} \\ \nu_{\sigma r} & -(\nu_r + js_i\omega_s) \end{bmatrix} \times \begin{bmatrix} \underline{\Psi}_{\alpha s(-)} \\ \underline{\Psi}_{xr(-)} \end{bmatrix} \tag{60}$$

We have defined the slip values for the direct (+) and respectively inverse (-) machines: $s_d = s = \dfrac{\omega_s - \omega_R}{\omega_s}; s_i = \dfrac{\omega_s + \omega_R}{\omega_s}$ with the interrelation expression: $s_i = 2 - s$.

The two machine-models create self-contained torques, which act simultaneously upon rotor. The resultant torque emerges from superposition effects procedure (Simion et al., 2009; Simion & Livadaru, 2010). The equation set (59), for M2D, gives two equations:

$$\underline{U}_{\alpha s(+)} = (\nu_s + j\omega_s)\underline{\Psi}_{\alpha s(+)} - \nu_{\sigma s}\underline{\Psi}_{xr(+)}; \quad 0 = \nu_{\sigma r}\underline{\Psi}_{\alpha s(+)} - (\nu_r + js\omega_s)\underline{\Psi}_{xr(+)} \tag{61}$$

which give further

$$\underline{\Psi}_{\alpha s(+)} = \frac{\left(v_r + js\omega_s\right)\underline{U}_{\alpha s(+)}}{\underline{\Delta}_{(+)}}; \underline{\Psi}_{xr(+)} = \frac{v_{\sigma r}\underline{U}_{\alpha s(+)}}{\underline{\Delta}_{(+)}}; \underline{\Delta}_{(+)} = \left(v_s + j\omega_s\right)\left(v_r + js\omega_s\right) - v_{\sigma s}v_{\sigma r} \tag{62}$$

Similarly, for M2I we obtain:

$$\underline{U}_{\alpha s(-)} = (v_s + j\omega_s)\underline{\Psi}_{\alpha s(-)} - v_{\sigma s}\underline{\Psi}_{xr(-)}; \quad 0 = v_{\sigma r}\underline{\Psi}_{\alpha s(-)} - \left[v_r + j(2-s)\omega_s\right]\underline{\Psi}_{xr(-)} \tag{63}$$

$$\underline{\Psi}_{\alpha s(-)} = \frac{\left[v_r + j(2-s)\omega_s\right]\underline{U}_{\alpha s(-)}}{\underline{\Delta}_{(-)}}; \underline{\Psi}_{xr(-)} = \frac{v_{\sigma r}\underline{U}_{\alpha s(-)}}{\underline{\Delta}_{(-)}}; \underline{\Delta}_{(-)} = \left(v_s + j\omega_s\right)\left[v_r + j(2-s)\omega_s\right] - v_{\sigma s}v_{\sigma r} \tag{64}$$

To determine the electromagnetic torque developed under unbalanced supply condition we use the symmetric components and the superposition effect. The *mean electromagnetic torque M2D* results from (25) but transformed in simplified complex quantities:

$$T_{e(+)} = -\frac{3p}{2}\Lambda_3 \cdot 2\mathrm{Re}\left(j\underline{\Psi}_{\alpha s(+)} \cdot \underline{\Psi}'^{*}_{xr(+)}\right) = \frac{3pv_{\sigma r}\Lambda_3}{2\omega_s} \cdot \frac{2U^2_{\alpha s(+)}s}{As^2 + 2Bs + C} \tag{65}$$

Similarly, the expression of the *mean electromagnetic torque M2D* is:

$$T_{e(-)} = -\frac{3p}{2}\Lambda_3 \cdot 2\mathrm{Re}\left(j\underline{\Psi}_{\alpha s(-)} \cdot \underline{\Psi}^{*}_{xr(-)}\right) = \frac{3pv_{\sigma r}\Lambda_3}{2\omega_s} \cdot \frac{2U^2_{\alpha s(-)}\left(2-s\right)}{A\left(2-s\right)^2 + 2B\left(2-s\right) + C} \tag{66}$$

The *mean resultant torque,* as a difference of the torques produced by M2D and M2I, can be written by using (65) and (66):

$$T_{erez} = \frac{3pv_{\sigma r}\Lambda_3}{2\omega_s}\left[\frac{s \cdot 2U^2_{\alpha s(+)}}{As^2 + 2Bs + C} - \frac{\left(2-s\right) \cdot 2U^2_{\alpha s(-)}}{A(2-s)^2 + 2B(2-s) + C}\right] \tag{67}$$

where we have defined the notations: $\omega_s^2 + v_s^2 = A; \quad v_{\sigma s}v_{\sigma r} = B; \quad v_r^2 + v_{tt}^2 = C;$ and

$$\sqrt{2}U_{\alpha s(+)} = U\sqrt{1 + k^2 + 2k\sin(\beta + \pi/6)}; \sqrt{2}U_{\alpha s(-)} = U\sqrt{1 + k^2 - 2k\sin(\beta - \pi/6)} \tag{68}$$

Finally, the expression of the mean resultant torque with the slip is:

$$T_{erez} = \frac{3pv_{\sigma r}\Lambda_3 U^2}{2\omega_s}\left[\frac{1 + k^2 + 2k\sin(\beta + \pi/6)}{As^2 + 2Bs + C}s - \frac{1 + k^2 - 2k\sin(\beta - \pi/6)}{A(2-s)^2 + 2B(2-s) + C}\left(2-s\right)\right] \tag{69}$$

The influence of the supply unbalances upon $T_e = f(s)$ characteristic are presented in Fig. 7. To this effect, let us take again into discussion the machine with the following parameters: supply voltages with the amplitude of 490 V (U_{as}=346.5V) and 2π/3 rad. shifted in phase; $R_s=R_r=2$; $L_{hs}=0{,}09$; $L_{\sigma s}= L_{\sigma r}=0{,}01$; J=0,05; p=2; $k_z=0{,}02$; $\omega_1=314{,}1$ (SI units). The characteristic corresponding to the three-phase symmetric machine is the curve A (the motoring pull-out torque is equal to 124 Nm and obviously $U_{as(-)} = 0$). If the voltage on phase *b* keeps the same amplitude as the voltage in phase *a*, for example, but the angle of phase difference changes with *π/24=7,5* degrees (from *2π/3=16π/24* to *17π/24* rad.) then the new characteristic is the B curve. The pull-out torque value decreases with approx. 12% but the pull-out slip keeps its value. Other two unbalance degrees are presented in Fig. 7 as well.

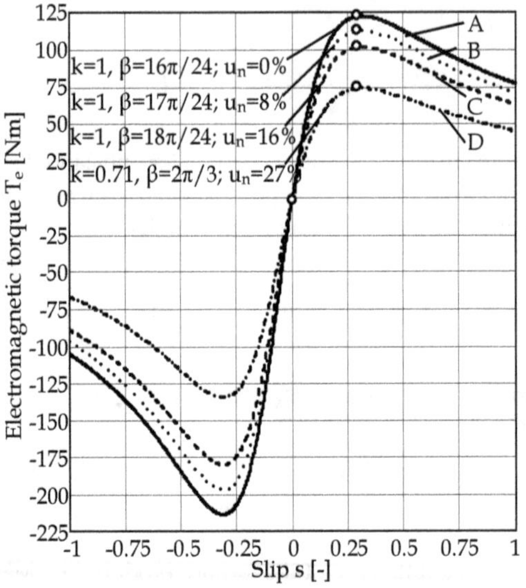

Figure 7. T_e=f(s) characteristic for different unbalance degrees

Usually, the *unbalance degree* of the supply voltage is defined as the ratio of inverse and direct components:

$$u_n = \frac{U_{as(-)}}{U_{as(+)}} = \frac{\sqrt{1+k^2-2k\sin(\beta-\pi/6)}}{\sqrt{1+k^2+2k\sin(\beta+\pi/6)}} \cdot 100[\%] \quad (70)$$

The curves A, B, C, and D from Fig. 7 correspond to the following values of the unbalance degree: u_n= 0; 8%; 16% and 27%. The highest unbalance degree (27% - curve D) causes a decrease of the pull-out torque by 40%.

b. The second approach takes into consideration the following reasoning. When the amplitudes of the three-phase supply system and/or the angles of the phase difference are not equal to 2π/3 then the *unbalanced system* can be replaced by two *balanced three-phase* systems that act in opposition. One of them is the *direct sequence* system and has higher voltages and the other is the *inverse sequence* system and has lower voltages. A transformation of the unbalanced voltages and total fluxes into two symmetric systems is again necessary. In other words, there is an unbalanced voltage system ($\underline{U}_{as}$, $\underline{U}_{bs}$, $\underline{U}_{cs}$), which is replaced by the *direct* and *inverse* symmetric systems. The mean resultant torque is the difference between the torques developed by the two symmetric machine-models. Taking into consideration their slip values ($s_d = s$ and $s_i = 2\text{-}s$) we can deduce the torque expression:

$$T_{erez} = -(3/2)p\Lambda_3 \cdot [3\mathrm{Re}\left(j\underline{\Psi}_{as1}\underline{\Psi}'^{*}_{ar1}\right) - 3\mathrm{Re}\left(j\underline{\Psi}_{as2}\underline{\Psi}'^{*}_{ar2}\right)] \tag{71}$$

$$T_{erez} = \frac{3pv_{\sigma r}\Lambda_3}{2\omega_s}\left[\frac{3sU^2_{as1}}{As^2 + 2Bs + C} - \frac{3(2-s)U^2_{as2}}{A(2-s)^2 + 2B(2-s) + C}\right] \tag{72}$$

and this is the same with (69) as we expected.

6. Simulation study upon some transient duties of the three-phase induction machine

6.1. Symmetric supply system

The mathematical model described by the equation system (26-1…8) allows a complete simulation study of the operation of the three-phase induction machine, which include start-up, any sudden change of the load and braking to stop eventually. To this end, the machine parameters (resistances, main and leakage phase inductances, moments of inertia corresponding to the rotor and the load, coefficients that characterize the variable speed and torque, etc.) have to be calculated or experimentally deduced. At the same time, the values of the load torque and the expressions of the instantaneous voltages applied to each stator phase winding are known, as well. The rotor winding is considered short-circuited. Using the above mentioned equation system, the structural diagram in the Matlab-Simulink environment can be carried out. Additionally, for a complete evaluation, virtual oscillographs for the visualization of the main physical parameters such as voltage, current, magnetic flux, torque, speed, rotation angle and current or specific characteristics (mechanical characteristic, angular characteristic or flux hodographs) fill out the structural diagram.

The study of the *symmetric three-phase* condition in the Matlab-Simulink environment takes into consideration the following parameter values: three identical supply voltages with the amplitude of 490 V (U_{as}=346.5V) and 2π/3 rad. shifted in phase; $u_{ar}=u_{br}=u_{cr}=0$ since the rotor winding is short-circuited; $R_s=R_r=2$; $L_{hs}=0{,}09$; $L_{\sigma s}= L_{\sigma r}=0{,}01$; J=0,05; p=2; $k_z=0{,}02$; $\omega_1=314{,}1$ (SI units). The equation system becomes:

$$\left(\overline{s} + 135{,}71\right)\overline{\psi}_{as} = \overline{u}_{as} - 32{,}14\left(\overline{\psi}_{bs} + \overline{\psi}_{cs}\right) + 32{,}14\left(2\overline{\psi}_{ar} - \overline{\psi}_{br} - \overline{\psi}_{cr}\right)\cos\theta_R + 55{,}67(\overline{\psi}_{cr} - \overline{\psi}_{br})\sin\theta_R$$

$$\left(\overline{s} + 135{,}71\right)\overline{\psi}_{bs} = \overline{u}_{bs} - 32{,}14\left(\overline{\psi}_{cs} + \overline{\psi}_{as}\right) + 32{,}14\left(2\overline{\psi}_{br} - \overline{\psi}_{cr} - \overline{\psi}_{ar}\right)\cos\theta_R + 55{,}67(\overline{\psi}_{ar} - \overline{\psi}_{cr})\sin\theta_R$$

$$\left(\bar{s}+135{,}71\right)\bar{\psi}_{cs}=\bar{u}_{cs}-32{,}14\left(\bar{\psi}_{as}+\bar{\psi}_{bs}\right)+32{,}14\left(2\bar{\psi}_{cr}-\bar{\psi}_{ar}-\bar{\psi}_{br}\right)\cos\theta_R+55{,}67(\bar{\psi}_{br}-\bar{\psi}_{ar})\sin\theta_R$$

$$\left(\bar{s}+135{,}71\right)\bar{\psi}_{ar}=0-32{,}14\left(\bar{\psi}_{br}+\bar{\psi}_{cr}\right)+32{,}14\left(2\bar{\psi}_{as}-\bar{\psi}_{bs}-\bar{\psi}_{cs}\right)\cos\theta_R+55{,}67(\bar{\psi}_{bs}-\bar{\psi}_{cs})\sin\theta_R$$

$$\left(\bar{s}+135{,}71\right)\bar{\psi}_{br}=0-32{,}14\left(\bar{\psi}_{cr}+\bar{\psi}_{ar}\right)+32{,}14\left(2\bar{\psi}_{bs}-\bar{\psi}_{cs}-\bar{\psi}_{as}\right)\cos\theta_R+55{,}67(\bar{\psi}_{cs}-\bar{\psi}_{as})\sin\theta_R$$

$$\left(\bar{s}+135{,}71\right)\bar{\psi}_{cr}=0-32{,}14\left(\bar{\psi}_{ar}+\bar{\psi}_{br}\right)+32{,}14\left(2\bar{\psi}_{cs}-\bar{\psi}_{as}-\bar{\psi}_{bs}\right)\cos\theta_R+55{,}67(\bar{\psi}_{as}-\bar{\psi}_{bs})\sin\theta_R$$

$$\begin{aligned}\dot{\theta}_R\left(\bar{s}+0{,}4\right)=(40)\Big\langle-(32{,}14)\Big\{\sin\theta_R\Big[\bar{\psi}_{as}\left(2\bar{\psi}_{ar}-\bar{\psi}_{br}-\bar{\psi}_{cr}\right)+\bar{\psi}_{bs}\left(2\bar{\psi}_{br}-\bar{\psi}_{cr}-\bar{\psi}_{ar}\right)+\\+\bar{\psi}_{cs}\left(2\bar{\psi}_{cr}-\bar{\psi}_{ar}-\bar{\psi}_{br}\right)\Big]+\sqrt{3}\cos\theta_R\cdot\Big[\bar{\psi}_{as}\left(\bar{\psi}_{br}-\bar{\psi}_{cr}\right)+\bar{\psi}_{bs}\left(\bar{\psi}_{cr}-\bar{\psi}_{ar}\right)+\bar{\psi}_{cs}\left(\bar{\psi}_{ar}-\bar{\psi}_{br}\right)\Big]\ \Big\}-T_{st}\Big\rangle\end{aligned} \quad (73\text{-}1\text{-}7)$$

$$\theta_R=\omega_R\frac{1}{s} \quad (73\text{-}8)$$

$$\begin{gathered}\bar{u}_{as}\leftrightarrow\frac{490}{\sqrt{2}}e^{j(314{,}1t)};\bar{u}_{bs}\leftrightarrow\frac{490}{\sqrt{2}}e^{j(314{,}1t-2{,}094)};\bar{u}_{cs}\leftrightarrow\frac{490}{\sqrt{2}}e^{j(314{,}1t-4{,}188)};\\U_{as\max}=U_{bs\max}=U_{cs\max}=490\end{gathered} \quad (73\text{-}9)$$

It has to be mentioned again that the above equation system allows the analysis of the three-phase induction machine under any condition, that is transients, steady state, symmetric or unbalanced, with one or both windings (from stator and rotor) connected to a supply system. Generally, a supplementary requirement upon the stator supply voltages is not mandatory. The case of short-circuited rotor winding, when the rotor supply voltages are zero, include the wound rotor machine under rated operation since the starting rheostat is short-circuited as well.

The presented simulation takes into discussion a varying duty, which consists in a *no-load* start-up (the load torque derives of frictions and ventilation and is proportional to the angular speed and have a steady state rated value of approx. 3 Nm) followed after 0,25 seconds by a sudden loading with a constant torque of 50 Nm. The simulation results are presented in Fig. 8, 10, 12, 14 and 15 and denoted by the symbol *RS-50.* A second simulation iterates the presented varying duty but with a load torque of 120 Nm, symbol *RS-120,* Fig. 9, 11 and 13. Finally, a third simulation takes into consideration a load torque of 125 Nm, which is a value over the pull-out torque. Consequently, the falling out and the stop of the motor in t≈0,8 seconds mark the varying duty (symbol *RS-125,* Fig. 16, 17, 18 and 19).

The *RS-50* simulation shows an upward variation of the angular speed to the no-load value (in t ≈ 0,1 seconds), which has a weak overshoot at the end, Fig. 8. The 50 Nm torque enforcement determines a decrease of the speed corresponding to a slip value of s ≈ 6,5%. In the case of the *RS-120* simulation, the start-up is obviously similar but the loading torque determines a much more significant decrease of the angular speed and the slip value gets to s ≈ 25%, Fig. 9.

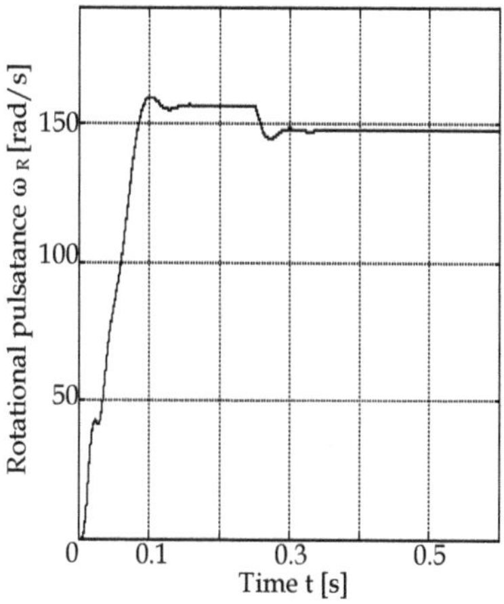

Figure 8. Time variation of rotational pulsatance – *RS-50*

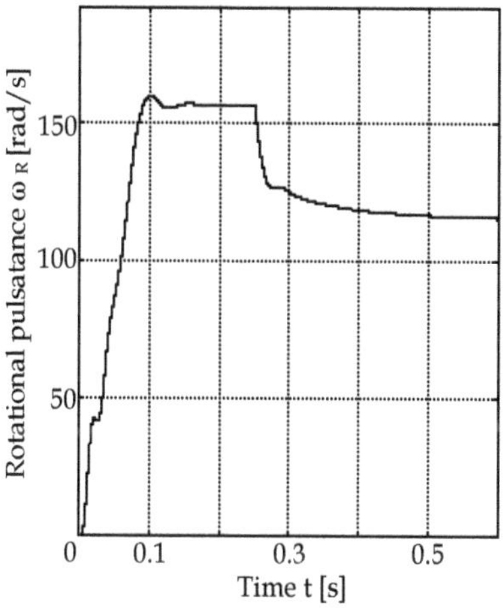

Figure 9. Time variation of rotational pulsatance – *RS-120*

In the first moments of the start-up, the electromagnetic torque oscillates around 100 Nm and after the load torque enforcement, it gets to approx. 53 Nm for *RS-50*, Fig. 10 and to approx. 122 Nm for *RS-120*, Fig. 11. The operation of the motor remains stable for the both duties.

The behavior of the machine is very interesting described by the hodograph of the resultant rotor flux (the locus of the head of the resultant rotor flux phasor), Fig. 12 and 13. With the connecting moment, the rotor fluxes start from 0 (O points on the hodograph) and track a corkscrew to the maximum value that corresponds to synchronism (ideal no-load operation), S points on the hodographs.

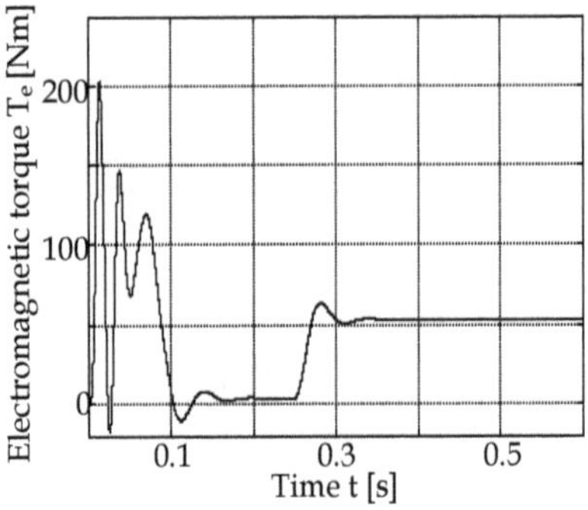

Figure 10. Time variation of electromagnetic torque – *RS-50*

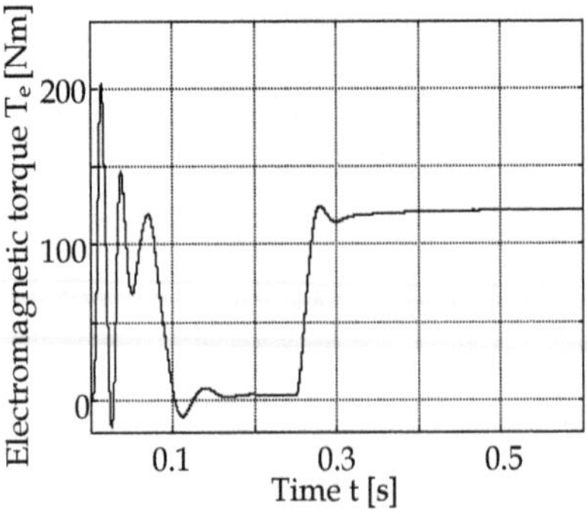

Figure 11. Time variation of electromagnetic torque – *RS-120*

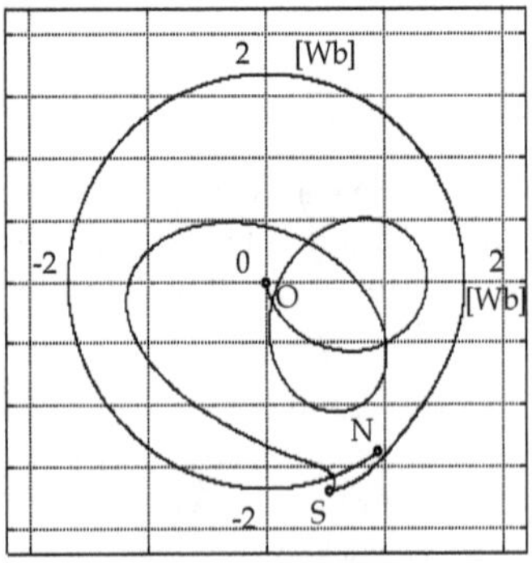

Figure 12. Hodograph of resultant rotor flux – *RS-50*

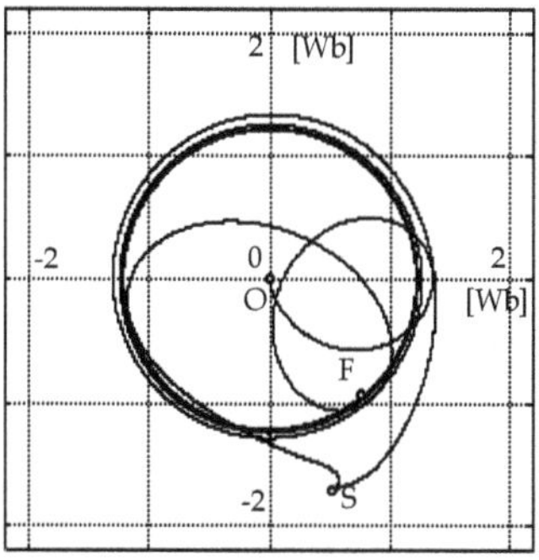

Figure 13. Hodograph of resultant rotor flux – *RS-120*

The enforcement of the load torque determines a decrease of the resultant rotor flux, which is proportional to the load degree, and is due to the rotor reaction. The locus of the head of the phasor becomes a circle whose radius is proportional to the amplitude of the resultant rotor flux. The speed on this circle is given by the rotor frequency that is by the slip value. It is interesting to notice that the load torque of 50 Nm causes a unique rotation of the rotor flux whose amplitude becomes equal to the segment ON (Fig. 12) whereas the 120 Nm torque causes approx. 4 rotations of the rotor flux and the amplitude OF is significantly smaller (Fig. 13).

If the expressions (1) and (2) are also used in the structural diagram then both stator and rotor phase currents can be plotted. The stator current corresponding to *as* phase has the frequency f_1=50 Hz and gets a start-up amplitude of approx. 70 A. This value decreases to approx. 6 A (no-load current) and after the torque enforcement (50 Nm) it rises to a stable value of approx. 14 A, Fig. 14. The rotor current on phase *ar*, which has a frequency value of $f_2 = s\cdot f_1$, gets a similar (approx. 70 A) start-up variation but in opposition to the stator current, i_{as}. Then, its value decrease and the frequency go close to zero. The loading of the machine has as result an increase of the rotor current up to 13 A and a frequency value of $f_2 \approx 3$Hz, Fig. 15. The fact that the current variations are sinusoidal and keep a constant frequency is an argument for a stable operation under symmetric supply conditions.

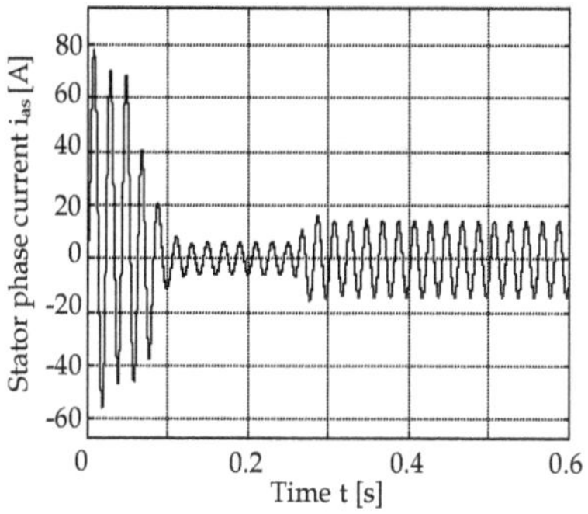

Figure 14. Time variation of stator phase current – *RS-50*

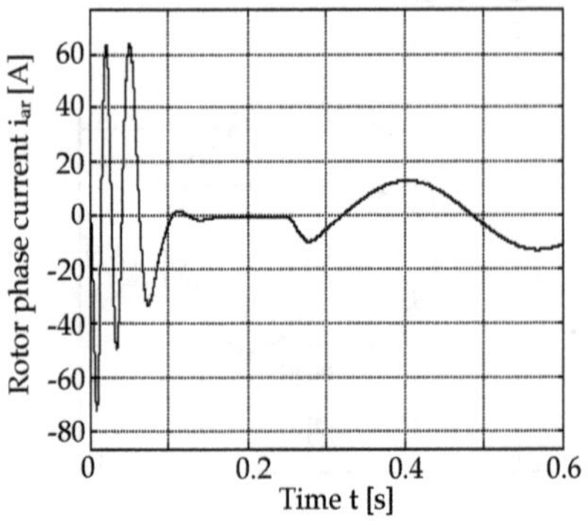

Figure 15. Time variation of rotor phase current – *RS-50*

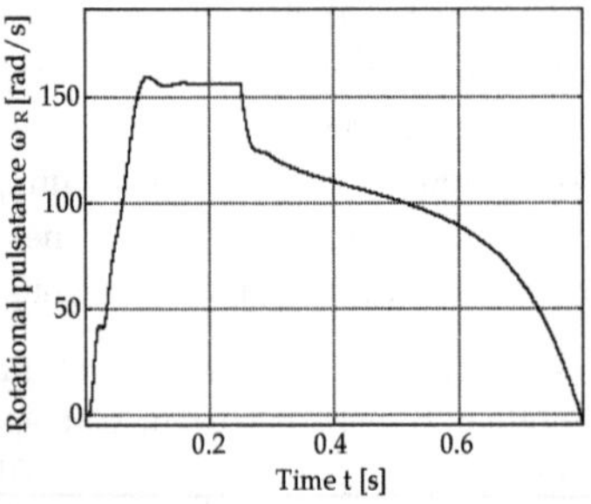

Figure 16. Time variation of rotational pulsatance – *RS-125 (start-up to locked-rotor)*

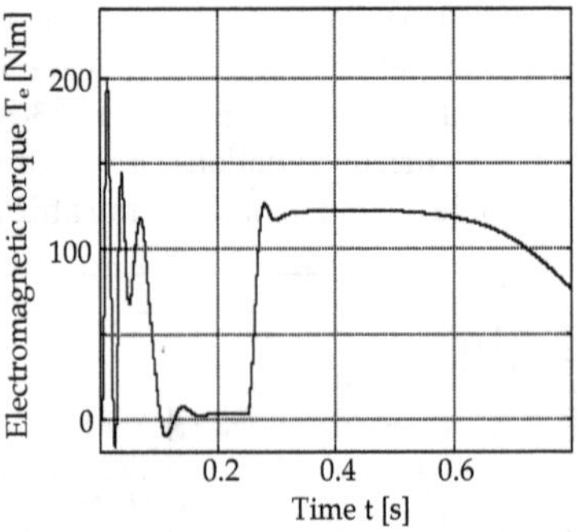

Figure 17. Time variation of electromagnetic torque – *RS-125*

The third simulation, *RS-125*, has a similar start-up but the enforcement of the load torque determines a fast deceleration of the rotor. The pull-out slip (s≈33%) happens in t≈0,5 seconds after which the machine falls out. The angular speed reaches the zero value in t≈0,8 seconds, Fig. 16, and the electromagnetic torque get a value of approx. 78 Nm. This value can be considered the locked-rotor (starting) torque of the machine, Fig. 17.

The described critical duty that involves no-load start-up and operation, overloading, falling out and stop is plotted in terms of resultant rotor flux and angular speed versus electromagnetic torque. The hodograph (Fig. 18) put in view a cuasi corkscrew section, corresponding to the start-up, characterized by its maximum value represented by the segment OS. The falling out tracks the corkscrew SP with a decrease of the amplitude, which is proportional to the deceleration of the rotor. The point P corresponds to the locked-rotor position (s=1). Fig. 19 presents the dynamic mechanical characteristic, which shows the variation of the electromagnetic torque under variable operation condition. During the no-load start-up, the operation point tracks successively the points O, M, L and S, that is from locked-rotor to synchronism with an oscillation of the electromagnetic torque inside certain limits (≈+200Nm to ≈-25Nm). The enforcement of the overload torque leads the operation point along the *downward* curve SK characterized by an *oscillation* section followed by the unstable falling out section, KP. The PKS curve, together with the marked points (Fig. 19) can be considered the *natural mechanical characteristic* under motoring duty.

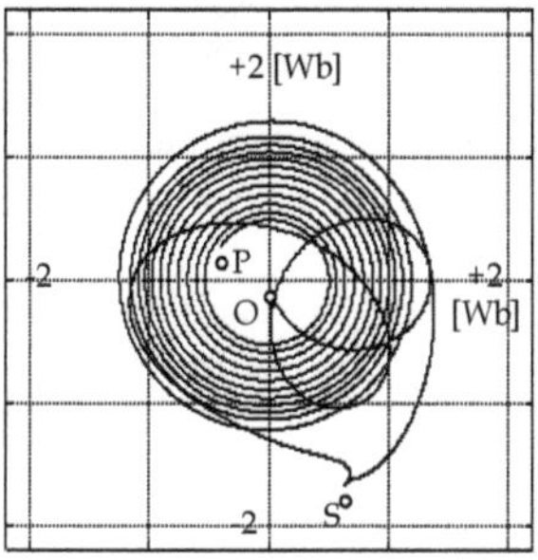

Figure 18. Hodograph of resultant rotor flux – *RS-125 (start-up to locked-rotor)*

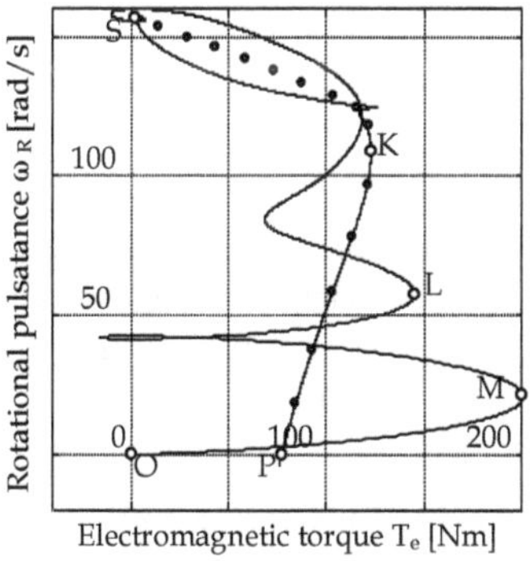

Figure 19. Rotational pulsatance vs. torque – *RS-125 (start-up to locked-rotor)*

6.2. Asymmetric supply system

A simulation study of the three-phase induction machine under unbalanced supply condition and varying duty (start-up, sudden torque enforcement and braking to stop eventually) is possible by using the same mathematical model described by the equation system (26-1...8). The values of the resistant torques and the expressions of the instantaneous phase voltages have to be stated. Since the rotor winding is short-circuited, the supply rotor voltages are $u_{ar}=u_{br}=u_{cr}=0$. On this basis, the structural diagram has been put into effect in the Matlab-Simulink environment. As regards the unbalanced three-phase supply system, it has to be mentioned that the phase voltages are no more equal in amplitude and the angles of phase difference may have other values than 2π/3 rad. In any event, the sum of the instantaneous values of the applied voltages must be zero, that is $u_{as}+u_{bs}+u_{cs}=0$. As an argument for this seemingly constraint stands the fact that the vast majority of the three-phase induction machines are connected to the industrial system via three supply leads (no neutral).

The simulation presented here takes into discussion an induction machine with the same parameters as above that is: $R_s=R_r=2$; $L_{hs}=0{,}09$; $L_{\sigma s}= L_{\sigma r}=0{,}01$; J=0,05; p=2; $k_z=0{,}02$; $\omega_1=314{,}1$ (SI units). Consequently, the equations (73-1) - (73-8) keep unchanged. The expressions (73-9) have to be modified in accordance with the asymmetry degree.

Two varying duties under unbalanced condition have been simulated. The first (denoted *RNS-1*) is characterized by an asymmetry degree, u_n = 16,5% and the following supply voltages:

$$\bar{u}_{as} \leftrightarrow \frac{490}{\sqrt{2}} e^{j(314{,}1t)}; \bar{u}_{bs} \leftrightarrow \frac{375}{\sqrt{2}} e^{j(314{,}1t-1{,}96)}; \bar{u}_{cs} \leftrightarrow \frac{490}{\sqrt{2}} e^{j(314{,}1t-3{,}927)}; u_n = 16{,}5\% \qquad (74)$$

The simulation results are presented in Fig. 20, 22, 24, 25 and 28. The second study simulation (denoted *RNS-2*) has an asymmetry degree of u_n = 27% given by the following stator voltages:

$$\bar{u}_{as} \leftrightarrow \frac{490}{\sqrt{2}} e^{j(314{,}1t)}; \bar{u}_{bs} \leftrightarrow \frac{346{,}43}{\sqrt{2}} e^{j(314{,}1t-2{,}357)}; \bar{u}_{cs} \leftrightarrow \frac{346{,}43}{\sqrt{2}} e^{j(314{,}1t-3{,}295)}; u_n = 27\% \qquad (75)$$

The simulation results are presented in Fig. 21, 23, 26, 27 and 29. The varying duties are similar to those discussed above and consist in a *no-load* start-up (the load torque derives of frictions and ventilation and is proportional to the angular speed and have a steady state rated value of approx. 3 Nm) followed after 0,25 seconds by a sudden loading with a constant torque of 50 Nm.

In comparison to symmetric supply, the unbalanced voltage system causes a longer start-up time with approx. 20% for *RNS-1* (Fig. 20) and with 50% for *RNS-2* (Fig. 21). Moreover, the higher asymmetry degree of *RNS-2* leads to the cancelation of the overshoot at the end of the start-up process. At the same time, significant speed oscillations are noticeable during the operation (no matter the load degree), which are higher with the increase of the asymmetry degree. These oscillations have a constant frequency, which is twice of the supply voltage frequency. They represent the main cause that determines the specific noise of the machines with unbalanced supply system.

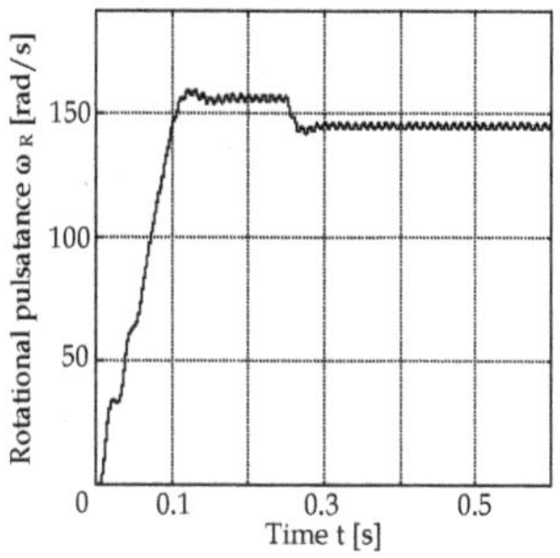

Figure 20. Time variation of rotational pulsatance – *RNS-1 (start-up + sudden load)*

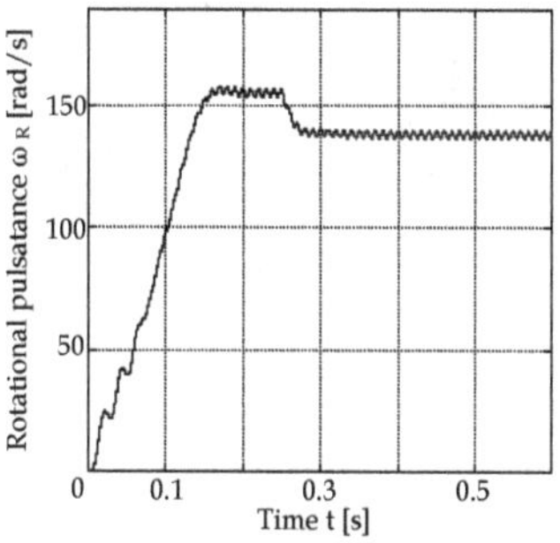

Figure 21. Time variation of rotational pulsatance – *RNS-2 (start-up + sudden load)*

The inspection of the electromagnetic torque variation (Fig. 22 and 23) shows the presence of a variable oscillating torque, whose frequency is twice the supply voltage frequency (in our case 100 Hz) and overlaps the average torque. *This oscillating component is demonstrated by the analytic expression of the instantaneous torque, which is written using nothing but total flux linkages* (25). The symmetric components theory, for example, is not capable to provide information about these oscillating torques. At the most, this theory evaluates the average torque, probably with inherent errors. Coming back to the torque variations, one can see that the amplitude oscillations increase with the asymmetry degree, but their frequency keeps unchanged.

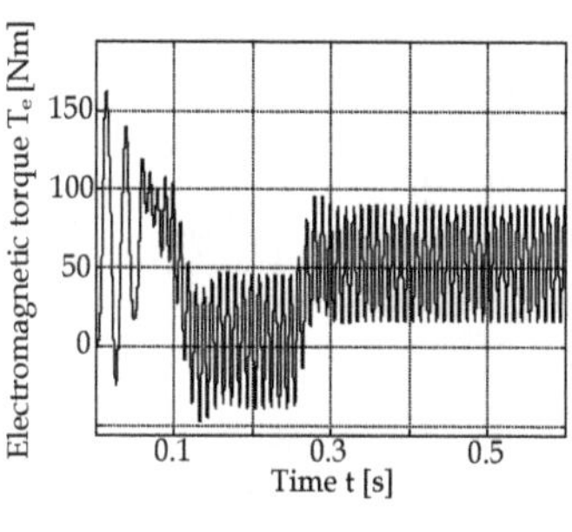

Figure 22. Time variation of electromagnetic torque – *RNS-1*

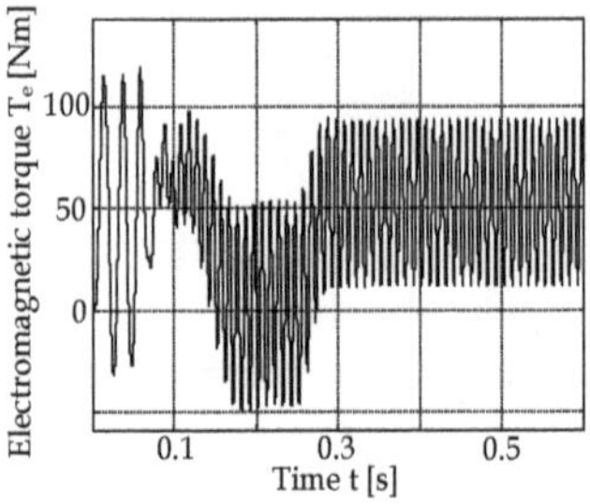

Figure 23. Time variation of electromagnetic torque – *RNS-2*

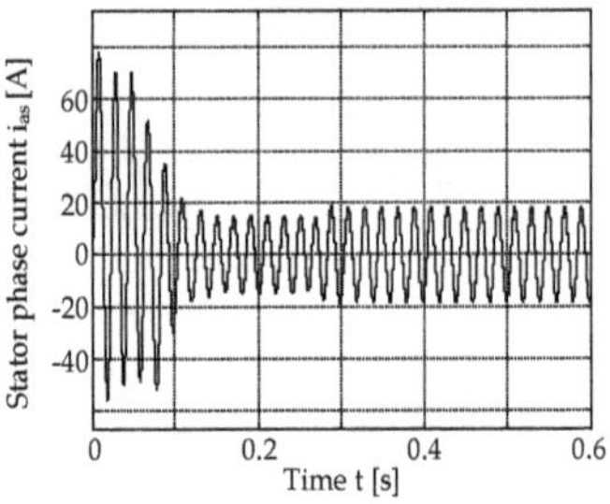

Figure 24. Time variation of stator phase current – *RNS-1*

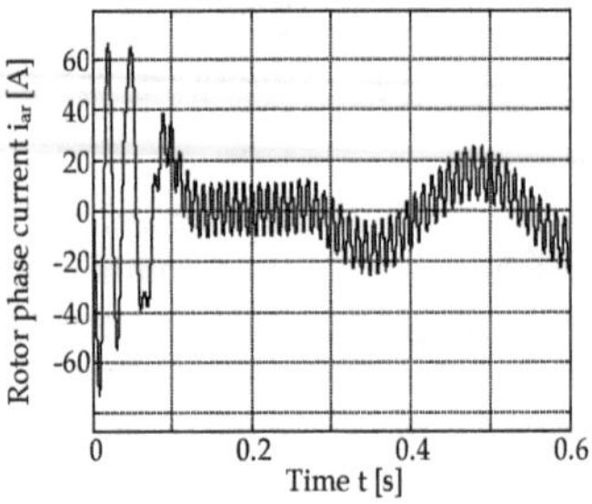

Figure 25. Time variation of rotor phase current – *RNS-1*

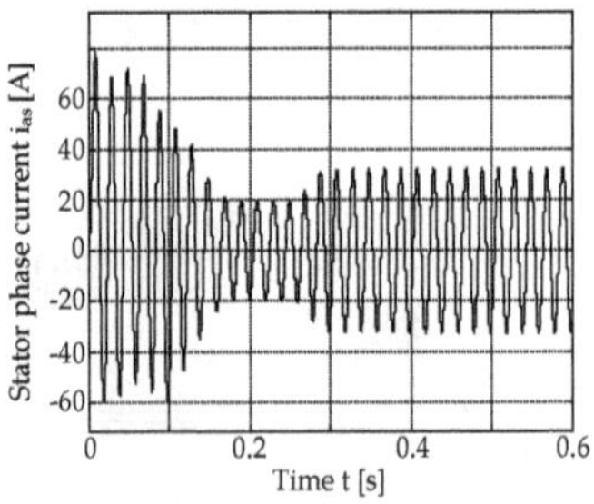

Figure 26. Time variation of stator phase current – *RNS-2*

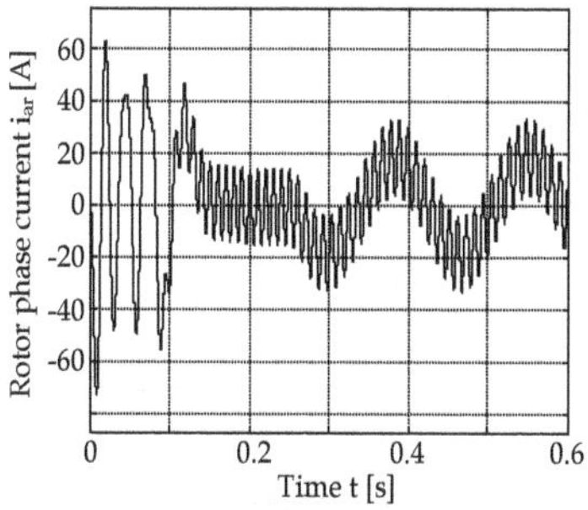

Figure 27. Time variation of rotor phase current – *RNS-2*

The stator currents variation, Fig. 24 and 26, have a sinusoidal shape and an unmodified frequency of 50 Hz. Their amplitude increases however with the asymmetry degree (approx. 18 A for *RNS-1* and approx. 32 A for *RNS-2*). As a consequence of this fact, both power factor and efficiency decrease. The rotor currents (Fig. 25 and 27) include besides the main component of f_2=s· f_1 frequency a second oscillating component of high frequency, $f'_2=(2-s)f_1$, which is responsible for parasitic torques and vibrations of the rotor. The amplitude of these oscillating currents increases with the asymmetry degree.

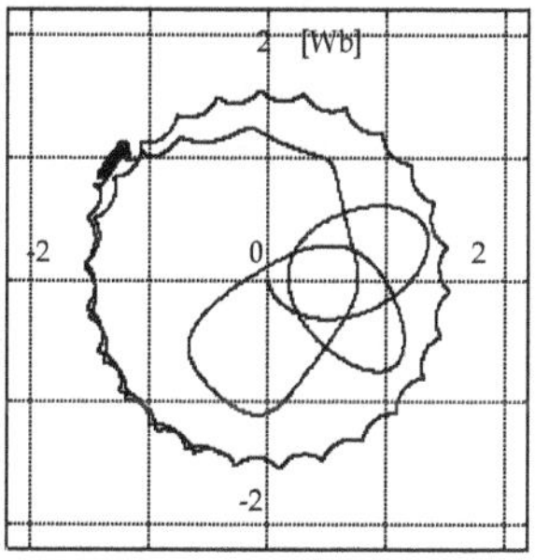

Figure 28. Hodograph of resultant rotor flux – *RNS-1*

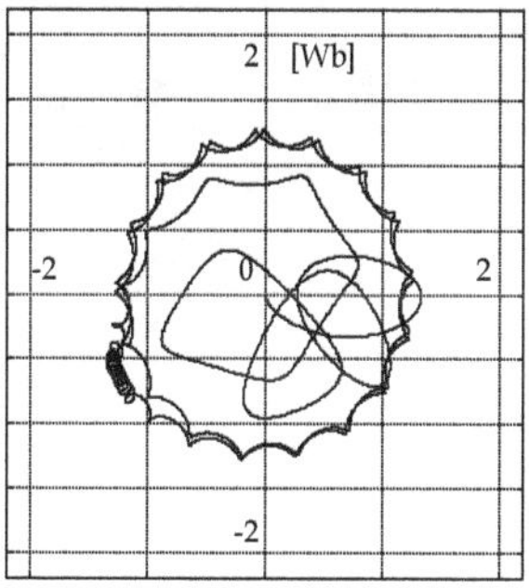

Figure 29. Hodograph of resultant rotor flux – *RNS-2*

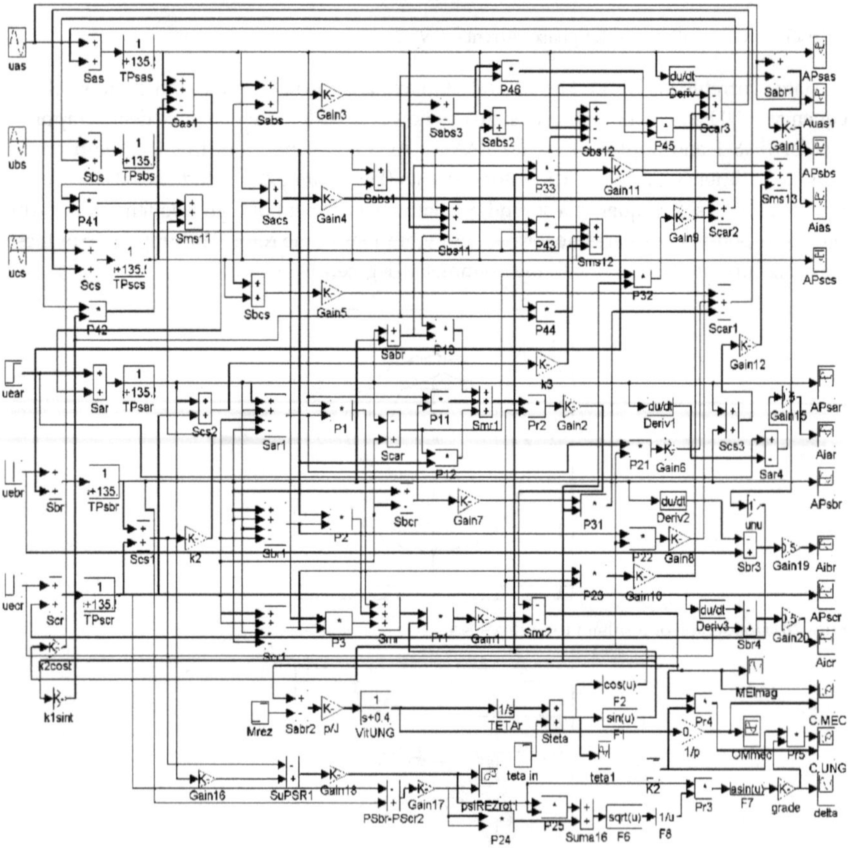

Figure 30. Structural diagram of the three-phase induction machine

The hodographs of the resultant rotor flux show a very interesting behavior of the unbalanced machines, Fig. 28 and 29. In comparison to the symmetric supply cases where the hodograph is a circle under steady state, the asymmetric system distort the curve into a „gear wheel" with a lot of teeth placed on a mean diameter whose magnitude depends inverse proportionally with the asymmetry degree. Generally, these curves do not overlap and prove that during the operation the interaction between stator and rotor fluxes is not constant in time since the rotor speed is not constant. Consequently, the rotor vibrations are usually propagated to the mechanical components and working machine.

In order to point out the superiority of the proposed mathematical model, Fig. 30 shows the structural diagram used in Simulink environment. The diagram is capable to simulate any steady-state and transient duty under balanced or unbalanced state of the induction machine including doubly-fed operation as generator or motor by simple modification of the input data. To prove this statement, a simulation of an unbalanced doubly-fed operation has been performed. The operation cycle involves: I. A no-load start-up (the wound rotor winding is short-circuited); II. Application of a supplementary output torque of (-70) Nm (at the moment t=0.4 sec.) which leads the induction machine to the generating duty (over synchronous speed); III. Supply of two series connected rotor phases with d.c. current (U_{ar}=+40V, U_{br}= −40V, U_{cr}=0V), at the moment time t=0.6 sec., which change the operation of the induction generator into a synchronized induction generator (SIG).

Fig. 31 and 32 show the dynamic mechanical characteristic, Te=f(Ω_R) and the hodograph of the resultant rotor flux respectively. The start-up corresponds to A-S1 curve, the over synchronous acceleration is modeled by S1-S curve and the operation under SIG duty corresponds to S-S2 curve. A few observations regarding Fig. 32 are necessary as well. The rotor flux hodograph is rotating in a *counterclockwise direction* corresponding to motoring duty, in a *clockwise direction* for generating duty and stands still at synchronism. The "in time" modification and the position of the hodograph corresponding to SIG duty depend on the moment of d.c. supply and the load angle of the machine.

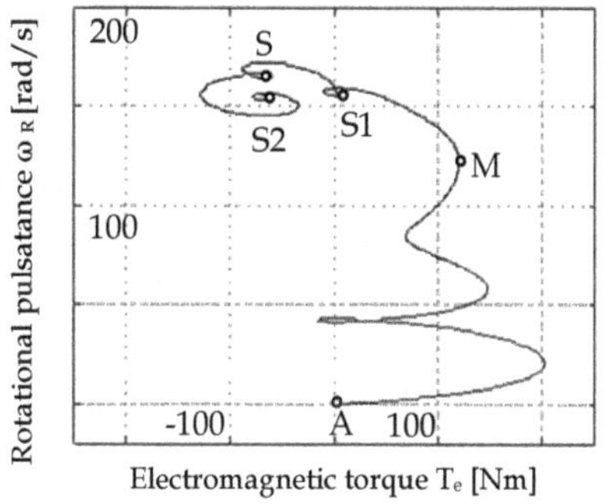

Figure 31. Dynamic mechanical characteristic

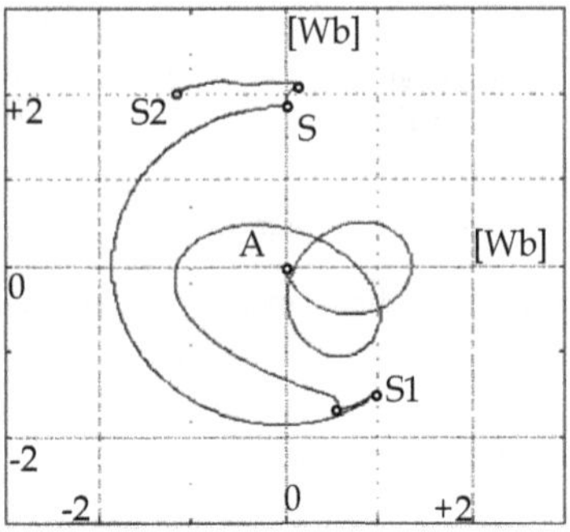

Figure 32. Hodograph of resultant rotor flux

7. Conclusion

The mathematical model presented in this contribution is characterized by the total lack of the winding currents and angular speed in the voltage equations. Since these parameters are differential quantities of other electric parameters, they usually bring supplementary calculus errors mainly for the dynamic duty analysis. Their removal assures a high accuracy of the results. If their variation is however necessary to be known then simple subsequent calculations can be performed.

The use of the mathematical model in total fluxes is appropriate for the study of the electric machines with permanent magnets where the definitive parameter is the magnetic flux and not the electric current.

The coefficients defined by (28.1-4), which depend on resistances and inductances, take into consideration the saturation. Consequently, the study of the induction machine covers more than the linear behavior of the magnetization phenomenon.

The most important advantage of the proposed mathematical model is its generality degree. Any operation duty, such as steady-state or transients, balanced or unbalanced, can be analyzed. In particular, the double feeding duty and the synchronized induction machine operation (feeding with D.C. current of a rotor phase while the other two are short-circuited) can be simulated as well.

The results obtained by simulation are based on the transformation of the equations in structural diagrams under Matlab-Simulink environment. They present the variation of electrical quantities (voltages and currents corresponding to stator and rotor windings), of mechanical quantities (expressed through rotational pulsatance) and of magnetic parameters (electromagnetic torque, resultant rotor and stator fluxes). They put in view the behavior of the induction machine for different transient duties. In particular, they prove that any unbalance of the supply system generates important variations of the electromagnetic torque and rotor speed. This fact causes vibrations and noise.

Author details

Alecsandru Simion, Leonard Livadaru and Adrian Munteanu
"Gh. Asachi" Technical University of Iaşi, Electrical Engineering Faculty, Romania

8. References

Ahmad, M. (2010). *High Performance AC Drives. Modeling Analysis and Control,* Springer, ISBN 978-3-642-13149-3, London, UK

Boldea, I. & Tutelea, L. (2010). *Electric Machines. Steady State, Transients and Design with MATLAB,* CRC Press, ISBN 978-1-4200-5572-6, Boca Raton, USA

Bose, B. (2006). *Power Electronics and Motor Drives,* Elsevier, ISBN 978-0-12-088405-6, San Diego, USA

Chiasson, J. (2005). *Modeling and High-Performance Control of Electrical Machines,* IEEE Press, Wiley Interscience, ISBN 0-471-68449-X, Hoboken, USA

De Doncker, R.; Pulle, D. & Veltman, A. (2011). *Advanced Electrical Drives. Analysis, Modeling, Control,* Springer, ISBN 978-94-007-0179-3, Dordrecht, Germany

Krause, P.; Wasynczuk, O. & Sudhoff, S. (2002). *Analysis of Electric Machinery and Drive Systems (sec. ed.),* IEEE Press, ISBN 0-471-14326-X, Piscataway, USA

Marino, R.; Tomei, P. & Verrelli, C. (2010). *Induction Motor Control Design,* Springer, ISBN 978-1-84996-283-4, London, UK

Ong, C-M. (1998). *Dynamic Simulation of Electric Machinery using Matlab/Simulink,* Prentice Hall, ISBN 0-13-723785-5, New Jersey, USA

Simion, Al.; Livadaru, L. & Lucache, D. (2009). Computer-Aided Simulation on the Reversing Operation of the Two-Phase Induction Machine. *International Journal of Mathematical Models and Methods in Applied Sciences,* Iss. 1, Vol. 3, pp. 37-47, ISSN 1998-0140

Simion, Al. (2010). Study of the Induction Machine Unsymmetrical Condition Using In Total Fluxes Equations. *Advances in Electrical and Computer Engineering,* Iss. 1 (February 2010), pp. 34-41, ISSN 1582-7445

Simion, Al. & Livadaru, L. (2010). On the Unsymmetrical Regime of Induction Machine. *Bul. Inst. Polit. Iaşi,* Tomul LVI(LX), Fasc.4, pp. 79-91, ISSN 1223-8139

Simion, Al.; Livadaru, L. & Munteanu, A. (2011). New Approach on the Study of the Steady-State and Transient Regimes of Three-Phase Induction Machine. *Buletinul AGIR,* Nr.4/2011, pp. 1-6, ISSN-L 1224-7928

Sul, S-K. (2011). *Control of Electric Machine Drive Systems,* IEEE Press, Wiley Interscience, ISBN 978-0-470-87655-8, Hoboken, USA

Wach, P. (2011). *Dynamics and Control of Electric Drives,* Springer, ISBN 978-3-642-20221-6, Berlin, Germany

The Behavior in Stationary Regime of an Induction Motor Powered by Static Frequency Converters

Sorin Muşuroi

Additional information is available at the end of the chapter

1. Introduction

Generally, the electric induction motors are designed for supply conditions from energy sources in which the supply voltage is a sinusoidal wave. The parameters and the functional sizes of the electric motors are guaranteed by designers only for it. If the electric motor is powered through an inverter, due to the presence in the input voltage waveform of superior time harmonics, both its parameters and its functional characteristic sizes will be more or less different from those in the case of the sinusoidal supply. The presence of these harmonics will result in the appearance of a deforming regime in the machine, generally with adverse effects in its operation. Under loading and speed conditions similar to those in the case of the sinusoidal supply, it is registered an amplification of the losses of the machine, of the electric power absorbed and thus a reduction in efficiency. There is also a greater heating of the machine and an electromagnetic torque that at a given load is not invariable, but pulsating, in rapport with the average value corresponding to the load. The occurrence of the deforming regime in the machine is inevitable, because any inverter produces voltages or printed currents containing, in addition to the fundamental harmonic, superior time harmonics of odd order. The deforming regime in the electric machine is unfortunately reflected in the supply power grid that powers the inverter. Generalizing, the output voltage harmonics are grouped into families centered on frequencies:

$$f_j = Jm_f f_c = Jm_f f_1 \quad (J = 1, 2, 3, \ldots), \tag{1}$$

and the various harmonic frequencies in a family are:

$$f_{(\nu)} = f_j \pm kf_c = (Jm_f \pm k)f_c = (Jm_f \pm k)f_1, \tag{2}$$

with

$$\nu = Jm_f \pm k \tag{3}$$

In the above relations, m_f represents the frequency modulation factor, f_1 is the fundamental's frequency and f_c is the frequency of the control modulating signal. Whereas the harmonic spectrum contains only ν order odd harmonics, in order that ($Jm_f \pm k$) is odd, an odd J determines an even k and vice versa. The present chapter aims to analyze the behavior of the induction motor when it is supplied through an inverter. The purpose of this study is to develop the theory of three-phase induction machine with a squirrel cage, under the conditions of the non-sinusoidal supply regime to serve as a starting point in improving the methodology of its constructive-technological design as advantageous economically as possible.

2. The mathematical model of the three-phase induction motor in the case of non-sinusoidal supply

In the literature there are known various mathematical models associated to induction machines fed by static frequency and voltage converters. The majority of these models are based on the association between an induction machine and an equivalent scheme corresponding to the fundamental and a lot of schemes corresponding to the various ν frequencies, corresponding to the Fourier series decomposition of the motor input voltage - see Fig. 1 (Murphy & Turnbull, 1988). In this model the skin effect is not considered.

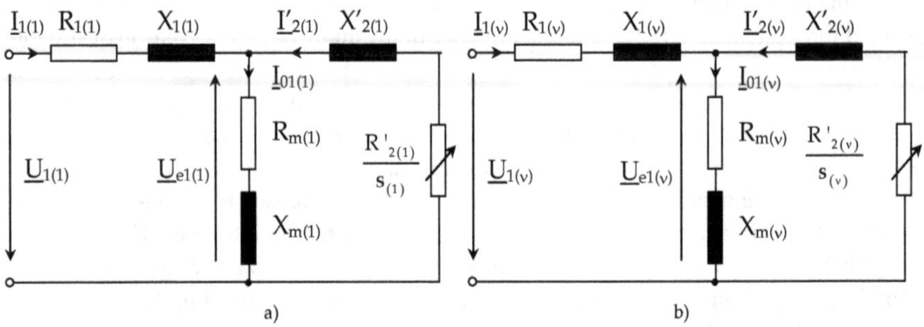

Figure 1. Equivalent scheme of the machine supplied through frequency converter: a) for the case of fundamental; b) for the ν order harmonics (positive or negative sequence).

For the equivalent scheme in Fig.1.a, corresponding to the fundamental, the electrical parameters are defined as:

$$R_{1(1)} = R_1 = R_{1n};\ X_{1(1)} = X_1 = aX_{1n};$$

$$R'_{2(1)} = R'_2 = R'_{2n};\ X'_{2(1)} = X'_2 = aX'_{2n};$$

$$R_{m(1)} = R_m = a^2R_{mn};\ X_{m(1)} = X_m = aX_{mn}; \tag{4}$$

$$\frac{R'_{2(1)}}{s_{(1)}} = \frac{R'_2}{s} = \frac{a}{c} R'_{2n}$$

In relations (4), R_{1n}, X_{1n}, R'_{2n}, X'_{2n}, R_{mn}, X_{mn} represents the values of the parameters R_1, X_1, R'_2, X'_2, R_m and X_m in nominal operating conditions (fed from a sinusoidal power supply, rated voltage frequency and load) and

$$a = \frac{f_1}{f_{1n}} = \frac{\omega_1}{\omega_{1n}} = \frac{n_1}{n_{1n}}; \; c = \frac{n_1 - n}{n_{1n}} = \frac{n_1 - n}{n_1} \cdot \frac{n_1}{n_{1n}} = s \cdot a \tag{5}$$

In the relations (5), f_1 and f_{1n} are random frequencies of the rotating magnetic field, and the nominal frequency of the rotating magnetic field respectively. For ν order harmonics, the scheme from Fig. 1.b is applicable. The slip $s_{(\nu)}$, corresponding to the ν order harmonic is:

$$s_{(\nu)} = \frac{\nu n_1 \mp n}{\nu n_1} = 1 \mp \frac{n}{\nu n_1} = 1 \mp \frac{1}{\nu} \pm \frac{c}{a} \frac{1}{\nu}, \tag{6}$$

where sign (-) (from the first equality) corresponds to the wave that rotates within the sense of the main wave and the sign (+) in the opposite one. For the case studied in this chapter - that of small and medium power machines – the resistances $R_{1(\nu)}$ and reactances $X_{1(\nu)}$ values are not practically affected by the skin effect. In this case we can write:

$$R_{1(\nu)} = R_{1(1)} = R_1 = R_{1n}, \tag{7}$$

$$X_{1(\nu)} = \omega_{1(\nu)} \cdot L_{1\sigma(\nu)} = \nu \omega_1 L_{1\sigma(\nu)}, \tag{8}$$

where $L_{1\sigma(\nu)}$ is the stator dispersion inductance corresponding to the ν order harmonic. If it is agreed that the machine cores are linear media (the machine is unsaturated), it results that the inductance can be considered constant, independently of the load (current) and flux, one can say that:

$$L_{1\sigma(\nu)} = L_{1\sigma(1)} = L_{1\sigma} \tag{9}$$

By replacing the inductance $L_{1\sigma(\nu)}$ expression from relation (9) in relation (8), we obtain:

$$X_{1(\nu)} = \nu \omega_1 L_{1\sigma} = \nu X_1 = \nu a X_{1n} \tag{10}$$

For the rotor resistance and rotor leakage reactance, corresponding to the ν order harmonic, both reduced to the stator the following expressions were established:

$$R'_{2(\nu)} = R'_{2(1)} = R'_2 = R'_{2n}, \tag{11}$$

$$X'_{2(\nu)} = \nu \cdot X'_2 = \nu \cdot a \cdot X'_{2n} \tag{12}$$

The magnetization resistance corresponding to the ν order harmonic, $R_{m\nu}$, is given by the relation:

$$R_{m(\nu)} = k_{K''} \cdot \nu^2 \cdot a^2 \cdot R_{mn} \tag{13}$$

$k_{K''}$ is a coefficient dependent on iron losses and on the magnetic field variation. The magnetization reluctance corresponding to the magnetic field produced by the ν order harmonic is:

$$X_{m(\nu)} = k_{K'} \nu \cdot a \cdot X_{mn} \tag{14}$$

Further the author intends to establish a single mathematical model associated to induction motors, supplied by static voltage and frequency converter, which consists of a single equivalent scheme and which describes the machine operation, according to the presence in the input power voltage of higher time harmonics. For this, the following simplifying assumptions are taken into account:

- the permeability of the magnetic core is considered infinitely large comparing to the air permeability and the magnetic field lines are straight perpendicular to the slot axis;
- both the ferromagnetic core and rotor cage (bar + short circuit rings) are homogeneous and isotropic media;
- the marginal effects are neglected, the slot is considered very long on the axial direction. The electromagnetic fields are considered, in this case plane-parallels;
- the skin effect is taken into account in the calculations only in bars that are in the transverse magnetic field of the slot. For the bar portions outside the slot and in short circuit rings, current density is considered as constant throughout the cross section of the bar;
- the passing from the constant density zone into the variable density zone occurs abruptly;
- in the real electric machines the skin effect is often influenced by the degree of saturation but the simultaneous coverage of both phenomena in a mathematical relationships, easily to be applied in practice is very difficult, even precarious. Therefore, the simplifying assumption of neglecting the effects of saturation is allowed as valid in establishing the relationships for equivalent parameters;
- the local variation of the magnetic induction and of current density is considered sinusoidal in time, both for the fundamental and for each ν harmonic;
- one should take into account only the fundamental space harmonic of the EMF.

Under these conditions of non-sinusoidal supply, the asynchronous motor may be associated to an equivalent scheme, corresponding to all harmonics. The scheme operates in the fundamental frequency $f_{1(1)}$ and it is represented in Fig. 2. According to this scheme, it can be formally considered that the motors, in the case of supplying through the power frequency converter (the corresponding parameters and the dimensions of this situation are marked with index "CSF") behave as if they were fed in sinusoidal regime at fundamental's frequency, $f_{1(1)}$ with the following voltages system:

$$u_A = \sqrt{2} \cdot U_{1(CSF)} \cdot \sin \omega_1 t \,;\, u_B = \sqrt{2} \cdot U_{1(CSF)} \cdot \sin\left(\omega_1 t - \frac{2\pi}{3}\right);\, u_C = \sqrt{2} \cdot U_{1(CSF)} \cdot \sin\left(\omega_1 t + \frac{2\pi}{3}\right), \quad (15)$$

where,
$$U_{1(CSF)} = \sqrt{U_{1(1)}^2 + \sum_{v \neq 1} U_{1(v)}^2} \quad (16)$$

$U_{1(v)}$ is the phase voltage supply corresponding to the v order harmonic. Corresponding to the system supply voltages, the current system which go through the stator phases is as follows:

$$\begin{cases} i_A = \sqrt{2} \cdot I_{1(CSF)} \cdot \sin\left(\omega_1 t - \varphi_{1(CSF)}\right) \\ i_B = \sqrt{2} \cdot I_{1(CSF)} \cdot \sin\left(\omega_1 t - \varphi_{1(CSF)} - \frac{2\pi}{3}\right) \\ i_C = \sqrt{2} \cdot I_{1(CSF)} \cdot \sin\left(\omega_1 t - \varphi_{1(CSF)} - \frac{4\pi}{3}\right) \end{cases}, \quad (17)$$

where $I_{1(CSF)}$ is given by:

$$I_{1(CSF)} = \sqrt{I_{1(1)}^2 + \sum_{v \neq 1} I_{1(v)}^2} \quad (18)$$

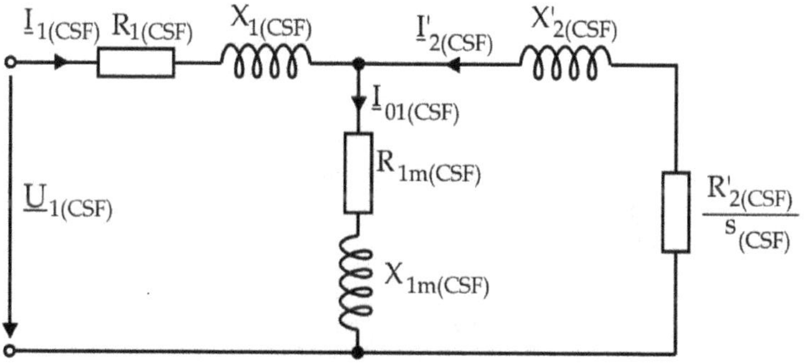

Figure 2. The equivalent scheme of the asynchronous motor powered by a static frequency converter.

Power factor in the deforming regime is defined as the ratio between the active power and the apparent power, as follows:

$$\Delta_{(CSF)} = \frac{P_{1(CSF)}}{S_{1(CSF)}} = \frac{P_{1(CSF)}}{U_{1(CSF)} I_{1(CSF)}} \quad (19)$$

If we consider the non-sinusoidal regime, the active power absorbed by the machine $P_{1(CSF)}$ is defined, as in the sinusoidal regime, as the average in a period of the instantaneous power. The following expression is obtained:

$$P_{1(CSF)} = \frac{1}{T}\int_0^T p \cdot dt = \sum_{v=1} U_{1(v)} I_{1(v)} \cos\varphi_{1(v)} = U_1 I_1 \cos\varphi_1 + \sum_{v\neq 1} U_{1(v)} I_{1(v)} \cos\varphi_{1(v)} \tag{20}$$

Therefore, the active power absorbed by the motor when it is supplied through a power static converter is equal to the sum of the active powers, corresponding to each harmonic (the principle of superposition effects is found). In relation (20), $\cos\varphi(1v)$ is the power factor corresponding to the v order harmonic having the expression:

$$\cos\varphi_{1(v)} = \frac{R_{1(v)} + \frac{R'_{2(v)}}{s_{(v)}}}{\sqrt{\left(R_{1(v)} + \frac{R'_{2(v)}}{s_{(v)}}\right)^2 + \left(X_{1(v)} + X'_{2(v)}\right)^2}} \tag{21}$$

The apparent power can be defined in the non-sinusoidal regime also as the product of the rated values of the applied voltage and current:

$$S_{1(CSF)} = U_{1(CSF)} \cdot I_{1(CSF)}, \tag{22}$$

Taken into account the relations (20), (21) and (22), the relation (19) becomes:

$$\Delta_{(CSF)} = \frac{U_1 I_1 \cos\varphi_1 + \sum_{v\neq 1} U_{1(v)} I_{1(v)} \cos\varphi_{1(v)}}{\sqrt{U_{1(1)}^2 + \sum_{v\neq 1} U_{1(v)}^2} \cdot \sqrt{I_{1(1)}^2 + \sum_{v\neq 1} I_{1(v)}^2}} \tag{23}$$

Because $\Delta_{(CSF)} \leq 1$, formally (the phase angle has meaning only in harmonic values) an angle $\varphi_{1(CSF)}$ can be associated to the power factor $\Delta_{(CSF)}$, as: $\cos\varphi_{1(CSF)} = \Delta_{(CSF)}$. With this, the relation (23) can be written:

$$\cos\varphi_{1(CSF)} = \frac{\cos\varphi_1 + \sum_{v\neq 1} \frac{U_{1(v)}}{U_{1(1)}} \frac{I_{1(v)}}{I_1} \cos\varphi_{1(v)}}{\sqrt{1 + \sum_{v\neq 1}\left(\frac{U_{1(v)}}{U_{1(1)}}\right)^2} \cdot \sqrt{1 + \sum_{v\neq 1}\left(\frac{I_{1(v)}}{I_{1(1)}}\right)^2}} \tag{24}$$

If one takes into account the relation (Murphy&Turnbull, 1988):

$$\frac{I_{1(v)}}{I_{1(1)}} = \frac{1}{v} \cdot \frac{1}{f_{1r} \cdot x^*_{sc}} \cdot \frac{U_{1(v)}}{U_{1(1)}}, \tag{25}$$

where x^*_{sc} is the reported short-circuit impedance, measured at the frequency $f_1 = f_{1n}$, relation (24) becomes:

$$\cos\varphi_{1(CSF)} = \frac{\cos\varphi_1 + \sum_{v\neq 1}\frac{1}{v}\cdot\frac{1}{f_{1r}\cdot x_{sc}^*}\cdot\left(\frac{U_{1(v)}}{U_{1(1)}}\right)^2 \cos\varphi_{1(v)}}{\sqrt{\left[1+\sum_{v\neq 1}\left(\frac{U_{1(v)}}{U_{1(1)}}\right)^2\right]\cdot\left[1+\sum_{v\neq 1}\left(\frac{1}{v}\cdot\frac{1}{f_{1r}\cdot x_{sc}^*}\cdot\frac{U_{1(v)}}{U_{1(1)}}\right)^2\right]}} \tag{26}$$

3. The determination of the equivalent parameters of the stator winding

The equivalent parameters of the scheme have been calculated at the fundamental's frequency, under the presence of all harmonics in the supply voltage. Under these conditions, we note by $p_{Cu1(CSF)}$ the losses that occur in the stator winding when the motor is supplied through a power frequency converter. These losses are in fact covered by some active power absorbed by the machine from the network, through the converter, $P_{1(CSF)}$. According to the principle of the superposition effects, it can be considered:

$$p_{Cu1(CSF)} = p_{Cu1(1)} + \sum_{v\neq 1} p_{Cu1(v)} = 3R_{1(1)}I_{1(1)}^2 + 3\sum_{v\neq 1} R_{1(v)}I_{1(v)}^2 \tag{27}$$

Further, the stator winding resistance corresponding to the fundamental, $R_{1(1)}$ and stator winding resistances corresponding to the all higher time harmonics $R_{1(v)}$, are replaced by a single equivalent resistance $R_{1(CSF)}$, corresponding to all harmonics, including the fundamental. The equalization is achieved under the condition that in this resistance the same loss $p_{Cu1(CSF)}$ occurs, given by relation (27), as if considering the "v" resistances $R_{1(v)}$, each of them crossed by the current $I_{1(v)}$. This equivalent resistance, $R_{1(CSF)}$, determined at the fundamental's frequency, is traversed by the current $I_{1(CSF)}$, with the expression given by (18). Therefore:

$$p_{Cu1(CSF)} = 3R_{1(CSF)}\cdot I_{1(CSF)}^2 = 3R_{1(CSF)}\left(I_{1(1)}^2 + \sum_{v\neq 1} I_{1(v)}^2\right) \tag{28}$$

Making the relations (27) and (28) equal, it results:

$$3R_{1(CSF)}\left(I_{1(1)}^2 + \sum_{v\neq 1} I_{1(v)}^2\right) = 3R_{1(1)}\left(I_{1(1)}^2 + \sum_{v\neq 1} I_{1(v)}^2\right) = 3R_{1(v)}\left(I_{1(1)}^2 + \sum_{v\neq 1} I_{1(v)}^2\right), \tag{29}$$

from which:

$$R_{1(CSF)} = R_{1(1)} = R_1 . \tag{30}$$

Applying the principle of the superposition effects to the reactive power absorbed by the stator winding $Q_{Cu1\,(CSF)}$, the following expression is obtained:

$$Q_{Cu1(CSF)} = Q_{Cu1(1)} + \sum_{v\neq 1} Q_{Cu1(v)} = 3\cdot X_{1(1)}I_{1(1)}^2 + 3\sum_{v\neq 1} X_{1(v)}I_{1(v)}^2 \tag{31}$$

As in the previous case, the stator winding reactance corresponding to the fundamental, $X_{1(1)}$ (determined at the fundamental's frequency $f_{1(1)}$) and the stator winding reactances, corresponding to all higher time harmonics $X_{1(\nu)}$ (determined at frequencies $f_{1(\nu)}=\nu\cdot f_1$ where Jmr±k) are replaced by an equivalent reactance, $X_{1(CSF)}$, determined at fundamental's frequency. This equivalent reactance, traversed by the current $I_{1(CSF)}$, conveys the same reactive power, $Q_{Cu1(CSF)}$ as in the case of considering "ν" reactances $X_{1(\nu)}$, (each of them determined at $f_{1(\nu)}$ frequency and traversed by the current $I_{1(\nu)}$). Following the equalization, the following expression can be written:

$$Q_{Cu1(CSF)} = 3X_{1(CSF)}I_{1(CSF)}^2 = 3X_{1(CSF)}\left(I_{1(1)}^2 + \sum_{\nu\neq 1} I_{1(\nu)}^2\right) \tag{32}$$

Making the relations (31) and (32) equal, it results:

$$X_{1(CSF)}\left(I_{1(1)}^2 + \sum_{\nu\neq 1} I_{1(\nu)}^2\right) = X_{1(1)}I_{1(1)}^2 + \sum_{\nu\neq 1} \nu X_1 I_{1(\nu)}^2 = X_1\left(I_{1(1)}^2 + \sum_{\nu\neq 1} \nu I_{1(\nu)}^2\right) \tag{33}$$

One can notice the following:

$$k_{X1} = \frac{X_{1(CSF)}}{X_1}$$

the factor that highlights the changes that the reactants of the stator phase value suffer in the case of a machine supplied through a power frequency converter, compared to sinusoidal supply, both calculated at the fundamental's frequency. From relations (25) and (33) it follows:

$$k_{X1} = \frac{X_{1(CSF)}}{X_1} = \frac{1+\sum_{\nu\neq 1}\nu\left(\frac{1}{f_{1r}X_{sc}^*}\right)^2\cdot\frac{1}{\nu^2}\left(\frac{U_{1(\nu)}}{U_{1(1)}}\right)^2}{1+\sum_{\nu\neq 1}\left(\frac{1}{f_{1r}X_{sc}^*}\right)^2\cdot\frac{1}{\nu^2}\left(\frac{U_{1(\nu)}}{U_{1(1)}}\right)^2} = \frac{1+\sum_{\nu\neq 1}\frac{1}{\nu}\left(\frac{1}{f_{1r}X_{sc}^*}\right)^2\left(\frac{U_{1(\nu)}}{U_{1(1)}}\right)^2}{1+\sum_{\nu\neq 1}\frac{1}{\nu^2}\left(\frac{1}{f_{1r}X_{sc}^*}\right)^2\left(\frac{U_{1(\nu)}}{U_{1(1)}}\right)^2} \tag{34}$$

where:

$$X_{sc}^* = \frac{X_{sc}}{Z_{(1)}}$$

- is the short circuit impedance reported, corresponding to the frequency $f_1=f_{1n}$ and f_{1r} is the reported frequency. One can notice that: $k_{X1}>1$. With the equivalent resistance given by (30) and the equivalent reactance resulting from the relationship (34) we can now write the relation for the equivalent impedance of the stator winding, $\underline{Z}_{1(CSF)}$ covering all frequency harmonics and including the fundamental:

$$\underline{Z}_{1(CSF)} = R_{1(CSF)} + jX_{1(CSF)} = R_{1(CSF)} + jk_{X1}X_1 \tag{35}$$

4. Determining the equivalent global change parameters for the power rotor fed by the static frequency converter

Further, it is considered a winding with multiple cages whose bars (in number of "c") are placed in the same notch of any form, electrically separated from each other (see Fig. 3). These bars are connected at the front by short-circuiting rings (one ring may correspond to several bars notch). This "generalized" approach, pure theoretically in fact, has the advantage that by its applying the relations of the two equivalent factors $k_{r(CSF)}$ and $k_{x(CSF)}$, valid for any notch type and multiple cages, are obtained. The rotor notch shown in Fig. 3 is the height h_c and it is divided into "n" layers (strips), each strip having a height $h_s = h_c/n$. The number of layers "n" is chosen so that the current density of each band should be considered constant throughout the height h_s (and therefore not manifesting the skin effect in the strip). The notch bars are numbered from 1 to c, from the bottom of the notch. The lower layer of each bar is identified by the index "i" and the top layer by the index "s". Thus, for a bar with index δ characterized by a specific resistance ρ_δ and an absolute magnetic permeabilityμ_δ, the lower layer is noted with $N_{\delta i}$ and the extremely high layer with $N_{\delta s}$. The current that flows through the bar δ is noted with $i_{c\delta}$ ($I_{c\delta}$ - rated value). The length of the bar, over which the skin effect occurs, is L. For the beginning, let us consider only the presence of the fundamental in the power supply, which corresponds to the supply pulsation, $\omega_{1(1)}=\omega_1=2\pi f_1$. In this case:

$$k_{r\delta(1)} = \frac{R_{\delta(1)\sim}}{R_{\delta\,-}} = \frac{1}{I^2_{c\delta(1)}} \cdot \sum_{\varepsilon=N_{\delta i}}^{N_{\delta s}} \frac{I^2_{\varepsilon(1)}}{b_\varepsilon} \cdot \sum_{\varepsilon=N_{\delta i}}^{N_{\delta s}} b_\varepsilon , \tag{36}$$

$$k_{x\delta(1)} = \frac{L_{\delta n\sigma(1)\sim}}{L_{\delta n\sigma\,-}} = \frac{\left|\mathrm{Re}\left[\underline{\Psi}_{\delta n\sigma(1)}\right]\right| \left(\sum_{\varepsilon=N_{\delta i}}^{N_{\delta s}} b_\varepsilon\right)^2}{\sqrt{2}\mu_\delta L h_s I_{c\delta(1)} \cdot \sum_{\lambda=N_{\delta i}}^{N_{\delta s}} \frac{1}{b_\lambda}\left[\left(\sum_{\varepsilon=N_{\delta i}}^{\lambda-1} b_\varepsilon\right)\left(\sum_{\varepsilon=N_{\delta i}}^{\lambda} b_\varepsilon\right) + \frac{b^2_\lambda}{3}\right]} \tag{37}$$

where b_λ and b_ε are the width of λ and ε order strips and $\Psi_{\delta n\sigma(1)}$ is the δ bar flux corresponding to the fundamental of the own magnetic field, assuming that for the λ order strip, the magnetic linkage corresponds to a constant repartition of the fundamental current density on the strip.

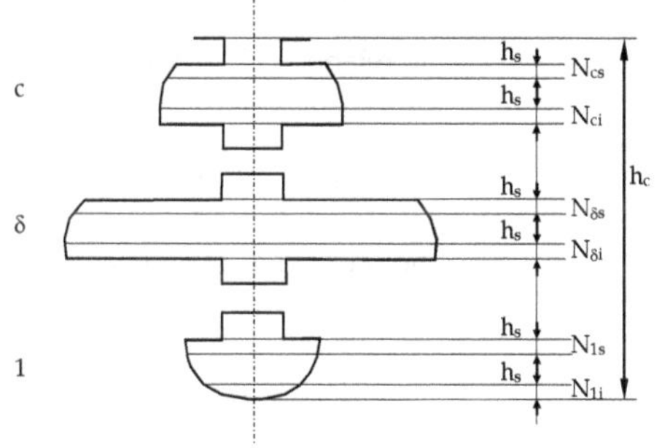

Figure 3. Notch generalized for multiple cages.

If in the motor power supply one considers only the ν order harmonic which corresponds to the supply pulsation $\omega_{1(\nu)}=\nu\omega_1$, the relations (36) and (37) remain valid with the following considerations: index "1" is replaced by index "ν" and the rotor phenomena are with the pulsation $\omega_{2(\nu)}$ given by the relation:

$$\omega_{2(\nu)} = s_{(\nu)} \cdot \omega_{1(\nu)} = \left(1 \mp \frac{1}{\nu} \pm \frac{s}{\nu}\right) \cdot \nu \cdot \omega_1 , \tag{38}$$

Subsequently we shall consider the real case, where in the δ bar both the fundamental and ν order time harmonics are present. For this, the equivalent d.c. global factor of the δ bar resistance modification is calculated with the relation:

$$k_{r\delta(CSF)} = \frac{p_{\delta(CSF)\sim}}{p_{\delta(CSF)-}} = \frac{R_{\delta(CSF)\sim}}{R_{\delta(CSF)-}} , \tag{39}$$

where $p_{\delta(CSF)\sim}$ represents the total a.c. losses in δ bar (considering the appropriate skin effect for all harmonics) and $p_{\delta(CSF)-}$ represents the bar δ total losses, without considering the repression phenomenon. The a.c. total losses in the δ bar are obtained by applying the effects superposition principle by adding all the δ bar a.c. losses caused by each ν order time, including the fundamental. Therefore one can obtain:

$$p_{\delta(CSF)\sim} = p_{\delta(1)} + \sum_{\nu \neq 1} p_{\delta(\nu)\sim} , \tag{40}$$

The a.c. loss in δ bar, corresponding to the fundamental, $p_{\delta(1)\sim}$, is calculated with the following relation:

$$p_{\delta(1)\sim} = I^2_{c\delta(1)} \cdot k_{r\delta(1)} \cdot R_{\delta -} \tag{41}$$

In the same way, the expression of the δ bar a.c. losses produced by some ν order time harmonic is obtained:

$$p_{\delta(\nu)\sim} = I^2_{c\delta(\nu)} \cdot R_{\delta(\nu)\sim} = I^2_{c\delta(\nu)} \cdot k_{r\delta(\nu)} \cdot R_{\delta -} \tag{42}$$

By replacing the relations (41) and (42) in relation (40), it results:

$$p_{\delta(CSF)\sim} = I^2_{c\delta(1)} \cdot k_{r\delta(1)} \cdot R_{\delta -} + \sum_{\nu \neq 1} I^2_{c\delta(\nu)} \cdot k_{r\delta(\nu)} \cdot R_{\delta -} = R_{\delta -}\left(I^2_{c\delta(1)} \cdot k_{r\delta(1)} + \sum_{\nu \neq 1} I^2_{c\delta(\nu)} \cdot k_{r\delta(\nu)}\right). \tag{43}$$

The δ bar losses without considering the repression phenomenon in the bar are calculated using the following relationship:

$$P_{\delta(CSF)-} = I^2_{c\delta(CSF)} \cdot R_{\delta -} , \tag{44}$$

where:

$$I_{c\delta(CSF)} = \sqrt{I^2_{c\delta(1)} + \sum_{v \neq 1} I^2_{c\delta(v)}} \tag{45}$$

is the rated value of the current which runs through the δ bar, in the case of a motor supplied by a frequency converter. By replacing the relation (45) in relation (44):

$$p_{\delta(CSF)-} = R_{\delta -} \left(I^2_{c\delta(1)} + \sum_{v \neq 1} I^2_{c\delta(v)} \right) \tag{46}$$

By replacing the relations (43) and (46) in (39) one obtains the expression for the global equivalent factor of the a.c. increasing resistance in the bar δ, $k_{r\delta\,(CSF)}$, in case of the presence of all harmonics in the motor power:

$$k_{r\delta(CSF)} = \frac{p_{\delta(CSF)\sim}}{p_{\delta(CSF)-}} = \frac{R_{\delta -} \left(I^2_{c\delta(1)} \cdot k_{r\delta(1)} + \sum_{v \neq 1} I^2_{c\delta(v)} \cdot k_{r\delta(v)} \right)}{R_{-} \left(I^2_{c\delta(1)} + \sum_{v \neq 1} I^2_{c\delta(v)} \right)} = \frac{k_{r\delta(1)} + \sum_{v \neq 1} k_{r\delta(v)} \left(\frac{I_{c\delta(v)}}{I_{c\delta(1)}} \right)^2}{1 + \sum_{v \neq 1} \left(\frac{I_{c\delta(v)}}{I_{c\delta(1)}} \right)^2} \tag{47}$$

The global equivalent change of a.c. δ bar inductance modification has the expression:

$$k_{x\delta(CSF)} = \frac{q_{\delta(CSF)\sim}}{q_{\delta(CSF)-}}, \tag{48}$$

where $q_{\delta(CSF)\sim}$ is the a.c. total reactive power, in the δ bar, and $q_{\delta(CSF)-}$ is the total reactive power for a uniform current distribution δ in the bar. Applying the superposition in the case of a.c. total reactive power, the following relationship is obtained:

$$q_{\delta(CSF)\sim} = q_{\delta(1)\sim} + \sum_{v \neq 1} q_{\delta(v)\sim}, \tag{49}$$

A.c. reactive power corresponding to the fundamental is calculated using the following relation:

$$q_{\delta(1)\sim} = \omega_1 \cdot k_{x\delta(1)} \cdot L_{\delta n\sigma -} \cdot I^2_{c\delta(1)} \tag{50}$$

In the same way, the expression of the a.c. reactive power in the δ bar corresponding to the ν order harmonic is obtained:

$$q_{\delta(v)\sim} = \omega_{1(v)} L_{\delta n\sigma(v)\sim} \cdot I^2_{c\delta(v)} = v \cdot \omega_1 \cdot k_{x\delta(v)} \cdot L_{\delta n\sigma -} \cdot I^2_{c\delta(v)} \tag{51}$$

By replacing the relations (50) and (51) in the relation (28), the expression for calculating the total a.c. reactive power in the δ bar is obtained:

$$q_{\delta(CSF)\sim} = \omega_1 \cdot k_{x\delta(1)} \cdot L_{\delta n\sigma\sim} \cdot I^2_{c\delta(1)} + \left(k_{x\delta(1)} \cdot I^2_{c\delta(1)} + \sum_{\nu \neq 1} \nu \cdot k_{x\delta(\nu)} \cdot I^2_{c\delta(\nu)} \right) = \\ = \omega_1 \cdot L_{\delta n\sigma\sim} \left(k_{x\delta(1)} \cdot I^2_{c\delta(1)} + \sum_{\nu \neq 1} \nu \cdot k_{x\delta(\nu)} \cdot I^2_{c\delta(\nu)} \right) \tag{52}$$

The total reactive power for an uniform current repartition in the δ bar, in the case of a motor supplied through a frequency converter, is calculated by the relation:

$$q_{\delta(CSF)\sim} = q_{\delta(1)\sim} + \sum_{\nu \neq 1} q_{\delta(\nu)\sim}\ , \tag{53}$$

where $q_{\delta(1)\sim}$ is the reactive power corresponding to the fundamental, in case of an uniform current distribution $I_{c\delta(1)}$ in the δ bar, while $q_{\delta(\nu)\sim}$ is the reactive power corresponding to the ν harmonic in case of a uniform current distribution $I_{c\delta(\nu)}$ in the δ bar:

$$q_{\delta(1)} = \omega_{1(1)} L_{\delta n\sigma\sim} \cdot I^2_{c\delta(1)} = \omega_1 \cdot L_{\delta n\sigma\sim} \cdot I^2_{c\delta(1)}\ . \tag{54}$$

Similarly, for the reactive power corresponding to the ν harmonic, in the case of an uniform current $I_{c\delta(\nu)}$ repartition in the δ bar, the following relation is obtained:

$$q_{\delta(\nu)\sim} = \omega_{1(\nu)} \cdot L_{\delta n\sigma\sim} \cdot I^2_{c\delta(\nu)} = \nu \cdot \omega_1 \cdot L_{\delta n\sigma\sim} \cdot I^2_{c\delta(\nu)} \tag{55}$$

By replacing the relations (54) and (55) in relation (53), the expression for the total reactive power for a uniform current distribution in the δ bar becomes:

$$q_{\delta(CSF)\sim} = \omega_1 \cdot L_{\delta n\sigma\sim} \cdot I^2_{c\delta(1)} + \sum_{\nu \neq 1} \nu \cdot \omega_1 \cdot L_{\delta n\sigma\sim} \cdot I^2_{c\delta(\nu)} = \omega_1 \cdot L_{\delta n\sigma\sim} \left(I^2_{c\delta(1)} + \sum_{\nu \neq 1} \nu \cdot I^2_{c\delta(\nu)} \right) \tag{56}$$

By replacing the relations (52) and (56) in relation (48), the expression for the global equivalent factor of the a.c. modifying inductance is obtained:

$$k_{x\delta(CSF)} = \frac{q_{\delta(CSF)\sim}}{q_{\delta(CSF)\sim}} = \frac{\omega_1 L_{\delta n\sigma\sim} \left(k_{x\delta(1)} \cdot I^2_{c\delta(1)} + \sum_{\nu \neq 1} \nu \cdot k_{x\delta(\nu)} \cdot I^2_{c\delta(\nu)} \right)}{\omega_1 L_{\delta n\sigma\sim} \left(I^2_{c\delta(1)} + \sum_{\nu \neq 1} \nu \cdot I^2_{c\delta(\nu)} \right)} = \frac{k_{x\delta(1)} + \sum_{\nu \neq 1} \left[\nu \cdot \left(\frac{I_{c\delta(\nu)}}{I_{c\delta(1)}} \right)^2 \cdot k_{x\delta(\nu)} \right]}{1 + \sum_{\nu \neq 1} \left[\nu \cdot \left(\frac{I_{c\delta(\nu)}}{I_{c\delta(1)}} \right)^2 \right]} \tag{57}$$

5. Determining the equivalent parameters of the winding rotor, considering the skin effect

The rotor winding's parameters are affected by the skin effect, at the start of the motor and also at the nominal operating regime. For establishing the relations that define these parameters, considering the skin effect, the expression of the rotor phase impedance reduced to the stator is used. For this, the rotor with multiple bars is replaced by a rotor with a single bar on the pole pitch. Initially only the fundamental present in the power supply of the motor is considered. The rotor impedance reduced to the stator has the equation:

$$Z'_{2(1)} = \frac{R'_{2(1)}}{s_{(1)}} + jX'_{2(1)} \tag{58}$$

Knowing that the induced EMF by the fundamental component of the main magnetic field from the machine in the pole pitch bars is:

$$\underline{U}_{e(1)} = \underline{I}'_{2(1)} \cdot \underline{Z}'_{2(1)} \; , \tag{59}$$

where, for the general case of multiple cages is valid the relation:

$$\underline{I}'_{2(1)} = \sum_{\delta=1}^{c} \underline{I}'_{c\delta(1)} = \frac{\underline{U}_{e(1)}}{\underline{\Delta}_{(1)}} \sum_{\delta=1}^{c} \underline{\Delta}_{\delta(1)} \tag{60}$$

In the relation (60), the number of the cages and respectively the rotor bars/ pole pitch is equal to "c". In the case of motors with the power up to 45 [kW], c=1 (simple cage or high bars) or c=2 (double cage). $\underline{\Delta}_{(1)}$ is the determinant corresponding to the equation system:

$$\underline{U}_{e(1)} = \sum_{\varepsilon=1}^{c} \underline{R}_{\delta\varepsilon(1)} \cdot \underline{I}_{c\varepsilon(1)} \; , \; \varepsilon=1, 2, \ldots, c \; , \tag{61}$$

having the expression:

$$\underline{\Delta}_{(1)} = \begin{vmatrix} \underline{R}_{11(1)} & \cdots & \underline{R}_{1n(1)} \\ \cdot & & \cdot \\ \cdot & & \cdot \\ \underline{R}_{n1(1)} & \cdots & \underline{R}_{nn(1)} \end{vmatrix} \tag{62}$$

$\underline{\Delta}_{\delta(1)}$ is the determinant corresponding to the fundamental obtained from $\underline{\Delta}_{(1)}$, where column δ is replaced by a column of 1:

$$\underline{\Delta}_{\delta(1)} = \begin{vmatrix} \underline{R}_{11(1)} & \cdots & \underline{R}_{1,\,\delta-1(1)} & 1 & \underline{R}_{1,\,\delta+1(1)} & \cdots & \underline{R}_{1n(1)} \\ & \cdot & & & & \cdot & \\ & \cdot & & & & \cdot & \\ \underline{R}_{n1(1)} & \cdots & \underline{R}_{n,\,\delta-1(1)} & 1 & \underline{R}_{n,\,\delta+1(1)} & \cdots & \underline{R}_{nn(1)} \end{vmatrix} \tag{63}$$

Because in the first phase the steady-state regime is under focus, the phenomenon in the rotor corresponding to the fundamental has the pulsation $\omega_{2(1)}=s\omega_1$, where s is the motor slip for the sinusoidal power supply in the steady-state regime. If the relation (63) is introduced in (60), the expression of the equivalent impedance of the rotor phase reduced to the stator, corresponding to the fundamental valid when considering the skin effect is obtained:

$$\underline{Z}'_{2(1)} = \frac{\underline{\Delta}_{(1)}}{\sum_{\delta=1}^{c} \underline{\Delta}_{\delta(1)}} \tag{64}$$

Thus, the expressions for the rotor phase resistance and inductance reduced to the stator, corresponding to the fundamental, both affected by the skin effect can be written.

$$\frac{R'_{2(1)}}{s_{(1)}} = \Re e\left[\underline{Z}'_{2(1)}\right], \tag{65}$$

$$X'_{2(1)} = \Im m\left[\underline{Z}'_{2(1)}\right] \tag{66}$$

By considering in the motor power supply the ν harmonic only, similar expressions are obtained for the corresponding rotor parameters. Thus:

$$\underline{Z}'_{2(\nu)} = \frac{\underline{\Delta}_{(\nu)}}{\sum_{\delta=1}^{c} \underline{\Delta}_{\delta(\nu)}}, \tag{67}$$

$$\frac{R'_{2(\nu)}}{s_{(\nu)}} = \Re e\left[\underline{Z}'_{2(\nu)}\right], \tag{68}$$

$$X'_{2(\nu)} = \Im m\left[\underline{Z}'_{2(\nu)}\right] \tag{69}$$

Further on we consider the real case of an electric induction machine fed by a frequency converter. For the beginning, the case of simple cage respectively high bars induction motors will be analyzed. Thus, a rotor phase resistance corresponding to the fundamental, $R'_{2(1)}$, and rotor phase resistance corresponding to higher order harmonics $R'_{2(\nu)}$ are replaced by an equivalent resistance $R'_{2(CSF)}$, which dissipates the same part of active power as in the case of "ν" resistances. This equivalent resistance is defined at the fundamental's frequency and it is traversed by the $I'_{2(CSF)}$ current:

$$I'_{2(CSF)} = \sqrt{I'^{2}_{2(1)} + \sum_{\nu \neq 1} I'^{2}_{2(\nu)}} \tag{70}$$

For the rotor phase equivalent resistance reduced to the stator, corresponding to all harmonics, defined at the fundamental's frequency, one can write:

$$R'_{2(CSF)} = k_{r(CSF)} \cdot R'_{2c} + R'_{2i}, \tag{71}$$

where: R'_{2c} is the resistance, considered at the fundamental's frequency of a part from the rotor phase winding from notches and reported to the stator, R'_{2i} is the resistance of a part of the rotoric winding, neglecting skin effect reported to the stator, $k_{r(CSF)}$ is the global modification factor of the rotor winding resistance, having the expression given by the relation (47). To track the changes that appear on the resistance of the rotor winding when the machine is supplied through a frequency converter, comparing to the case when the machine is fed in the sinusoidal regime, the $k_{R'2}$ factor is introduced:

$$k_{R_2'} = \frac{R'_{2(CSF)}}{R'_2}, \tag{72}$$

where R'_2 is the rotor winding resistance reported to the stator, when the machine is fed in the sinusoidal regime:

$$R'_2 = k_r R'_{2c} + R'_{2i}, \tag{73}$$

where k_r is the modification factor of the a.c. rotor resistance, in the case of sinusoidal: $k_r \cong k_{r(1)}$. It is obtained:

$$k_{R_2'} = \frac{k_{r(CSF)} R'_{2c} + R'_{2i}}{k_r R'_{2c} + R'_{2i}} \tag{74}$$

If both the nominator and the denominator of the second member on the relation (74) are divided by k_r and then by R'_{2c}, the following expression is obtained:

$$k_{R_2'} = \frac{\frac{k_{r(CSF)}}{k_r} + \frac{R'_{2i}}{R'_{2c}} \cdot \frac{1}{k_r}}{1 + \frac{R'_{2i}}{R'_{2c}} \cdot \frac{1}{k_r}} = \frac{k_{kr} + r_2 \cdot \frac{1}{k_r}}{1 + r_2 \cdot \frac{1}{k_r}}, \tag{75}$$

where:

$$r_2 = \frac{R'_{2i}}{R'_{2c}} \cong \text{const.},$$

which is constant for the same motor, at a given fundamental's frequency. For c=1, $k_{kr}>1$, it results that $k_{R'2} > 1$, which means that $R'_{2(CSF)} > R'_2$ also. The procedure is similar for the reactance. The rotor phase reactance, corresponding to the fundamental, $X'_{2(1)}$, and also the reactance corresponding to the higher harmonics, $X'_{2(\nu)}$, are replaced by an equivalent reactance $X'_{2(CSF)}$. As in the case of the rotor resistance, we can write:

$$k_{X_2'} = \frac{X'_{2(CSF)}}{X'_2}, \tag{76}$$

where $X'_{2(CSF)}$ is the equivalent reactance of the rotor phase, reduced to the stator, corresponding to all harmonics, including the fundamental, on the fundamental's frequency:

$$X'_{2(CSF)} = k_{X(CSF)} X'_{2c} + X'_{2i} , \tag{77}$$

and X'_2 is the reactance of the rotor phase reduced to the stator which characterizes the machine when it is fed in the sinusoidal regime:

$$X'_2 = k_X X'_{2c} + X'_{2i} \tag{78}$$

In relation (77) and (78), we noted: X'_{2c} -the reactance of the rotor winding part from the notches, reduced to the stator, in which the skin effect is present, X'_{2i}- the reactance of the rotor winding phase where the skin effect can be neglected. $k_{X(CSF)}$ is defined in relation (57), where $c \cong 1$. Taking into account the relations (77) and (78), the relation (76) becomes:

$$k_{X'_2} = \frac{k_{X(CSF)} X'_{2c} + X'_{2i}}{k_X X'_{2c} + X'_{2i}} = \frac{\frac{k_{X(CSF)}}{k_X} + \frac{X'_{2i}}{X'_{2c}} \cdot \frac{1}{k_X}}{1 + \frac{X'_{2i}}{X'_{2c}} \cdot \frac{1}{k_X}} = \frac{k_{k_X} + x_2 \frac{1}{k_X}}{1 + x_2 \frac{1}{k_X}} , \tag{79}$$

where:

$$x_2 = \frac{X'_{2i}}{X'_{2c}} ,$$

is a constant for the same motor at a given fundamental's frequency $k_{kX}<1$, with the consequences $k_{X'2}<1$ and $X'_{2(CSF)}<X'_2$. With this, the impedance of a rotor phase reported to the stator in the case of a machine supplied by a power converter, receives the form:

$$\underline{Z}'_{2(CSF)} = \frac{R'_{2(CSF)}}{s_{(CSF)}} + jX'_{2(CSF)} , \tag{80}$$

where:

$$s_{(CSF)} = \frac{R'_{2(CSF)} I'_{2(CSF)}}{U_{e1(CSF)}} \tag{81}$$

and:

$$U_{e1(CSF)} = \sqrt{U^2_{e1(1)} + \sum_{v \neq 1} U^2_{e1(v)}} \tag{82}$$

In the case of double cage induction motors, the rotor parameters are necessary to be determined for both cages. The principle of calculation keeps its validity from the above presented case, the induction motors with simple cage, respectively cage with high bars, with one remark: in the relations for determining $k_{r(CSF)}$ respectively $k_{x(CSF)}$,it is considered that c=2 (for δ=1 the working work cage results and for δ=c=2 the startup cage results). The complex structure of the used algorithm and its component computing relations synthetically presented in the paper, request a very high volume of calculation. Therefore the presence of a computer in solving this problem is absolutely necessary. In the Laboratory of Systems dedicated to control the electrical servomotors from the Polytechnic University of Timişoara the software calculation CALCMOT has been designed. It allows the determination and the analysis of the factors $k_{r(CSF)}$, $k_{x(CSF)}$ and the parameters of the equivalent winding machine induction in the non-sinusoidal regime. Further on, the expressions of the equivalent parameters for the magnetic circuit will be set (corresponding to all harmonics). Thus, to determine the equivalent resistance of magnetization $R_{1m(CSF)}$, we have to take into account that this is determined only by the ferromagnetic stator core losses which are covered directly by the stator power without making the transition through the stereo-mechanical power. By approximating that $I_{01(CSF)} \approx I_{\mu(CSF)}$, for $R_{1m(CSF)}$ it is obtained:

$$R_{1m(CSF)} = \frac{p_{z1(CSF)} + p_{j1(CSF)}}{3I^2_{\mu(CSF)}}, \tag{83}$$

where $p_{z1(CSF)}$ and $p_{j1(CSF)}$ are global losses occurring respectively in the stator teeth and in the yoke due to the supplying of the motor through the frequency converter. In determining the total magnetization current $I_{\mu(CSF)}$, the principle of the superposition effects is applied:

$$I_{\mu(CSF)} = \sqrt{I^2_{\mu(1)} + \sum_{v \neq 1} I^2_{\mu(v)}} \tag{84}$$

For the equivalent magnetizing reactance, corresponding to all harmonics, determined at the fundamental's magnetization frequency $f_{1(1)}$, we obtain:

$$X_{1m(CSF)} \cong \sqrt{\left(\frac{U_{1(CSF)}}{I_{\mu(CSF)}}\right)^2 - \left(R_{1(CSF)} + R_{1m(CSF)}\right)^2} \tag{85}$$

For the equivalent impedance of the magnetization circuit it can be written:

$$\underline{Z}_{1m(CSF)} = R_{1m(CSF)} + j \cdot X_{1m(CSF)} \tag{86}$$

Given these assumptions and considering that the equivalent parameters were calculated reduced to the fundamental's frequency (in the conditions of a sinusoidal regime), one may formally accept the calculation in complex quantities. Corresponding to the unique scheme shown in Fig. 2, the motor equations are:

$$\underline{U}_{1(CSF)} = \underline{Z}_{1(CSF)} \cdot \underline{I}_{1(CSF)} - \underline{U}_{e1(CSF)};$$

$$\underline{U}'_{e2(CSF)} = \underline{Z}'_{2(CSF)} \cdot \underline{I}'_{2(CSF)} = \underline{U}_{e1(CSF)};$$

$$\underline{U}_{e1(CSF)} = -\underline{Z}_{m(CSF)} \cdot \underline{I}_{01(CSF)}; \tag{87}$$

$$\underline{I}_{01(CSF)} = \underline{I}_{1(CSF)} + \underline{I}'_{2(CSF)}$$

6. Experimental validation

The induction machines which have been tested are: MAS 0,37 [kW] x 1500 [rpm] and MAS 1,1 [kW] x 1500 [rpm]. To validate the experimental studies of the theoretical work, tests were made both for the operation of motors supplied by a system of sinusoidal voltages, and for the operation in case of static frequency converter supply. In Tables 1 and 2 are presented theoretical values (obtained by running the calculation program) and the results of measurements, for $k_{R'2}$ and $k_{X'2}$, factors, respectively the calculation errors of, for both motors tested.

Nr.	$f_{1(1)}$ [Hz]	$k_{R'2} = \frac{R'_{2(CSF)}}{R'_2}$ (calculated)	$k_{R'2}$ (measured)	$\varepsilon_{kR'2}$ [%]	$k_{X'2} = \frac{X'_{2(CSF)}}{X'_2}$ (calculated)	$k_{X'2}$ (measured)	$\varepsilon_{kX'2}$ [%]
1.	25	1,048	1,11	5,58	0,863	0,894	3,6
2.	30	1,026	1,077	4,97	0,912	0,857	-6,03
3.	40	1,021	1,061	3,77	0,944	0,884	-6,35
4.	50	1,014	1,075	6,01	0,967	0,897	-7,23
5.	60	1,011	1,079	6,82	0,975	0,914	-6,25

Table 1. The theoretical and experimental values of factors $k_{R'2}$ and $k_{X'2}$, respectively the errors of calculation, corresponding to 0.37 [kW] x 1500 [rpm] MAS.

Nr.	$f_{1(1)}$ [Hz]	$k_{R'2} = \frac{R'_{2(CSF)}}{R'_2}$ (calculated)	$k_{R'2}$ (measured)	$\varepsilon_{kR'2}$ [%]	$k_{X'2} = \frac{X'_{2(CSF)}}{X'_2}$ (calculated)	$k_{X'2}$ (measured)	$\varepsilon_{kX'2}$ [%]
1.	20	1,098	1,185	7,92	0,812	0,821	1,108
2.	30	1,041	1,120	7,58	0,886	0,916	3,386
3.	40	1,034	1,106	6,96	0,926	0,891	-3,77
4.	50	1,023	1,089	6,45	0,956	0,863	-9,72
5.	60	1,018	1,082	6,28	0,966	0,871	-9,83

Table 2. The theoretical and experimental values of factors $k_{R'2}$ şi $k_{X'2}$, respectively the errors of calculation, corresponding to 1.1 [kW] x 1500 [rpm] MAS.

Parameters of the winding machine supplied by the power converter can be calculated with errors less than 10 [%]. The main cause of errors is the assumption of saturation neglect. Even in this case the results can be considered satisfactory, which leads to validate the theoretical study carried out in the paper.

7. Theoretical analysis of the magnetic losses

7.1. Statoric iron losses

7.1.1. The main stator iron losses

A. The main stator teeth losses

In the teeth, the magnetic field is alternant and generates this type of losses. In the case of the direct supplying system the total losses from the stator teeth p_{zl} are being composed by the magnetic hysteresis losses, p_{zlh} and the eddy currents losses, p_{zlw}:

$$p_{z1} = \left(k_{zh} \cdot \sigma_h \cdot f_1 + k_{zw} \cdot \sigma_w \cdot f_1^2 \cdot \Delta^2\right) \cdot B_{z1m}^2 \cdot G_{z1}, \tag{88}$$

where: σ_h is a material constant depending on the thickness and the quality of the steel sheet, f_1 is the supplying frequency, B_{zlm} represents the magnetic induction in the middle of the stator tooth, G_{zl} represent the weight of the stator teeth, σ_w is a material constant similar to σ_h, depending on the sheet thickness and quality and Δ represents the thickness of the sheet. k_{zh} and k_{zw} are two factors which have the mission of underlining respectively the hysteresis losses increment and the eddy currents losses increment due to the mechanical modifications of the stator's sheets. In the case of converters-mode supplying system, at the total losses from the stators teeth caused by the fundamental the losses induced by the higher time harmonics must be taken into account. For an exact analytic expression in the following it is proposed an analysis method of the iron losses based upon the equalization of the hysteresis losses with the eddy currents ones. For the start, only the fundamental is considered present in the supplying system. Distinct from the sine-mode supplying system, when in most cases the supplying frequency is $f_1=f_{1n}=50$ [Hz], is the fact that in the case of the inverter based supplying system the fundamental frequency can take values higher than 50 [Hz]. At very high magnetization frequencies the influence of the skin effect must be taken in consideration. In the following, the minimum value of the magnetization frequency is being determined and for that the skin effect must be considered. The computing relation for the magnetization frequency f_1 is the following:

$$f_1 = \left(\frac{\xi}{\Delta}\right)^2 \cdot \frac{\rho}{\mu\pi}, \tag{89}$$

where ξ is the refulation factor.

The minimum magnetization frequency f_{min}, computed with the relation (89), from which the skin effect must be considered is 140[Hz]. Consequently, in the fundamental - wave supplying mode, at which usually we have $f1 \le 120$ [Hz], the principal losses from the stators teeth, can be written as following:

$$p_{z1(1)} = \left(k_{zh} \cdot \sigma_h \cdot f_1 + k_{zw} \cdot \sigma_w \cdot f_1^2 \cdot \Delta^2\right)^2 \cdot B_{z1m(1)}^2 \cdot G_{z1} \ , \tag{90}$$

where $B_{z1m(1)}$ represents the magnetic induction from the middle of the tooth, $B_{z1m(1)} = B_{z1m}$. In order to be able to apply the principle of over position effects, the machine is being considered as being ideal; therefore we neglect the hysteresis phenomenon. For this, we proposed the equalization of the hysteresis losses with the eddy current losses, an assumption that allows the linearization of the machines' equations. Through this equalization, the real machine – that is practically non-linear and in which the principal losses are made of a sum of two components: the one of eddy currents losses and the one of hysteresis losses - is being replaced with a theoretical linear machine, characterized only by its eddy currents losses. Energetically speaking, the two machines must be equivalent. As a following, if we take $p^*_{z1w(1)}$ as the eddy currents losses corresponding to the fundamental, which appear in the theoretical model of the machine adopted, than these losses must be equal to the main losses from the stator teeth characteristic to the real machine, losses given through the relation:

$$p^*_{z1w(1)} = p_{z1(1)} \tag{91}$$

We consider these equivalent losses, $p^*_{z1w(1)}$, equal to the real losses through the eddy currents corresponding to the fundamental, $p_{z1w(1)}$, multiplied with a $k_{z1e(1)}$ factor. This is an equalization factor of the real losses from the stators teeth with losses resulted only from "$p_{z1w(1)}$" – fundamental-mode supplying state:

$$p^*_{z1w(1)} = k_{z1e(1)} \cdot p_{z1w(1)} \tag{92}$$

We consider that through this equalization factor a covering value of the principal stator teeth losses is obtained. The relation (91) made explicit becomes:

$$\left(k_{zh} \cdot \sigma_h \cdot f_1 + k_{zw} \cdot \sigma_w \cdot f_1^2 \cdot \Delta^2\right) \cdot B_{z1m(1)}^2 \cdot G_{z1} = k_{z1e(1)} \cdot k_{zw} \cdot \sigma_w \cdot f_1^2 \cdot \Delta^2 \cdot B_{z1m(1)}^2 \cdot G_{z1} \ . \tag{93}$$

Because of the fact that the usually used sheets have the thickness $\Delta = 0.5$ [mm]=const, one can consider that:

$$k_{z1e(1)} = 1 + \frac{K_{z\Delta}}{f_1} \tag{94}$$

where we have

$$K_{z\Delta} = K_z / \Delta^2 \text{ with } K_z = \frac{\sigma_h \cdot k_{zh}}{\sigma_w \cdot k_{zw}}$$

In the following part we consider that only the ν order harmonic is present in the supplying wave, characterized by the magnetization frequency $f_{1(\nu)} = \nu \cdot f_1$. Therefore, the principal losses in the stator teeth occurring in the real machine corresponding to the ν order time harmonic must be corrected through the two factors $k_{h(\nu)}$ and $k_{w(\nu)}$, which are a function of the reaction of the eddy currents:

$$p_{z1(\nu)} = \left(k_{zh} \cdot k_{h(\nu)} \cdot \sigma_h \cdot \nu \cdot f_1 + k_{zw} \cdot k_{w(\nu)} \cdot \sigma_w \cdot \nu^2 \cdot f_1^2 \cdot \Delta^2\right) \cdot B_{z1m(\nu)}^2 \cdot G_{z1} \tag{95}$$

In the relation (95), $B_{z1m(\nu)}$ represents the magnetic induction according to the ν order time harmonic from the middle of the tooth. The factors $k_{h(\nu)}$ and $k_{w(\nu)}$ have the expressions:

$$k_{h(\nu)} = \frac{\xi_{(\nu)}}{2} \cdot \frac{sh\xi_{(\nu)} + \sin\xi_{(\nu)}}{ch\xi_{(\nu)} - \cos\xi_{(\nu)}}; \; k_{w(\nu)} = \frac{3}{\xi_{(\nu)}} \cdot \frac{sh\xi_{(\nu)} - \sin\xi_{(\nu)}}{ch\xi_{(\nu)} - \cos\xi_{(\nu)}}; \tag{96}$$

As in the case of the fundamental-wave supplying case, the real machine is replaced by a theoretical linear machine which has only losses given by the eddy currents. Reasoning as in the case of the fundamental, we obtain:

$$k_{z1e(\nu)} = 1 + \frac{K_z}{\Delta^2} \cdot \frac{1}{\nu \cdot f_1} \cdot \frac{k_{h(\nu)}}{k_{w(\nu)}} == 1 + \frac{K_{z\Delta}}{\nu \cdot f_1} \cdot \frac{k_{h(\nu)}}{k_{w(\nu)}}, \tag{97}$$

$$p_{z1(\nu)} = p_{z1w(\nu)}^* = k_{z1e(\nu)} \cdot p_{z1w(\nu)} = k_{z1e(\nu)} \cdot k_{zw} \cdot k_{w(\nu)} \cdot \sigma_w \cdot \nu^2 \cdot f_1^2 \cdot \Delta^2 \cdot B_{z1m(\nu)}^2 \cdot G_{z1} \tag{98}$$

where $p^*_{z1w(\nu)}$ are the equivalent losses corresponding to the ν harmonic. If we have $p_{z1(CSF)}$ for the losses from the stators teeth with the machine supplied by inverters, by applying the principle of over position effects for the theoretical linear model of the machine, it will be written:

$$p_{z1(CSF)} = k_{zw} \cdot \sigma_w \cdot f_1^2 \cdot \Delta^2 \cdot B_{z1m(1)}^2 \cdot G_{z1} \left[k_{z1e(1)} + \sum_{\nu \neq 1} k_{z1e(\nu)} \cdot k_{w(\nu)} \cdot \nu^2 \left(\frac{B_{z1m(\nu)}}{B_{z1m(1)}} \right)^2 \right] \tag{99}$$

In order to analyze the modifications suffered by the main losses in the stators teeth while the motor is supplied by an inverter versus the sine-mode supplying system, we analyze the ratio between the relations (99) and (88). After making the intermediary computations in which the relations (93), (94) and (99) are taken into account we obtain:

$$k_{pz1} = \frac{p_{z1(CSF)}}{p_{z1}} = 1 + \sum_{\nu \neq 1} \left(\frac{k_{z1e(\nu)}}{k_{z1e(1)}} \cdot k_{w(\nu)} \cdot \nu^2 \cdot k_{Bz1(\nu,1)}^2 \right), \tag{100}$$

where $k_{Bz(\nu,1)} = B_{z1m(\nu)} / B_{z1m(1)}$.

B. The principal losses in the stator yoke

In the case of the direct – mode supplying system of the machine, the principal yoke losses consist of the hysteresis losses, p_{j1h} and eddy currents losses, p_{j1w}:

$$p_{j1} = \left(\sigma_h \cdot f_1 \cdot k_{j1h} + \sigma_w \cdot \Delta^2 \cdot f_1^2 \cdot k_{j1w}\right) \cdot B_{j1}^2 \cdot G_{j1} \quad (101)$$

where: B_{jl} is the magnetic induction in the stator yoke, G_{jl} represents the weight of the stator yoke, $k_{j1w} = k_{j1w1} \cdot k_{j1w2}$, where k_{j1w1} is a coefficient that corresponds to the non uniform repartition of the magnetic induction in the yoke and k_{j1w2} is a coefficient that corresponds to the currents closing perpendicular to the sheets, through the places with imperfections in the sheets isolation layer and also in the wholes made in the cutting process. In the case on an inverter supplying system at the total losses from the stator yoke caused by the fundamental, the superior time harmonics losses must be added. In order to apply the principle of over-position effect the method is similar to the one used in the case of the principal losses in the teeth. We equalize energetically the real machine with the linear theoretical one where we consider only the eddy currents losses. As a following, for the fundamental supplying mode, the principal losses in the stator yoke for a real machine, $p_{j1(1)}$ are:

$$p_{j1(1)} = \left(\sigma_h \cdot f_1 \cdot k_{j1h} + \sigma_w \cdot \Delta^2 \cdot f_1^2 \cdot k_{j1w}\right) \cdot B_{j1(1)}^2 \cdot G_{j1} \quad (102)$$

If we have $p^*_{j1w(1)}$ as losses in eddy currents, than these must be equalized with the principal losses from the stator yoke described with the relation (102):

$$p^*_{j1w(1)} = p_{j1(1)} \quad (103)$$

These equivalent losses, $p^*_{j1w(1)}$ are considered equal to the real eddy currents losses $p_{j1w(1)}$, multiplied with an equalizing factor of the real yoke losses with "$p_{j1w(1)}$" type losses, $k_{j1e(1)}$:

$$p^*_{j1w(1)} = k_{j1e(1)} \cdot p_{j1w(1)} \quad (104)$$

Similarly to point A, as a following of the equalization we obtain the relation:

$$k_{j1e(1)} = 1 + \frac{K_w}{\Delta^2 \cdot f_1} = 1 + \frac{K_{w\Delta}}{f_1}, \quad (105)$$

where we have:

$$K_w = \frac{\sigma_h \cdot k_{j1h}}{\sigma_w \cdot k_{j1w}} \text{ and } K_{w\Delta} = \frac{K_w}{\Delta^2}$$

As a following we consider present in the supplying system of the machine only the ν order superior time harmonic. Because of the fact that the magnetization frequency $f_{1(\nu)}$ is the fundamental one multiplied with ν, the principal losses from the stator yoke which appear in the fundamental must be adjusted with the two coefficients: $k_{h(\nu)}$ and $k_{w(\nu)}$. These factors take into account respectively the skin effect and the eddy currents reaction.

$$p_{j1(\nu)} = \left(k_{h(\nu)} \cdot \sigma_h \cdot \nu \cdot f_1 \cdot k_{j1h} + k_{w(\nu)} \cdot \sigma_w \cdot \Delta^2 \cdot \nu^2 \cdot f_1^2 \cdot k_{j1w}\right) \cdot B_{j1(\nu)}^2 \cdot G_{j1} \quad (106)$$

In the relation (106), $B_{j1(v)}$ represents the magnetic induction accordingly to the ν order harmonic. Through the energetically equalization realized from the replacement of the real machine with the linear model, we obtain the equalizing factor of the stator yoke losses, with the "$p_{j1w(v)}$" type losses:

$$k_{j1e(v)} = 1 + \frac{K_w}{\Delta^2} \cdot \frac{1}{v \cdot f_1} \cdot \frac{k_{h(v)}}{k_{w(v)}} = 1 + \frac{K_{w\Delta}}{v \cdot f_1} \cdot \frac{k_{h(v)}}{k_{w(v)}} \tag{107}$$

In conclusion, the principal losses in the stator yoke, corresponding to the ν order time harmonic can be written by equalizing as:

$$p_{j1(v)} = p^*_{j1w(v)} = k_{j1e(v)} \cdot p_{j1w(v)}, \tag{108}$$

where:

$$p_{j1w(v)} = k_{w(v)} \cdot \sigma_w \cdot \Delta^2 \cdot v^2 \cdot f_1^2 \cdot k_{j1w} \cdot B^2_{j1(v)} \cdot G_{j1} \tag{109}$$

As a following we have considered the situation of the machine supplied by the fundamental and the superior time harmonics as well. Taking $p_{j1(CSF)}$ as the global losses occurring in the stator yoke due to the converter supplying mode, by applying the over position effect principle on the theoretical linear model we can write:

$$p_{j1(CSF)} = \sigma_w \cdot f_1^2 \cdot \Delta^2 \cdot k_{j1w} \cdot B^2_{j1(1)} \cdot G_{j1} \left[k_{j1e(1)} + \sum_{v \neq 1} k_{j1e(v)} \cdot k_{w(v)} \cdot v^2 \left(\frac{B_{j1(v)}}{B_{j1(1)}} \right)^2 \right] \tag{110}$$

In order to analyze the changes that the principal losses from the stator yoke suffer when the machine is being supplied through an inverter versus the sine-mode supplying case, we divide the relation (110) at (101). After finishing the computations we have:

$$k_{pj1} = \frac{p_{j1(CSF)}}{p_{j1}} = 1 + \sum_{v \neq 1} \left(\frac{k_{j1e(v)}}{k_{j1e(1)}} \cdot k_{w(v)} \cdot v^2 \cdot k^2_{Bj1(v,1)} \right), \tag{111}$$

where: $k_{Bj(v,1)} = B_{j1(v)} / B_{j1(1)}$.

7.1.2. The supplementary stator iron losses

A. Surface supplementary losses

In the case of a network supplying mode, the magnetic induction distribution curve over the polar step is not very different from a sine-curve. The surface stator losses are given by the expression:

$$P_{\sigma 1} = \frac{1}{2} \cdot l \cdot \pi \cdot D \cdot \frac{\tau_{c1} - b_{41}}{\tau_{c1}} \cdot k_o \cdot (N_{c2} \cdot n)^{1,5} \cdot (\tau_{c2} \cdot \beta_2 \cdot k_{\delta 2} \cdot B_\delta)^2 \tag{112}$$

In the relation (112) the significance of the sizes is the following: D is the inner diameter of the stator, τ_{c1} is the step of the stator slot and τ_{c2} is the step of the rotor slot, b_{41} is the opening of the stator slot, N_{c2} is the number of stator slots, n is the rotation speed, β_2 is a factor dependent on the ratio b_{42}/δ (b_{42} is the opening of the rotor slot), $k_{\delta 2}$ is an air gap factor, k_o is an adjustment factor which depends on the materials resistivity and its magnetic permeability. In the case of the inverter supplying method, due to the deforming state at the supplementary losses produced by the fundamental, the surface losses produced by the superior time harmonics must be considered. Because of the fact that the surface losses in the polar pieces are treated as the eddy current losses developed in the inductor sheets, we can apply the over position effect principle without any further parallelism. Therefore, the surface supplementary losses in the stator in the case of a machine supplied by inverters can be computed with the relation:

$$P_{\sigma 1(CSF)} = \frac{1}{2} \cdot 1 \cdot \pi \cdot D \cdot \frac{\tau_{c1} - b_{41}}{\tau_{c1}} \cdot k_o \cdot (N_{c2} \cdot n)^{1,5} \cdot (\tau_{c2} \cdot \beta_2 \cdot k_{\delta 2} \cdot B_{\delta(1)})^2 \left[1 + \sum_{v \neq 1} \left(\frac{B_{\delta(v)}}{B_{\delta(1)}} \right)^2 \right] \quad (113)$$

Dividing the supplementary losses in the stator surface when having an inverter supplying system for the machine, $P_{\sigma 1(CSF)}$, by the supplementary losses in the stator surface when we have the sine-mode supplying system for the machine, $P_{\sigma 1}$, and making the intermediary computations we obtain the increment factor of the supplementary stator surface losses in the inverter versus the sine-mode supplying case, $k_{P\sigma 1}$, as following:

$$k_{P\sigma 1} = \frac{P_{\sigma 1(CSF)}}{P_{\sigma 1}} = 1 + \sum_{v \neq 1} \left(\frac{B_{\delta(v)}}{B_{\delta(1)}} \right)^2 = 1 + \sum_{v \neq 1} k^2_{B\delta(v,1)} > 1 \ , \quad (114)$$

where $k_{B\delta(v,1)} = B_{\delta(v)} / B_{\delta(1)}$. By analyzing the relation (114) one can notice the fact that the $k_{P\sigma 1}$ factor tends to 1 because of the fact that the value is practically very low. Consequently, the surface supplementary losses increase due to the inverter supplying system to an extent that is not to be taken into consideration.

B. The pulsation supplementary losses

In the case of the sine-mode supplying system, the pulsation supplementary losses in the stator, provided that the magnetic field along the polar step is not much different from a sine-wave, has the following expression:

$$P_{P1} = \frac{1}{2} \cdot \sigma_w \cdot k_{wP1} \cdot (\Delta N_{c2} n)^2 \cdot \left(\frac{\gamma_2 \delta k_\delta}{2\tau_{c1}} \right)^2 \cdot G_{z1} \cdot B^2_{z1m} \ , \quad (115)$$

where k_{wP1} is an increment coefficient of the stator losses by eddy currents due to processing, k_δ is the total air gap factor and γ_2 is constant for the one and the same machine, depended on the opening of the stator slot and the air gap dimension. In the situation in which the machine is supplied by inverters, by applying the over position effect principle, the following expression for the supplementary pulsation losses in the stator $P_{P1(CSF)}$ is obtained:

$$P_{P1(CSF)} = \frac{1}{2} \cdot \sigma_w \cdot k_{wP1} \cdot \left(\Delta N_{c2} n\right)^2 \cdot \left(\frac{\gamma_2 \delta k_\delta}{2\tau_{c1}}\right)^2 \cdot G_{z1} \cdot B^2_{z1m(1)} \left[1 + \sum_{v \neq 1} \left(\frac{B_{z1m(v)}}{B_{z1m(1)}}\right)^2\right] \quad (116)$$

Dividing the pulsation stator losses in the case of the inverter supplying system $P_{P1(CSF)}$, by the pulsation stator losses in the case of sine-mode supplying system P_{P1}, we obtain the increment factor of the supplementary pulsation losses in the inverter versus sine-wave supplying system, k_{Pp1}:

$$k_{Pp1} = \frac{P_{P1(CSF)}}{P_{P1}} = 1 + \sum_{v \neq 1} \left(\frac{B_{z1m(v)}}{B_{z1m(1)}}\right)^2 = 1 + \sum_{v \neq 1} k^2_{Bz1(v,1)} > 1 \quad (117)$$

By analyzing the relation (117) we can state that in the case of an inverter supplied machine we have not obtained a significant increment of the pulsation losses in the stator due to the small value of the $k^2_{Bz1(v,1)}$.

7.2. Rotor iron losses

7.2.1. Principal losses in the rotor iron

A. The principal losses in the rotor's teeth

Firstly, only one superior time harmonic is considered present in the supplying system of the machine, of an average order v. The real losses that this harmonic produces in the rotor teeth have the expression:

$$p_{z2(v)} = \left(k_{zh} \cdot k_{h(v)} \cdot \sigma_h \cdot s_{(v)} \cdot v \cdot f_1 + k_{zw} \cdot k_{w(v)} \cdot \sigma_w \cdot s^2_{(v)} \cdot v^2 \cdot f_1^2 \cdot \Delta^2\right) B^2_{z2m(v)} \cdot G_{z2} \quad (118)$$

In the relation (118), $B_{z2m(v)}$ represents the magnetic induction corresponding to the v order harmonic from the middle of the rotor tooth. In the theoretical model adopted, these losses given by the relation (118) are produced only by eddy currents:

$$p_{z2(v)} = p^*_{z2w(v)} = k_{z2e(v)} \cdot p_{z2w(v)}\ , \quad (119)$$

where $k_{z2e(v)}$ is an equalizing factor of the real losses from the rotor teeth, only with the losses of "$p_{z2w(v)}$" type, corresponding to the v order time harmonic. Developing the relation (119) by using the relation (118), after finishing the intermediary computations we obtain:

$$k_{z2e(v)} = 1 + \frac{K_z}{\Delta^2} \cdot \frac{1}{s_{(v)} \cdot v \cdot f_1} \cdot \frac{k_{h(v)}}{k_{w(v)}} = 1 + \frac{K_{z\Delta}}{s_{(v)} \cdot v \cdot f_1} \cdot \frac{k_{h(v)}}{k_{w(v)}} \quad (120)$$

Therefore, the principal losses from the rotor teeth, corresponding to the ν order time harmonic can be written by equalization as it follows:

$$p_{z2(\nu)} = k_{z2e(\nu)} \cdot k_{zw} \cdot k_{w(\nu)} \cdot \sigma_w \cdot s_{(\nu)}^2 \cdot \nu^2 \cdot f_1^2 \cdot \Delta^2 \cdot B_{z2m(\nu)}^2 \cdot G_{z2} \tag{121}$$

In the conditions in which in the supplying system of the machine all the superior time harmonics are present, the principal losses in the rotor teeth can be written as:

$$p_{z2(CSF)} = \sum_{\nu \neq 1} p_{z2(\nu)} \tag{122}$$

B. The principal losses from the rotor's yoke

In the hypotheses in which in the supplying system only the ν order harmonic is present, the real principal losses induced by it in the rotor yoke have the expression:

$$p_{j2(\nu)} = \left(k_{h(\nu)} \cdot \sigma_h \cdot s_{(\nu)} \cdot \nu \cdot f_1 \cdot k_{j1h} + k_{w(\nu)} \cdot \sigma_w \cdot s_{(\nu)}^2 \cdot \nu^2 \cdot f_1^2 \cdot \Delta^2 \cdot k_{j2w}\right) \cdot B_{j2(\nu)}^2 \cdot G_{j2} \tag{123}$$

Through the energetic equalization, due to the replacement of the real machine by a theoretical linear model we can obtain the equality:

$$p_{j2(\nu)} = p_{j2w(\nu)}^{*} = k_{j2e(\nu)} \cdot p_{j2w(\nu)} \tag{124}$$

Reasoning as in the previous cases, we can determine the equalizing factor of the real losses in the rotor yoke, only with losses of the type "$p_{j2w(\nu)}$" type as it follows:

$$k_{j2e(\nu)} = 1 + \frac{K_W}{\Delta^2} \cdot \frac{1}{s_{(\nu)} \cdot \nu \cdot f_1} \cdot \frac{k_{h(\nu)}}{k_{w(\nu)}} = 1 + \frac{K_{W\Delta}}{s_{(\nu)} \cdot \nu \cdot f_1} \cdot \frac{k_{h(\nu)}}{k_{w(\nu)}} \tag{125}$$

Consequently, the principal rotor yoke losses corresponding to the ν order harmonic can be written by equalization in the form:

$$p_{j2(\nu)} = k_{j2e(\nu)} \cdot k_{j2w} \cdot k_{w(\nu)} \cdot \sigma_w \cdot s_{(\nu)}^2 \cdot \nu^2 \cdot f_1^2 \cdot \Delta^2 \cdot B_{j2(\nu)}^2 \cdot G_{j2} \tag{126}$$

Disregarding all these, in the case of the inverter supplying system the total principal losses in the rotor yoke, $p_{j2(CSF)}$, are computed with the relation:

$$p_{j2(CSF)} = \sum_{\nu \neq 1} p_{j2(\nu)} \tag{127}$$

7.2.2. The supplementary losses in the rotor iron

A. The surface supplementary losses

If the machine is directly supplied from the power supply, the surface supplementary rotor losses are calculated with the relation:

$$P_{\sigma 2} = \frac{1}{2} \cdot p_{\sigma 2} \cdot l \cdot \pi \cdot (\Delta - 2\delta) \cdot \frac{\tau_{c2} - b_{42}}{\tau_{c2}}, \tag{128}$$

where the specific rotor surface losses $p_{\sigma 2}$ have the expression:

$$p_{\sigma 2} = k_o (N_{c1} \cdot n)^{1,5} \cdot (\tau_{c1} \cdot \beta_1 \cdot k_{\delta 1} \cdot B_\delta)^2 \tag{129}$$

In the relations (128) and (129) we noted by b_{42} the opening of the rotor slot, N_{c1} the number of rotor slots, β_1 a factor dependent on the b_{41}/δ ratio and $k_{\delta 1}$ the air gap factor. Proceeding similarly we can obtain the expression of the increment factor of the supplementary losses in the rotor surface while the machine is being supplied by inverters versus the sine-mode supplying system, $k_{P\delta 2}$:

$$k_{P\sigma 2} = \frac{P_{\sigma 2(CSF)}}{P_{\sigma 2}} = 1 + \sum_{v \neq 1} \left(\frac{B_{\delta(v)}}{B_{\delta(1)}} \right)^2 = 1 + \sum_{v \neq 1} k^2_{B\delta(v,1)} = k_{P\sigma 1} > 1 \tag{130}$$

B. The supplementary pulsation losses

The supplementary pulsation rotor losses, in the sine-mode supplying system have the following expression:

$$P_{P2} = \frac{1}{2} \sigma_w \cdot k_{wP2} (\Delta \cdot N_{c1} \cdot n \cdot B_{P2})^2 \cdot G_{z2} \tag{131}$$

B_{P2} represents the pulsation induction in the rotor teeth. Consequently, taking into account the fact that:

$$\frac{B_{z2m(v)}}{B_{z2m(1)}} = \frac{B_{\delta(v)}}{B_{\delta(1)}} = k_{B\delta(v,1)}, \tag{132}$$

we obtain:

$$k_{Pp2} = \frac{P_{P2(CSF)}}{P_{P2}} = 1 + \sum_{v \neq 1} k^2_{B\delta(v,1)} > 1 \tag{133}$$

8. Conclusions

This paper aims to study the theoretical behavior of asynchronous three-phase motor in the case of supplying through a power frequency converter. This study has aimed to develop the theory of the asynchronous three-phase motor in non-sinusoidal periodic regime to serve as a starting point in optimizing the design methodology. Given that the asynchronous three-phase motor is fed through a static frequency converter, the machine operation in the presence of higher time harmonics in the supply voltage can be described by a single mathematical model. The model consists of a single equivalent scheme corresponding to all harmonics and it is defined at the fundamental frequency.

Author details

Muşuroi Sorin
Politehnica University of Timişoara, Romania

9. References

[1] Muşuroi, S.; Vătău, D.; Andea, P.; Şurianu, F.D.; Frigură, F. & Bărbulescu, C. (2007). Analysis of the Magnetic Losses from the Induction Machines Supplied by Inverters, *Proceedings of EUROCON 2007 International Conference on Computer as a Tool*, pp. 1800-1809, ISBN 978-1-4244-0812-2, Warsaw, Poland, September 9-12, 2007

[2] Muşuroi, S.; Olărescu, N.V.; Vătău, D. & Şorândaru, C. (2009). Theoretical and Experimental Determination of Equivalent Parameters of Three-Phase Induction Motor Windings in Case of Power Electronic Converters Supply, *WSEAS Transactions on Systems*, Vol.8, Issue10, (October 2009), pp. 1115-1124, ISSN 1109-2777

[3] Muşuroi, S.; Şorândaru, C.; Olărescu, N.V. & Svoboda, M. (2009) Mathematical Model Associated to Three-Phase Induction Servomotors in the Case of Scalar Control, *WSEAS Transactions on Systems*, Vol.8, Issue 10, (October 2009), pp. 1125-1134, ISSN 1109-2777

[4] Mohan, N.; Undeland, T.M. & Robbins, W.P. (1995). *Power Electronics: Converters, Applications and Design*, (2nd edition), John Wiley & Sons, ISBN 0-471-58408-8, Inc., New-York, USA

[5] Murphy, J. & Turnbull, F. (1988). *Power Electronic Control of AC Motors*, Pergarmon Press, ISBN 0-08-022683, Oxford, 1988

[6] Dordea, T. & Dordea, P.T. (1984). Ersatzläuferimpedanz einer induktionsmachine mit vielfachem käfig und in den selben nuten untergebrachten stäben, *Revue Roumaine des Sciences Techniques*, Vol.29, No.2, (December 1984), pp. 151-159, ISSN 0035-4066

Modelling and Analysis of Squirrel Cage Induction Motor with Leading Reactive Power Injection

Adisa A. Jimoh, Pierre-Jac Venter and Edward K. Appiah

Additional information is available at the end of the chapter

1. Introduction

Induction motors are by far the most used electro-mechanical device in industry today. Induction motors hold many advantages over other types of motors. They are cheap, rugged, easily maintainable and can be used in hazardous locations. Despite its advantages it has one major disadvantage. It draws reactive power from the source to be able to operate and therefore the power factor of the motor is inherently poor especially under starting conditions and under light load (Jimoh and Nicolae, 2007). Poor power factor adversely affects the economics of distribution and transmission systems and therefore may lead to higher electricity charges (Muljadi et al., 1989). At starting, power drawn by the motor is mainly reactive and it can draw up to 8 times its rated current at a power factor of about 0.2 until it reaches rated speed after which the power factor will increase to more than 0.6 if the motor is properly loaded and depending on the size of the motor.

To improve the power factor, reactive power compensation is needed where reactive power is injected. Several techniques have been suggested including synchronous compensation which is complex and expensive. Switched capacitor banks which requires expensive switchgear and may cause voltage regeneration, over voltage and high inrush currents (El-Sharkawi et al., 1985).

In this chapter another approach for power factor correction is explored where the stator of an induction motor has two sets of three phase windings which is electrically isolated but magnetically coupled. The main winding is connected to the three phase supply and the auxiliary winding connected to fixed capacitors for reactive power injection.

The first part of this chapter focuses on the development of a mathematical model for a normal three phase induction motor, the second part of the chapter focuses on the development of the mathematical model for a dual winding three phase induction motor with reactive power injection, where the derived mathematical model is simulated using Matlab/Simulink environment and the third part of the chapter focuses on the performance analysis of both theoretical and experimental results.

2. Arbitrary reference frame theory

Arbitrary reference frame theory is mainly used in the dynamic analysis of electrical machines. Because of the highly coupled nature of the machine, especially the inductances within the winding make it rather impossible to perform dynamic simulations and analysis on electrical machines.

Arbitrary reference frame theory was discovered by Blondel, Dreyfus, Doherty and Nickle as mentioned in the classical paper (Park, 1929). This newly found theory was generalised by Park on synchronous machines and this method was later extended by Stanley to the application of dynamic analysis of induction machines (Stanley, 1938).

By using this method a poly-phase machine is transformed to a two-phase machine with their magnetic axis in quadrature as illustrated in Figure 1. This method is also commonly referred to as the *dq* method in balanced systems and to the *dq0* method in unbalanced systems with the '*0*'relating to the zero sequence or homopolar component in the Fortescue Transformation.

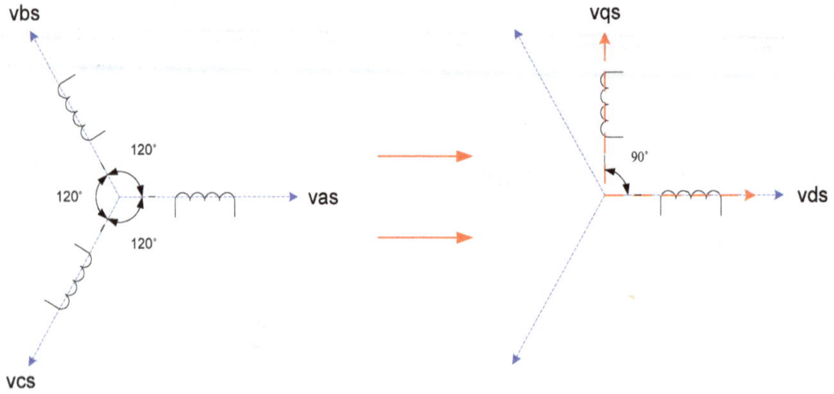

Figure 1. Park's Transform

This transformation eliminates mutual magnetic coupling between the phases and therefore makes the magnetic flux linkage of one winding independent of the current of another winding.

The transformation is done by applying a transformation matrix, Equation (1) while the inverse transformation matrix, Equation (2) will transform back to the natural reference frame. Equations (1)&(2) applies to a three phase system but can be modified to accommodate a system with any number of phases which might be useful in the case of the machine having an auxiliary winding as proposed in this work.

$$[C] = \frac{2}{3}\begin{bmatrix} \cos\theta & \cos\left(\theta - \frac{2\pi}{3}\right) & \cos\left(\theta - \frac{4\pi}{3}\right) \\ \sin\theta & \sin\left(\theta - \frac{2\pi}{3}\right) & \sin\left(\theta - \frac{4\pi}{3}\right) \\ \frac{1}{2} & \frac{1}{2} & \frac{1}{2} \end{bmatrix} \quad (1)$$

$$[C]^{-1} = \begin{bmatrix} \cos\theta & \sin\theta & 1 \\ \cos\left(\theta - \frac{2\pi}{3}\right) & \sin\left(\theta - \frac{2\pi}{3}\right) & 1 \\ \cos\left(\theta - \frac{4\pi}{3}\right) & \sin\left(\theta - \frac{4\pi}{3}\right) & 1 \end{bmatrix} \quad (2)$$

3. Modelling of three-phase induction motor

The winding arrangement of a symmetrical induction machine is shown in Figure 2. The stator windings are identical and sinusoidally distributed, displaced 120° apart, with N_s equivalent turns and resistance r_s per winding per phase. Similarly the rotor windings are also considered as three identical sinusoidally distributed windings, displaced 120° apart, with N_r equivalent turns and resistance of r_r per winding per phase.

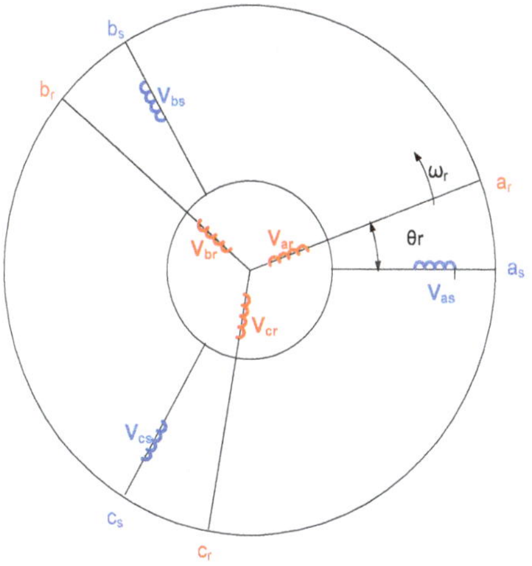

Figure 2. Three-Phase Winding Arrangement

In developing the equations which describe the behaviour of the induction machine the following assumptions are made:

1. The airgap is uniform.
2. Eddy currents, friction and windage losses and saturation are neglected.
3. The windings are distributed sinusoidally around the air gap.
4. The windings are identical

3.1. Voltage equations

Using Kirchoff's Voltage Law, the voltage equations for each winding on the stator and rotor can be determined.

Stator Windings:

$$v_{as} = r_s i_{as} + \frac{d\lambda_{as}}{dt} \tag{3}$$

$$v_{bs} = r_s i_{bs} + \frac{d\lambda_{bs}}{dt} \tag{4}$$

$$v_{cs} = r_s i_{cs} + \frac{d\lambda_{cs}}{dt} \tag{5}$$

Rotor Windings:

$$v_{ar} = r_r i_{ar} + \frac{d\lambda_{ar}}{dt} \tag{6}$$

$$v_{br} = r_r i_{br} + \frac{d\lambda_{br}}{dt} \tag{7}$$

$$v_{cr} = r_r i_{cr} + \frac{d\lambda_{cr}}{dt} \tag{8}$$

With subscript 'a','b','c' referring to the phases, subscript 's' referring to stator variables, subscript 'r' referring to rotor variables, 'v' referring to instantaneous voltage, 'i' referring to instantaneous current and 'λ' referring to flux linkage.

After obtaining the voltage equations in the natural reference frame, the transformation to the arbitrary reference frame can be done. It is very convenient to first refer all rotor variables to the stator by applying the appropriate turns ratio. Equations (9)-(11) represents all rotor variables and is expressed in a simplified way including the variables of all the rotor phases in one equation.

$$i'_{abcr} = \frac{N_r}{N_s} i_{abcr} \tag{9}$$

$$v'_{abcr} = \frac{N_s}{N_r} v_{abcr} \tag{10}$$

$$\lambda'_{abcr} = \frac{N_s}{N_r} \lambda_{abcr} \tag{11}$$

The transformation of the voltage equations to the arbitrary reference frame are dealt with in Section 3.3. It is important to determine the different inductances which will influence the flux linkage in Equations (3)-(8) and also transform it to the arbitrary reference frame.

3.2. Inductances

The flux linkages as seen in the voltage equations are functions of inductance and therefore the inductances within the motor must be determined.

The inductances within the motor consist of self inductance, leakage inductance, magnetizing inductance and mutual inductance. The flux linkage equation is shown in Equation (12) and contains the inductance matrix [L].

$$\begin{bmatrix} \lambda_{as} \\ \lambda_{bs} \\ \lambda_{cs} \\ \lambda_{ar} \\ \lambda_{br} \\ \lambda_{cr} \end{bmatrix} = \begin{bmatrix} L_{asas} & L_{asbs} & L_{ascs} & L_{asar} & L_{asbr} & L_{ascr} \\ L_{bsas} & L_{bsbs} & L_{bscs} & L_{bsar} & L_{bsbr} & L_{bscr} \\ L_{csas} & L_{csbs} & L_{cscs} & L_{csar} & L_{csbr} & L_{cscr} \\ L_{aras} & L_{arbs} & L_{arcs} & L_{arar} & L_{arbr} & L_{arcr} \\ L_{bras} & L_{brbs} & L_{brcs} & L_{brar} & L_{brbr} & L_{brcr} \\ L_{cras} & L_{crbs} & L_{crcs} & L_{crar} & L_{crbr} & L_{crcr} \end{bmatrix} \times \begin{bmatrix} i_{as} \\ i_{bs} \\ i_{cs} \\ i_{ar} \\ i_{br} \\ i_{cr} \end{bmatrix} \tag{12}$$

Where the inductance is defined by the subscript, for example L_{asas} refers to the inductance between winding *as* and winding *as*, meaning that this is self inductance in winding *as*; and L_{asbr} refers to the inductance between winding *as* and winding *br*, meaning that this is a mutual inductance.

3.2.1. Self inductance

The self inductance in the stator windings consists of magnetizing and leakage inductance. The windings are identical and therefore the self inductance of all stator windings will be identical.

$$L_{asas} = L_{bsbs} = L_{cscs} = L_{ms} + L_{ls} \tag{13}$$

The magnetizing inductance (L_{ms}) can be expressed as in Equation (14) (Lipo and Novotny, 1996).

$$L_{ms} = \frac{\mu_0 \ell r N_s^2 \pi}{4g} \tag{14}$$

The self inductance in the rotor windings is similar to that of the stator windings.

$$L_{arar} = L_{brbr} = L_{crcr} = L_{mr} + L_{lr} \tag{15}$$

And,

$$L_{mr} = \frac{\mu_0 \ell r N_r^2 \pi}{4g} \tag{16}$$

Where N_s and N_r is the effective number of turns of the stator and rotor windings, r is the radius of the motor cross-section, ℓ is the length of the motor and g is the airgap radial length.

3.2.2. Mutual inductance

Mutual inductance exists between all windings of both the stator and the rotor. There are four different types of mutual inductance which are stator-stator (mutual inductance between two different stator windings), rotor-rotor (mutual inductance between two different rotor windings), stator-rotor (mutual inductance between a stator and rotor winding) and rotor-stator (mutual inductance between a rotor and stator winding).

Stator-stator mutual inductance can be expressed as Equation (17) (Lipo and Novotny, 1996).

$$L_{xsys} = \frac{\mu_0 \ell r N_s^2 \pi}{4g} cos\theta_{xsys} \tag{17}$$

Where L_{xsys} is the inductance between any stator winding 'x' and any other stator winding 'y' and θ_{xsys} is the angle between stator winding 'x' and 'y'.

Using Equation (14), Equation (17) can be modified to be:

$$L_{xsys} = L_{ms} cos\theta_{xsys} \tag{18}$$

When considering the winding distribution in Figure 2, it can be seen that the only possible displacement between two stator windings are 120° and 240° in both directions. This implies that $cos\theta_{xsys}$ in Equation (18) can be evaluated as follows:

$$cos\theta_{xsys} = \cos(\pm 120^{\circ}) = \cos(\pm 240^{\circ}) = -\frac{1}{2} \tag{19}$$

From Equations (18)&(19) the expression describing the mutual inductance between any two stator windings can be simplified to Equation (20).

$$L_{asbs} = L_{ascs} = L_{bscs} = L_{bsas} = L_{csas} = L_{csbs} = -\frac{1}{2} L_{ms} \tag{20}$$

The rotor-rotor mutual inductances are similar to that of the stator-stator mutual inductances and can be expressed as:

$$L_{arbr} = L_{arcr} = L_{brcr} = L_{brar} = L_{crar} = L_{crbr} = -\frac{1}{2} L_{mr} \tag{21}$$

The stator-rotor mutual inductances depend on the position of the rotor according to the following relationship.

$$L_{xsyr} = L_{sr} cos\theta_{xsyr} \tag{22}$$

Where L_{xsyr} is the mutual inductance between any stator winding 'x' and any rotor winding 'y'; and θ_{xsyr} is the angle between them.

The expression for L_{sr} in Equation (22) is given by Equation (23).

$$L_{sr} = \left(\frac{N_s}{2}\right)\left(\frac{N_r}{2}\right)\frac{\mu_0 \pi r \ell}{g} \tag{23}$$

(Krause, 1986, Lipo and Novotny, 1996)

Now, using Equation (22) and Figure 2, the expressions for the stator-rotor mutual inductances can be deduced.

$$L_{asar} = L_{bsbr} = L_{cscr} = L_{sr} cos\theta_r \tag{24}$$

$$L_{asbr} = L_{bscr} = L_{csar} = L_{sr} cos\left(\theta_r + \frac{2\pi}{3}\right) \tag{25}$$

$$L_{ascr} = L_{bsar} = L_{csbr} = L_{sr}cos\left(\theta_r - \frac{2\pi}{3}\right) \tag{26}$$

Likewise it can be shown that the rotor-stator mutual inductances are:

$$L_{aras} = L_{brbs} = L_{crcs} = L_{sr}\cos(-\theta_r) \tag{27}$$

$$L_{arbs} = L_{brcs} = L_{cras} = L_{sr}cos\left(\frac{2\pi}{3} - \theta_r\right) \tag{28}$$

$$L_{arcs} = L_{bras} = L_{crbs} = L_{sr}cos\left(\frac{4\pi}{3} - \theta_r\right) \tag{29}$$

All inductances have now been quantified. The complete inductance matrix can now be constructed but to simplify the work the inductance matrix in Equation (12) is first divided into sub-matrices. The inductance matrix as in Equation (12) is repeated as Equation (30).

$$L = \begin{bmatrix} L_{asas} & L_{asbs} & L_{ascs} & L_{asar} & L_{asbr} & L_{ascr} \\ L_{bsas} & L_{bsbs} & L_{bscs} & L_{bsar} & L_{bsbr} & L_{bscr} \\ L_{csas} & L_{csbs} & L_{cscs} & L_{csar} & L_{csbr} & L_{cscr} \\ L_{aras} & L_{arbs} & L_{arcs} & L_{arar} & L_{arbr} & L_{arcr} \\ L_{bras} & L_{brbs} & L_{brcs} & L_{brar} & L_{brbr} & L_{brcr} \\ L_{cras} & L_{crbs} & L_{crcs} & L_{crar} & L_{crbr} & L_{crcr} \end{bmatrix} \tag{30}$$

The inductance matrix is divided into four sub-matrices.

$$L = \begin{bmatrix} L_s & L_{SR} \\ L_{RS} & L_r \end{bmatrix} \tag{31}$$

Where L_s is the inductance within the stator windings, L_r is the inductances within the rotor windings, L_{SR} is the inductances between stator and rotor windings and L_{RS} is the inductances between the rotor and stator windings.

Using Equation (30) and dividing according to Equation (31) and substituting inductances yield the following:

$$L_s = \begin{bmatrix} L_{ms} + L_{ls} & -\frac{1}{2}L_{ms} & -\frac{1}{2}L_{ms} \\ -\frac{1}{2}L_{ms} & L_{ms} + L_{ls} & -\frac{1}{2}L_{ms} \\ -\frac{1}{2}L_{ms} & -\frac{1}{2}L_{ms} & L_{ms} + L_{ls} \end{bmatrix} \tag{32}$$

$$L_r = \begin{bmatrix} L_{mr} + L_{lr} & -\frac{1}{2}L_{mr} & -\frac{1}{2}L_{mr} \\ -\frac{1}{2}L_{mr} & L_{mr} + L_{lr} & -\frac{1}{2}L_{mr} \\ -\frac{1}{2}L_{mr} & -\frac{1}{2}L_{mr} & L_{mr} + L_{lr} \end{bmatrix} \tag{33}$$

$$L_{SR} = L_{sr}\begin{bmatrix} cos\theta_r & \cos(\theta_r + \frac{2\pi}{3}) & \cos(\theta_r - \frac{2\pi}{3}) \\ \cos(\theta_r - \frac{2\pi}{3}) & cos\theta_r & \cos(\theta_r + \frac{2\pi}{3}) \\ \cos(\theta_r + \frac{2\pi}{3}) & \cos(\theta_r - \frac{2\pi}{3}) & cos\theta_r \end{bmatrix} \tag{34}$$

$$L_{RS=}(L_{SR})^T = L_{sr}\begin{bmatrix} cos\theta_r & \cos(\theta_r - \frac{2\pi}{3}) & \cos(\theta_r + \frac{2\pi}{3}) \\ \cos(\theta_r + \frac{2\pi}{3}) & cos\theta_r & \cos(\theta_r - \frac{2\pi}{3}) \\ \cos(\theta_r - \frac{2\pi}{3}) & \cos(\theta_r + \frac{2\pi}{3}) & cos\theta_r \end{bmatrix} \quad (35)$$

It can be proven that$L_{RS=}(L_{SR})^T$.

It is evident that θ_r exists in Equations (34)&(35) where θ_r relates to rotor position, this rotor position changes continuously which means that in the natural state the inductances are varying with time. To be able to derive a rigorous dynamic model de-coupling has to be done by using arbitrary reference frame theory as mentioned in Section 2.

Before transforming the inductances to the arbitrary reference frame all rotor parameters must be referred to the stator.

The magnetizing inductances (L_{ms}, L_{mr}) and mutual inductances (L_{sr}, L_{rs}) are from the same magnetic flux path and are therefore related. From Equations (14)&(23) it can be deduced that:

$$L_{ms} = \left(\frac{N_s}{N_r}\right) L_{sr} \quad (36)$$

By using the effective turns ratio L_{sr} can be referred to the stator.

$$L'_{sr} = \frac{N_s}{N_r}(L_{sr}) \quad (37)$$

And therefore

$$L'_{sr} = L_{ms} \quad (38)$$

The mutual inductance matrix referred to the stator can now be expressed as:

$$L'_{SR} = L_{ms}\begin{bmatrix} cos\theta_r & \cos(\theta_r + \frac{2\pi}{3}) & \cos(\theta_r - \frac{2\pi}{3}) \\ \cos(\theta_r - \frac{2\pi}{3}) & cos\theta_r & \cos(\theta_r + \frac{2\pi}{3}) \\ \cos(\theta_r + \frac{2\pi}{3}) & \cos(\theta_r - \frac{2\pi}{3}) & cos\theta_r \end{bmatrix} \quad (39)$$

Using the same approach the inductances within the rotor windings can also be simplified as:

$$L'_r = \begin{bmatrix} L'_{lr} + L_{ms} & -\frac{1}{2}L_{ms} & -\frac{1}{2}L_{ms} \\ -\frac{1}{2}L_{ms} & L'_{lr} + L_{ms} & -\frac{1}{2}L_{ms} \\ -\frac{1}{2}L_{ms} & -\frac{1}{2}L_{ms} & L'_{lr} + L_{ms} \end{bmatrix} \quad (40)$$

The flux linkage can now be expressed as:

$$\begin{bmatrix} \lambda_{abcs} \\ \lambda'_{abcr} \end{bmatrix} = \begin{bmatrix} L_s & L'_{SR} \\ (L'_{SR})^T & L'_r \end{bmatrix} \times \begin{bmatrix} i_{abcs} \\ i'_{abcr} \end{bmatrix} \quad (41)$$

Equation (41) can be transformed to the arbitrary reference frame as indicated in Equation (42) (Krause, 1986).

$$\begin{bmatrix} \lambda_{qd0s} \\ \lambda'_{qdor} \end{bmatrix} = \begin{bmatrix} K_s L_s K_s^{-1} & K_s L'_{SR} K_r^{-1} \\ K_r (L'_{SR})^T K_s^{-1} & K_r L'_r K_r^{-1} \end{bmatrix} \times \begin{bmatrix} i_{qd0s} \\ i'_{qd0r} \end{bmatrix} \quad (42)$$

Where;

$$K_s = \frac{2}{3} \begin{bmatrix} cos\theta & cos\left(\theta - \frac{2\pi}{3}\right) & cos\left(\theta - \frac{4\pi}{3}\right) \\ sin\theta & sin\left(\theta - \frac{2\pi}{3}\right) & sin\left(\theta - \frac{4\pi}{3}\right) \\ \frac{1}{2} & \frac{1}{2} & \frac{1}{2} \end{bmatrix} \quad (43)$$

$$K_s^{-1} = \begin{bmatrix} cos\theta & sin\theta & 1 \\ cos\left(\theta - \frac{2\pi}{3}\right) & sin\left(\theta - \frac{2\pi}{3}\right) & 1 \\ cos\left(\theta - \frac{4\pi}{3}\right) & sin\left(\theta - \frac{4\pi}{3}\right) & 1 \end{bmatrix} \quad (44)$$

$$K_r = \frac{2}{3} \begin{bmatrix} cos\beta & cos\left(\beta - \frac{2\pi}{3}\right) & cos\left(\beta - \frac{4\pi}{3}\right) \\ sin\beta & sin\left(\beta - \frac{2\pi}{3}\right) & sin\left(\beta - \frac{4\pi}{3}\right) \\ \frac{1}{2} & \frac{1}{2} & \frac{1}{2} \end{bmatrix} \quad (45)$$

$$K_r^{-1} = \begin{bmatrix} cos\beta & sin\beta & 1 \\ cos\left(\beta - \frac{2\pi}{3}\right) & sin\left(\beta - \frac{2\pi}{3}\right) & 1 \\ cos\left(\beta - \frac{4\pi}{3}\right) & sin\left(\beta - \frac{4\pi}{3}\right) & 1 \end{bmatrix} \quad (46)$$

Where;

$$\beta = \theta - \theta_r \quad (47)$$

K_s is the transformation matrix and K_s^{-1} the inverse transformation matrix for the stator parameters, K_r is the transformation matrix and K_r^{-1} the inverse transformation matrix for the rotor parameters.

Evaluating Equation (42) with Equations (43)-(46) yield the flux linkage in the arbitrary reference frame as shown in Equation (48).

Comparing Equation (48) with Equations (32)-(35), it is clear that θ_r has been eliminated from the flux linkage equations by using the arbitrary reference frame transformation. It means that the flux linkage is no longer a function of rotor position.

$$\begin{bmatrix} \lambda_{qs} \\ \lambda_{ds} \\ \lambda_{0s} \\ \lambda'_{qr} \\ \lambda'_{dr} \\ \lambda'_{0r} \end{bmatrix} = \begin{bmatrix} L_{ls} + \frac{3}{2}L_{ms} & 0 & 0 & \frac{3}{2}L_{ms} & 0 & 0 \\ 0 & L_{ls} + \frac{3}{2}L_{ms} & 0 & 0 & \frac{3}{2}L_{ms} & 0 \\ 0 & 0 & L_{ls} & 0 & 0 & 0 \\ \frac{3}{2}L_{ms} & 0 & 0 & L'_{lr} + \frac{3}{2}L_{ms} & 0 & 0 \\ 0 & \frac{3}{2}L_{ms} & 0 & 0 & L'_{lr} + \frac{3}{2}L_{ms} & 0 \\ 0 & 0 & 0 & 0 & 0 & L'_{lr} \end{bmatrix} \times \begin{bmatrix} i_{qs} \\ i_{ds} \\ i_{0s} \\ i'_{qr} \\ i'_{dr} \\ i'_{0r} \end{bmatrix} \quad (48)$$

$$L_m = \frac{3}{2} L_{ms} \tag{49}$$

3.3. Voltage equations in arbitrary reference frame

Recalling and repeating the voltage equations in the natural reference frame in section 3.1, for ease of reading only, now it is presented in matrix format.

$$v_{abcs} = r_s i_{abcs} + \frac{d\lambda_{abcs}}{dt} \tag{50}$$

$$v'_{abcr} = r'_r i'_{abcr} + \frac{d\lambda'_{abcr}}{dt} \tag{51}$$

Taking only the stator voltage equations as in Equation (50) and only considering the resistive part, it can be transformed to the arbitrary reference frame as follows:

$$v_{qd0s}^{res} = K_s r_s K_s^{-1} i_{qd0s} \tag{52}$$

$$K_s r_s K_s^{-1} = \begin{bmatrix} r_s & 0 & 0 \\ 0 & r_s & 0 \\ 0 & 0 & r_s \end{bmatrix} \tag{53}$$

Therefore;

$$v_{qd0s}^{res} = r_s i_{qd0s} \tag{54}$$

Where;

$$r_s = \begin{bmatrix} r_s & 0 & 0 \\ 0 & r_s & 0 \\ 0 & 0 & r_s \end{bmatrix} \tag{55}$$

The superscript *'res'* refers to the resistive part of the voltage equation.

Now considering only the inductive part of the of the voltage equation as in Equation (50) which can be transformed to the arbitrary reference frame as follows:

$$v_{qd0s}^{ind} = K_s \frac{d}{dt}\left[K_s^{-1} \lambda_{qd0s}\right] \tag{56}$$

Expanding Equation (56) using the product rule:

$$v_{qd0s}^{ind} = K_s \frac{d}{dt}[K_s^{-1}] \lambda_{qd0s} + K_s K_s^{-1} \frac{d}{dt} \lambda_{qd0s} \tag{57}$$

Now, evaluating parts of the terms in Equation (57) separately;

Knowing that;

$$\theta = \int \omega(\xi)\, d\xi + \theta(0) \tag{58}$$

Where ξ is a dummy variable for integration.

$$\frac{d}{dt}[K_s^{-1}] = \omega \begin{bmatrix} -sin\theta & cos\theta & 0 \\ -sin\left(\theta - \frac{2\pi}{3}\right) & cos\left(\theta - \frac{2\pi}{3}\right) & 0 \\ -sin\left(\theta + \frac{2\pi}{3}\right) & cos\left(\theta + \frac{2\pi}{3}\right) & 0 \end{bmatrix} \tag{59}$$

$$K_s \frac{d}{dt}[K_s^{-1}] = \begin{bmatrix} 0 & \omega & 0 \\ -\omega & 0 & 0 \\ 0 & 0 & 0 \end{bmatrix} \tag{60}$$

$$K_s K_s^{-1} = \begin{bmatrix} 1 & 0 & 0 \\ 0 & 1 & 0 \\ 0 & 0 & 1 \end{bmatrix} \tag{61}$$

Using Equations (58)-(61) to evaluate Equation (57) yields;

$$v_{qd0s}^{ind} = \omega \begin{bmatrix} \lambda_{ds} \\ -\lambda_{qs} \\ 0 \end{bmatrix} + \frac{d}{dt} \begin{bmatrix} \lambda_{qs} \\ \lambda_{ds} \\ \lambda_{0s} \end{bmatrix} \tag{62}$$

The superscript *'ind'* refers to the inductive part of the voltage equation.

Now adding the voltage equations for the resistive and inductive parts gives the full stator voltage equations in the arbitrary reference frame.

$$v_{qd0s} = \begin{bmatrix} r_s & 0 & 0 \\ 0 & r_s & 0 \\ 0 & 0 & r_s \end{bmatrix} \begin{bmatrix} i_{qs} \\ i_{ds} \\ i_0 \end{bmatrix} + \omega \begin{bmatrix} \lambda_{ds} \\ -\lambda_{qs} \\ 0 \end{bmatrix} + \frac{d}{dt} \begin{bmatrix} \lambda_{qs} \\ \lambda_{ds} \\ \lambda_{0s} \end{bmatrix} \tag{63}$$

Using the same method the rotor voltage equation can be determined.

$$v'_{qd0r} = \begin{bmatrix} r'_r & 0 & 0 \\ 0 & r'_r & 0 \\ 0 & 0 & r'_r \end{bmatrix} \begin{bmatrix} i'_{qr} \\ i'_{dr} \\ i'_{0r} \end{bmatrix} + (\omega - \omega_r) \begin{bmatrix} \lambda'_{dr} \\ -\lambda'_{qr} \\ 0 \end{bmatrix} + \frac{d}{dt} \begin{bmatrix} \lambda'_{qr} \\ \lambda'_{dr} \\ \lambda_{0r} \end{bmatrix} \tag{64}$$

Where ω is the rotational speed of the reference frame and ω_r is the rotational speed of the rotor.

The model developed up to now is a general model in the arbitrary reference frame which means that this model can take the form of any reference frame depending the value substituted for ω , therefore called arbitrary.

Different reference frames are obtained by substituting the appropriate value of the reference frame speed into ω. Three different reference frames are commonly used, the stationary reference frame where $\omega = 0$, the synchronous reference frame where ω is set to the angular velocity of the supply voltage ($\omega = \omega_e$) and the rotor reference frame where ω is set to the angular velocity of the rotor ($\omega = \omega_r$). The use of reference frames depends on the nature of the problem to be solved.

3.4. Electromagnetic torque

The torque equation for a three phase induction machine is well known and is not derived in this section. The torque equation of the three phase machine with auxiliary winding is derived in Section 4.

$$T_e = \left(\frac{3}{2}\right)\left(\frac{P}{2}\right)\left(\lambda_{ds} i_{qs} - \lambda_{qs} i_{ds}\right) \tag{65}$$

An electric motor is an electro-mechanical device and needs an equation that couples the electrical and mechanical systems.

$$T_{em} = J\left(\frac{2}{P}\right)\frac{d\omega_r}{dt} + T_L \tag{66}$$

Where P is number of poles, J is moment of inertia, T_L is torque connected to the shaft and ω_r is the angular rotational speed of the rotor.

3.5. Equivalent circuit

The full mathematical model of the three phase induction machine is given by Equations (48),(63)&(64). These equations are used to develop the equivalent circuits for the three phase induction machine as in Figure 3

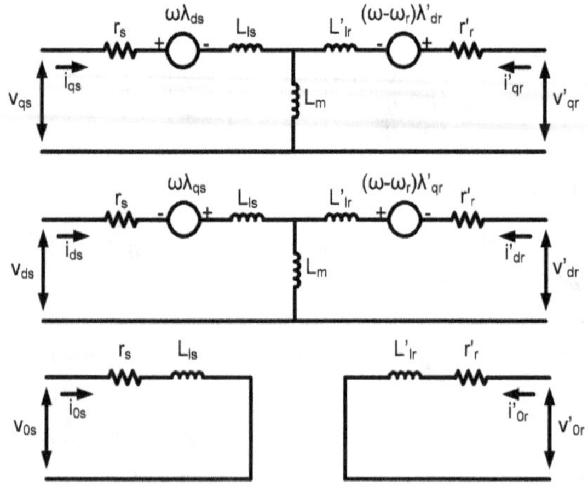

Figure 3. Equivalent Circuits

4. Modelling of three-phase with auxiliary winding

This machine consists of two three phase windings arranged on top of each other in the same slots. This means that there is no displacement between the two windings. These two windings are electrically isolated but magnetically connected. One of these windings is treated as the main winding and will be supplied with a three phase voltage. The main winding is labelled with the subscript *'abc'*. The remaining winding is treated as the auxiliary winding. The auxiliary winding is connected to static capacitors for reactive power injection. The injection of reactive power will improve the power factor of the machine. The auxiliary winding is labelled with the subscript *'xyz'*. The winding arrangement is as shown in Figure 4.

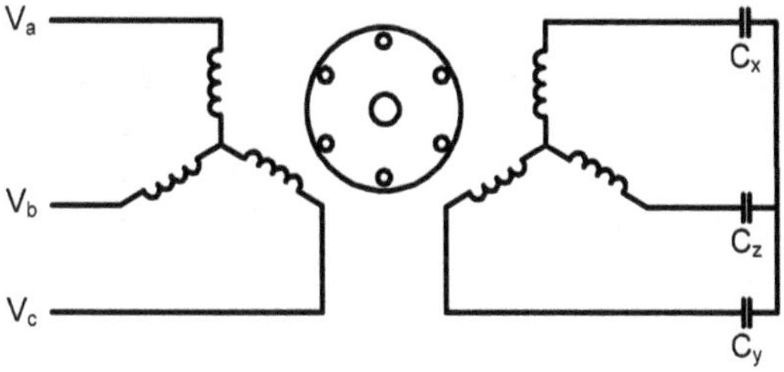

Figure 4. Winding arrangement of Main and Auxiliary Windings

The assumptions in developing the equations which describe the behaviour of this machine are the same as the assumptions mentioned in Section 3 with one addition. It is assumed that the main and auxiliary winding is identical. It has the same conductor cross section and the same number of turns.

4.1. Voltage equations

The voltage equations for this machine are developed in the same way as described in Section 3.1. There are three additional voltage equations because of the additional set of three phase windings. For simplicity the voltage equations are represented in matrix format. It is assumed that the main and auxiliary windings are identical and will therefore have the same resistance.

Main Stator Windings:

$$v_{abcs} = r_s i_{abcs} + \frac{d\lambda_{abcs}}{dt} \tag{67}$$

Auxiliary Stator Windings:

$$v_{xyzs} = r_s i_{xyzs} + \frac{d\lambda_{xyzs}}{dt} \tag{68}$$

Rotor Windings:

$$v'_{abcr} = r'_r i'_{abcr} + \frac{d\lambda'_{abcr}}{dt} \tag{69}$$

4.2. Inductances

Because of the addition of the auxiliary winding the dimensions of the inductance matrix will increase. The dimension of the inductance matrix is equal to the number of windings, in this case a [9 x 9] matrix. The inductance matrix can again be divided into smaller more manageable sub-matrices as in Equation (70).

$$\begin{bmatrix} L_{abc\_s} \\ L_{xyz\_s} \\ L_{abc\_r} \end{bmatrix} = \begin{bmatrix} L_{abcs} & L_{abcsxyzs} & L_{abcsabcr} \\ L_{xyzsabcs} & L_{xyzs} & L_{xyzsabcr} \\ L_{abcrabcs} & L_{abcrxyzs} & L_{abcr} \end{bmatrix} \tag{70}$$

where L_{abcs}, L_{xyzs}, L_{abcr} are the inductances within the different three phase sets of windings. The other sub-matrices refer to mutual inductances between different sets of windings. This includes stator and rotor windings, for example $L_{abcsxyzs}$ refer to the mutual inductance between the main winding and the auxiliary winding on the stator. Many of the inductances in this structure are very similar to that of the conventional three phase machine. Therefore, many of the developed inductances in Section 3 can be used.

$$L_{abcs} = L_{xyzs} = \begin{bmatrix} L_{ms} + L_{ls} & -\frac{1}{2}L_{ms} & -\frac{1}{2}L_{ms} \\ -\frac{1}{2}L_{ms} & L_{ms} + L_{ls} & -\frac{1}{2}L_{ms} \\ -\frac{1}{2}L_{ms} & -\frac{1}{2}L_{ms} & L_{ms} + L_{ls} \end{bmatrix} \tag{71}$$

$$L_{abcr} = \begin{bmatrix} L_{mr} + L_{lr} & -\frac{1}{2}L_{mr} & -\frac{1}{2}L_{mr} \\ -\frac{1}{2}L_{mr} & L_{mr} + L_{lr} & -\frac{1}{2}L_{mr} \\ -\frac{1}{2}L_{mr} & -\frac{1}{2}L_{mr} & L_{mr} + L_{lr} \end{bmatrix} \tag{72}$$

$$L_{abcsabcr} = L_{xyzsabcr} = L_{ms} \begin{bmatrix} \cos\theta_r & \cos(\theta_r + \frac{2\pi}{3}) & \cos(\theta_r - \frac{2\pi}{3}) \\ \cos(\theta_r - \frac{2\pi}{3}) & \cos\theta_r & \cos(\theta_r + \frac{2\pi}{3}) \\ \cos(\theta_r + \frac{2\pi}{3}) & \cos(\theta_r - \frac{2\pi}{3}) & \cos\theta_r \end{bmatrix} \tag{73}$$

$$L_{abcrabcs} = L_{abcrxyzs} = L_{abcsabcr}^{T} \tag{74}$$

All of the above was taken from Section 3 because of similar relationships between the windings. The only sub-matrices remaining are the two describing the inductances between the main and auxiliary windings on the stator. For determining the expressions of all the elements of the sub-matrices, equation (18) is used.

$$L_{abcsxyzs} = \begin{bmatrix} L_{asxs} & L_{asys} & L_{aszs} \\ L_{bsxs} & L_{bsys} & L_{bszs} \\ L_{csxs} & L_{csys} & L_{cszs} \end{bmatrix} \tag{75}$$

Where;

$$L_{asxs} = L_{bsys} = L_{cszs} = L_{ms}\cos(0^\circ) = L_{ms} \tag{76}$$

$$L_{asys} = L_{aszs} = L_{bsxs} = L_{bszs} = L_{csxs} = L_{cszs} = L_{ms}\cos(\pm 120^\circ) = -\frac{1}{2}L_{ms} \tag{77}$$

Now;

$$L_{abcsxyzs} = \begin{bmatrix} L_{ms} & -\frac{1}{2}L_{ms} & -\frac{1}{2}L_{ms} \\ -\frac{1}{2}L_{ms} & L_{ms} & -\frac{1}{2}L_{ms} \\ -\frac{1}{2}L_{ms} & -\frac{1}{2}L_{ms} & L_{ms} \end{bmatrix} \quad (78)$$

It can be shown that;

$$L_{xyzsabcs} = L_{abcsxyzs} \quad (79)$$

Mutual leakage inductance exists between the main and auxiliary windings because they are sharing the same stator slots. This mutual leakage inductance should be reflected in Equation (78) but are neglected for the purpose of this study.

Transforming the inductance matrix in the natural reference frame to the arbitrary reference frame as explained in Section 3.2 yields the following inductance matrix in the arbitrary reference frame.

$$\begin{bmatrix} L_{q1s} \\ L_{d1s} \\ L_{01s} \\ L_{q2s} \\ L_{d2s} \\ L_{02s} \\ L_{qr} \\ L_{dr} \\ L_{0r} \end{bmatrix} = \begin{bmatrix} L_{ss} & 0 & 0 & L_m & 0 & 0 & L_m & 0 & 0 \\ 0 & L_{ss} & 0 & 0 & L_m & 0 & 0 & L_m & 0 \\ 0 & 0 & L_{ls} & 0 & 0 & 0 & 0 & 0 & 0 \\ L_m & 0 & 0 & L_{ss} & 0 & 0 & L_m & 0 & 0 \\ 0 & L_m & 0 & 0 & L_{ss} & 0 & 0 & L_m & 0 \\ 0 & 0 & 0 & 0 & 0 & L_{ls} & 0 & 0 & 0 \\ L_m & 0 & 0 & L_m & 0 & 0 & L_{rr} & 0 & 0 \\ 0 & L_m & 0 & 0 & L_m & 0 & 0 & L_{rr} & 0 \\ 0 & 0 & 0 & 0 & 0 & 0 & 0 & 0 & L'_{lr} \end{bmatrix} \quad (80)$$

Where;

$$L_{ss} = L_{ls} + L_m \quad (81)$$

$$L_{rr} = L'_{lr} + L_m \quad (82)$$

Where the subscripts *'1'* refers to the main stator winding, *'2'* refers to the auxiliary stator winding and *'r'* refers to the rotor winding.

Because flux linkage appears in the voltage equations it is very convenient to represent flux linkage instead of inductance.

$$\boldsymbol{\lambda} = \boldsymbol{LI} \quad (83)$$

Where Equation (83) is in matrix format and refers to any flux linkage in the system.

4.3. Voltage equation in the arbitrary reference frame

The voltage equations as developed in Section 4.1 can be transformed to the arbitrary reference frame. This is quite an easy task when the developments in Section 3 are used. The equation for the main winding on the stator and the rotor winding remains the same as that for the conventional three-phase machine as shown in Equations (84)-(85).

$$v_{qd01s} = \begin{bmatrix} r_s & 0 & 0 \\ 0 & r_s & 0 \\ 0 & 0 & r_s \end{bmatrix} \begin{bmatrix} i_{q1s} \\ i_{d1s} \\ i_{01s} \end{bmatrix} + \omega \begin{bmatrix} \lambda_{d1s} \\ -\lambda_{q1s} \\ 0 \end{bmatrix} + \frac{d}{dt} \begin{bmatrix} \lambda_{q1s} \\ \lambda_{d1s} \\ \lambda_{01s} \end{bmatrix} \quad (84)$$

$$v'_{qd0r} = \begin{bmatrix} r'_r & 0 & 0 \\ 0 & r'_r & 0 \\ 0 & 0 & r'_r \end{bmatrix} \begin{bmatrix} i'_{qr} \\ i'_{dr} \\ i'_{0r} \end{bmatrix} + (\omega - \omega_r) \begin{bmatrix} \lambda'_{dr} \\ -\lambda'_{qr} \\ 0 \end{bmatrix} + \frac{d}{dt} \begin{bmatrix} \lambda'_{qr} \\ \lambda'_{dr} \\ \lambda_{0r} \end{bmatrix} \quad (85)$$

The voltage equations of the auxiliary winding are different from the main winding because no voltage is applied directly to the winding. Capacitors are rather connected to the auxiliary winding as in Figure 4. It is therefore important to develop the voltage equation for a capacitor in the arbitrary reference frame. This is given in Equation (86).

$$Vc_{qd02s} = \frac{1}{C} \int [i_{qd0s2}] dt + \omega \begin{bmatrix} Vc_{d02} \\ -Vc_{q02} \\ 0 \end{bmatrix} \quad (86)$$

Now, the voltage equation for the auxiliary winding becomes;

$$v_{qd02s} = 0 = \begin{bmatrix} r_s & 0 & 0 \\ 0 & r_s & 0 \\ 0 & 0 & r_s \end{bmatrix} \begin{bmatrix} i_{q2s} \\ i_{d2s} \\ i_{02s} \end{bmatrix} + \omega \begin{bmatrix} \lambda_{d2s} \\ -\lambda_{q2s} \\ 0 \end{bmatrix} + \frac{d}{dt} \begin{bmatrix} \lambda_{q2s} \\ \lambda_{d2s} \\ \lambda_{02s} \end{bmatrix} + \begin{bmatrix} Vc_{q2} \\ Vc_{d2} \\ 0 \end{bmatrix} \quad (87)$$

4.4. Electromagnetic torque

The electromagnetic torque can be derived from the energy stored in the coupling system. The stored energy for a normal three-phase can be written as Equation (88) (Krause, 1986).

$$W_f = \frac{1}{2}(i_{abcs})^T (L_s - L_{ls} I) i_{abcs} + (i_{abcs})^T L'_{sr} i'_{abcr} + \frac{1}{2}(i'_{abcr})^T (L'_r - L'_{lr} I) i'_{abcr} \quad (88)$$

Where I is the identity matrix. The machine is assumed to be magnetically linear and therefore, the field energy W_f is equal to the co-energy W_c.

The change in mechanical energy in a rotational system delivering mechanical power is given as:

$$dW_m = T_{em} d\theta_{rm} \quad (89)$$

Where T_{em} is electromagnetic torque and θ_{rm} is the actual angular displacement of the rotor. The flux linkages, currents, W_f and W_c are all expressed as a function of the electrical angular displacement θ_r.

$$\theta_r = \left(\frac{P}{2}\right) \theta_{rm} \quad (90)$$

Substituting Equation (90) into Equation (89) yields:

$$dW_m = T_{em} \frac{2}{P} d\theta_r \quad (91)$$

Because W_f=W_c the electromagnetic torque can be evaluated with:

$$T_{em}(i_j, \theta_r) = \frac{P}{2}\frac{\partial W_{c(i_j,\theta_r)}}{\partial \theta_r} \tag{92}$$

Substituting Equation (88) into Equation (92) gives the electromagnetic torque of the stator side as

$$\frac{P}{2} i_s^T \frac{\partial [L_{ss}]}{\partial \theta_r} is \tag{93}$$

Where,

$$i_s = [\ iabcs \quad ixyzs] \tag{94}$$

$$L_s = \begin{bmatrix} L_{abcs} & L_{abcxyzs} \\ L_{xyzabcs} & L_{xyzs} \end{bmatrix} \tag{95}$$

$$L_{ss} = (L_s - L_{ls}) \tag{96}$$

Substituting Equation (95) into Equation (93) gives the torque for each of the stator currents.

$$T_{em} = \frac{P}{2}\langle(\ iabcs)^T \frac{\partial [L_{abcs}]}{\partial \theta_r} \quad iabcs$$

$$+(\ iabcs)^T \frac{\partial [L_{abcxyzs}]}{\partial \theta_r} \quad ixyzs + (\ ixyzs)^T \frac{\partial [L_{xyzs}]}{\partial \theta_r} \quad ixyzs +$$

$$+(\ ixyzs)^T \frac{\partial [L_{xyzabcs}]}{\partial \theta_r} \quad iabcs\rangle \tag{97}$$

Applying the transformation from *'abc'* to *'dq0'* yields the torque equation:

$$T_{em} = \frac{3}{2}\frac{P}{2}(\lambda_{d1s}i_{q1s} - \lambda_{q1s}i_{d1s} + \lambda_{d2s}i_{q1s} - \lambda_{q2s}i_{d2s}) \tag{98}$$

4.5. Equivalent circuit

The electrical system of this machine as described by Equations (80)&(84)-(87) can be represented with an equivalent circuit. The zero sequence circuit diagram is omitted because the system is assumed to be balanced.

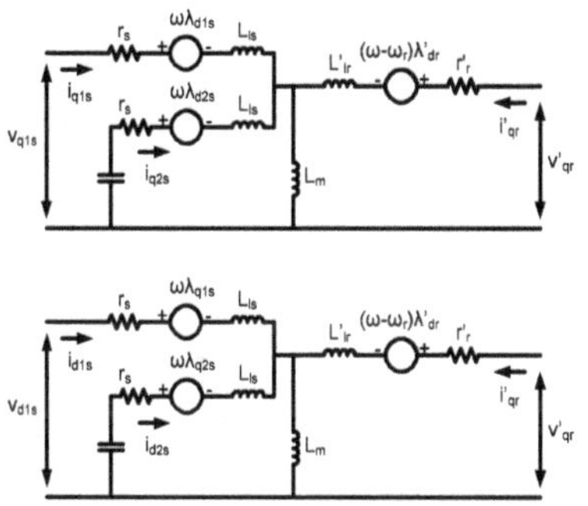

Figure 5. Equivalent *'qd'* circuit diagrams

5. Simulation results

The derived mathematical models implemented in the Matlab/Simulink environment can be used to generate steady-state and dynamic simulation results. The machine without compensation is used as reference. Capacitance is added to the auxiliary winding and compared with the behaviour of the reference machine. This will show the effect of the capacitors connected to the auxiliary winding on the performance of the modified machine. The dynamic model can be used for steady state analysis by taking readings after the transient.

5.1. Steady state analysis

The main objective of the modifications done in terms of the addition of the auxiliary winding is to improve the poor power factor an induction motor has. It is therefore important to focus on the behaviour of the machine parameters that involves power factor when introducing reactive power injection.

Figure 6 shows that the injection of reactive power in the auxiliary winding improves the power factor of the motor. The bigger the size of the capacitor, the more reactive power is injected and hence the better the power factor. For this specific machine, capacitors of 30μF connected per phase as in Figure 4 leads to a power factor very close to unity.

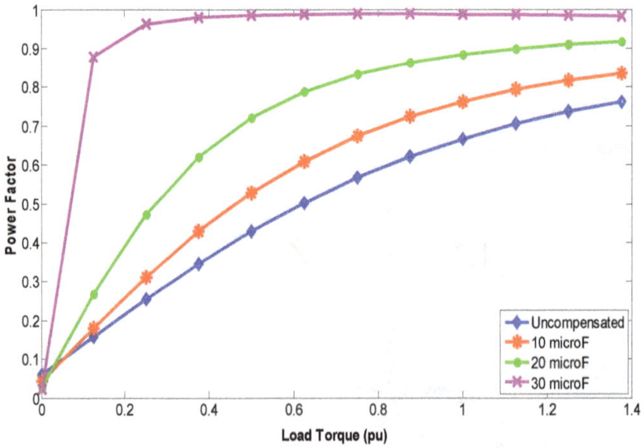

Figure 6. Torque – Power Factor waveform

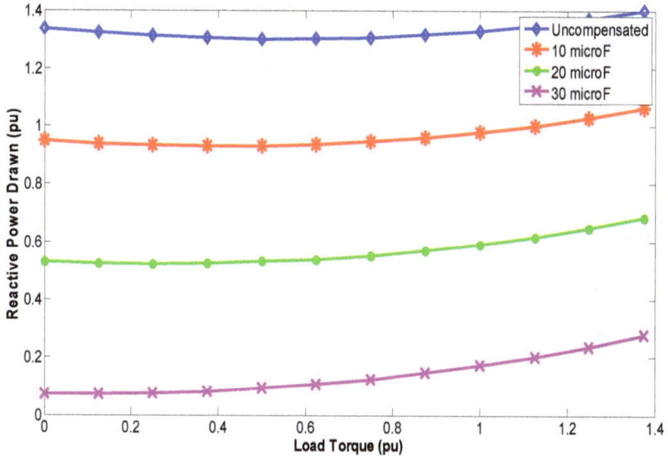

Figure 7. Torque – Reactive Power waveform

With the increase in power factor as seen in Figure 6 it is expected that less reactive power will be drawn from the source with the addition of capacitors to the auxiliary winding. Figure 7 supports this expectation. In Figure 7 the reactive power drawn from the source reduces with increasing capacitor size.

Because the reactive component of the supply current decreases with the reactive power injection, the magnitude of the supply current therefore decreases. This is shown in Figure 8.

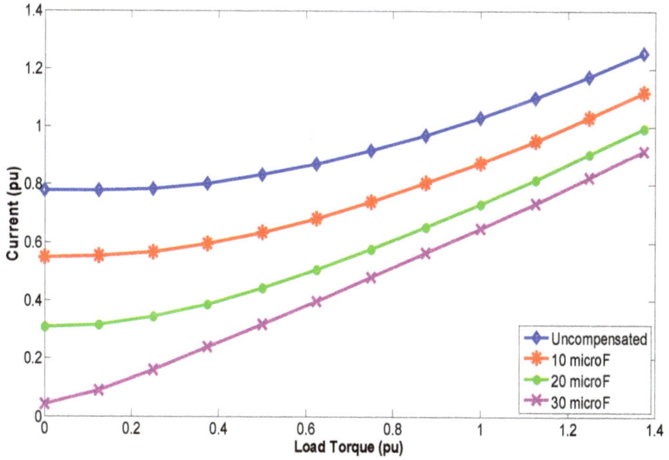

Figure 8. Torque–Current waveform

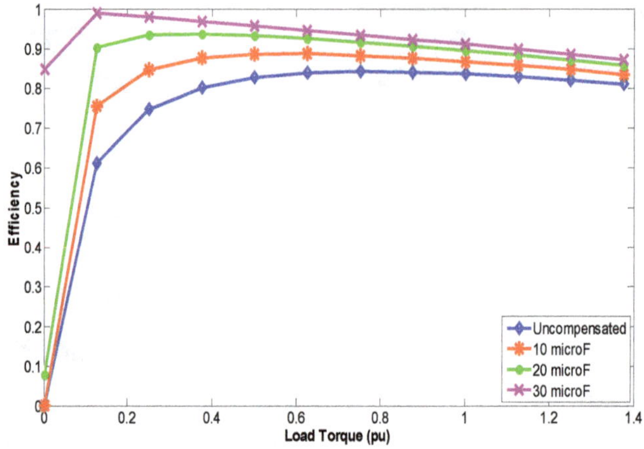

Figure 9. Torque Efficiency waveform

The active power drawn from the source consists of different components of which one is copper losses (I^2R losses). With the decrease of current shown in Figure 8, it is logical that the copper losses of the main stator winding will also decrease. This will lead to a decrease in active power drawn from the source without a change in output power and hence the improvement in the efficiency of the motor as seen in Figure 9.

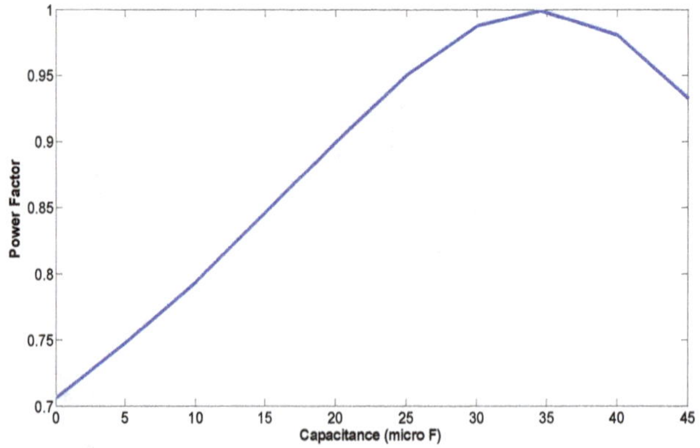

Figure 10. Capacitance-Power Factor waveform

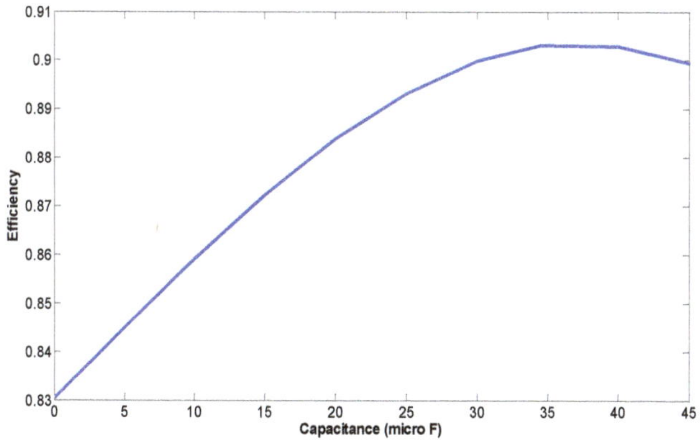

Figure 11. Capacitance-Efficiency waveform

The effect that the change of capacitance has on the performance of the machine is studied in Figures 10 and 11. Figure 10 shows that as the capacitance increase the power factor also increases. It also shows that it is possible to over-compensate the machine which will lead to a decreasing power factor. With the current machine at the current load it can be seen that the optimum value for the capacitor is slightly less than 35μF and will lead to a power factor close to unity. Figure 11 shows the improvement in efficiency as capacitance increases. The efficiency of this machine at current load can be improved with about 0.07 as seen in Figure 11.

5.2. Dynamic analysis

Steady-state analysis is not always sufficient in determining the behaviour of an electrical machine. Transient and dynamic periods are the most likely periods for harming an electrical machine. The dynamic model will show the exact behaviour of the machine during transient and or dynamic periods.

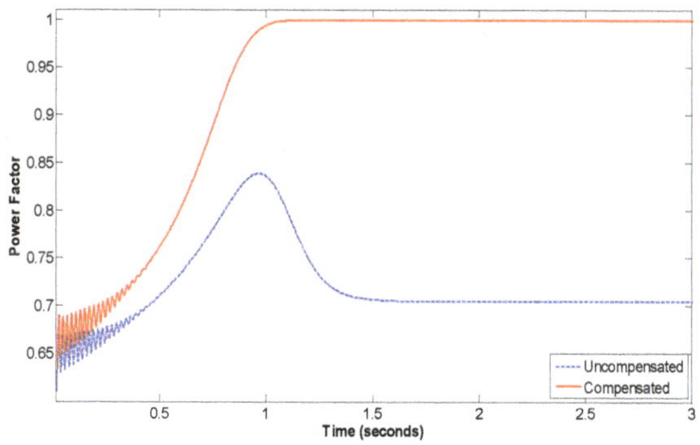

Figure 12. Power Factor

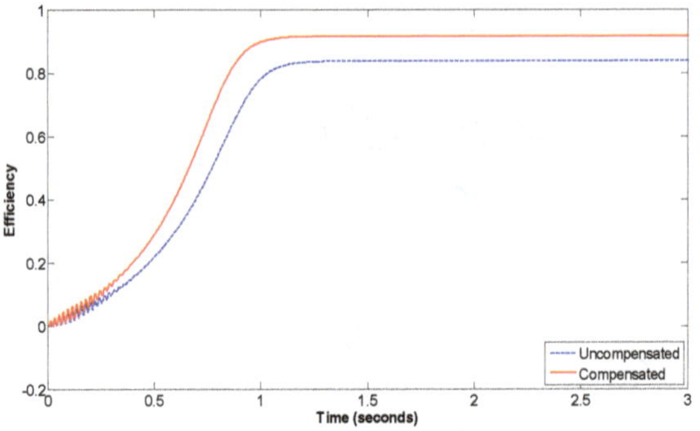

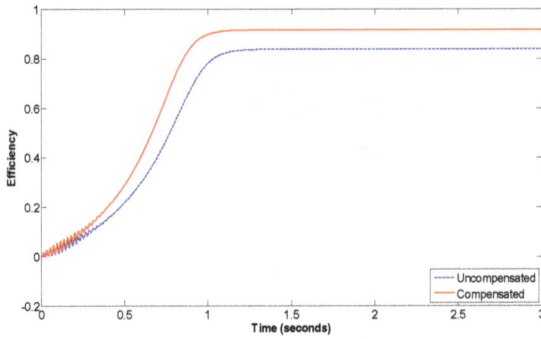

Figure 13. Efficiency

The dynamic behaviour of a compensated and uncompensated induction machine is compared in Figure 12. The uncompensated machine (dashed waveform) has a low power factor when starting and settles at a power factor of just more than 0.7. The compensated machine (solid waveform) has a higher power factor when starting and settles at a power factor close to unity. This shows how effective this concept is in power factor correction.

The earlier statement that the improvement in power factor will improve the efficiency is supported in Figure 13.

The inrush current of the machine is shown in Figure 14. This machine has a transient state when starting where the current can reach eight times rated current.

The current of the auxiliary winding is shown in Figure 15.

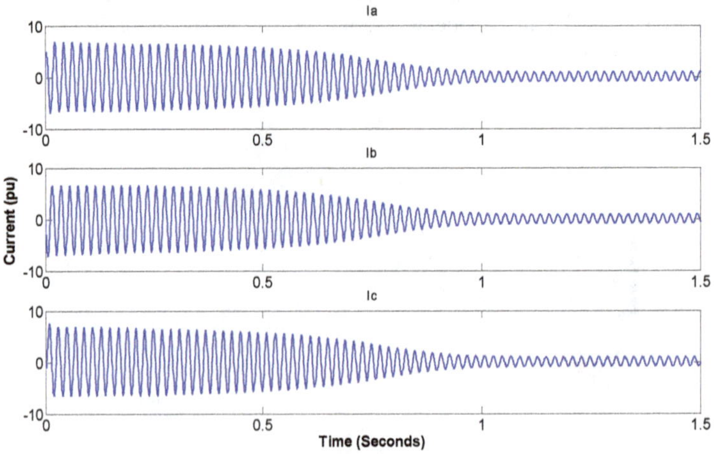

Figure 14. Phase Currents – Main Stator Winding

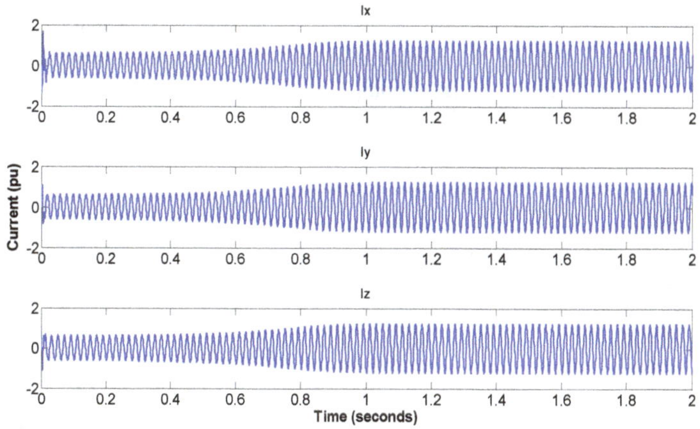

Figure 15. Phase Currents – Auxiliary Stator Winding

6. Experimental validation

In order to validate the theoretical model with the practical model, three capacitor values of 10, 20 and 30μF are used for the three phase auxiliary winding.

Figure 16. The Experimental set up

The stator current of the motor is observed for both uncompensated and compensated windings. It is seen that the starting current for the uncompensated winding is high as compared to the compensated. The current at steady state also identifies the stator current for the uncompensated to be lower as compared to the compensated. These results are shown in figures 17 and 18.

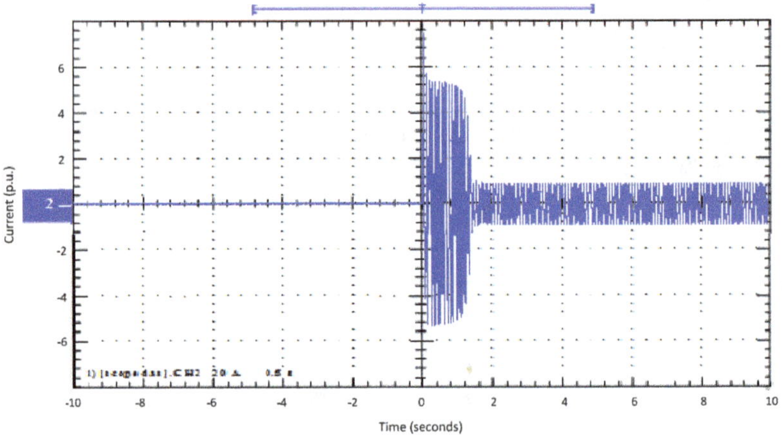

Figure 17. Stator current of uncompensated winding

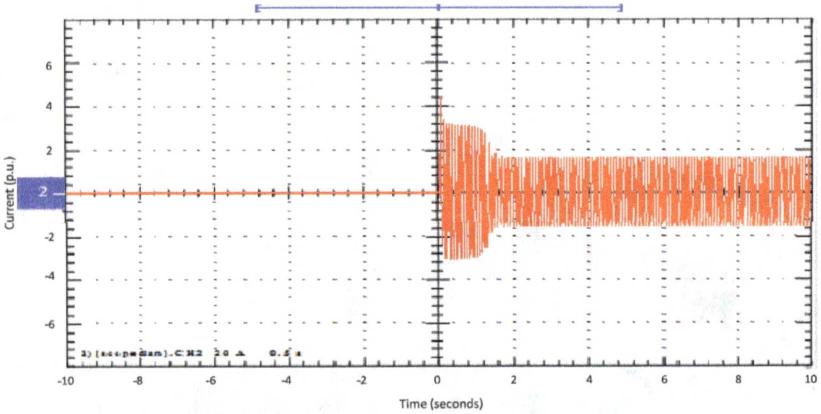

Figure 18. Phase Currents-Stator current of the compensated winding

Other experimental results such as the active power versus the capacitance and power factor versus capacitance are shown in figure 19. These results conform to the theoretical simulations.

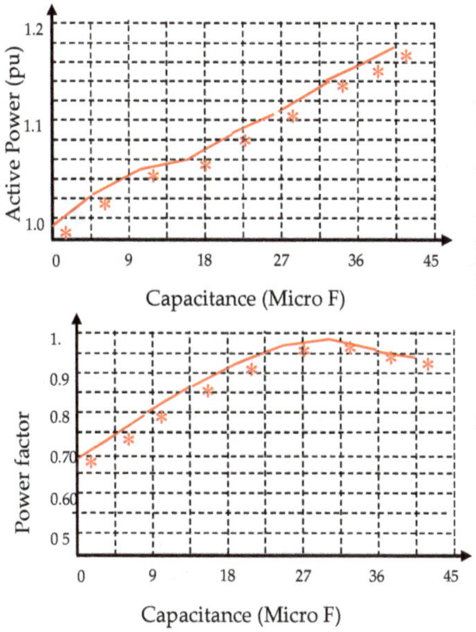

Figure 19. Experimental results of active power and power factor versus capacitance values

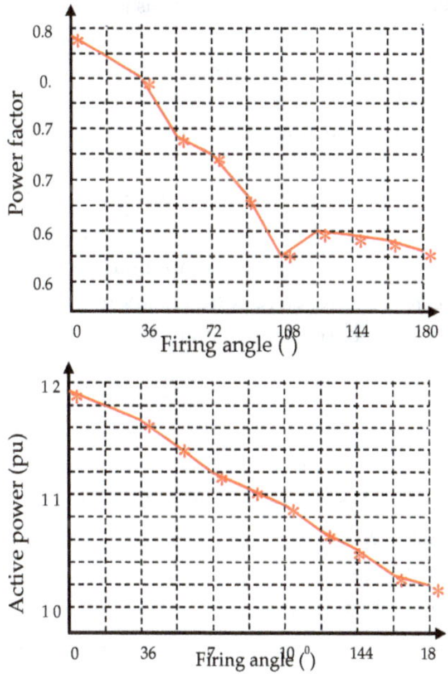

Figure 20. Experimental results of power factor and active power versus firing angle

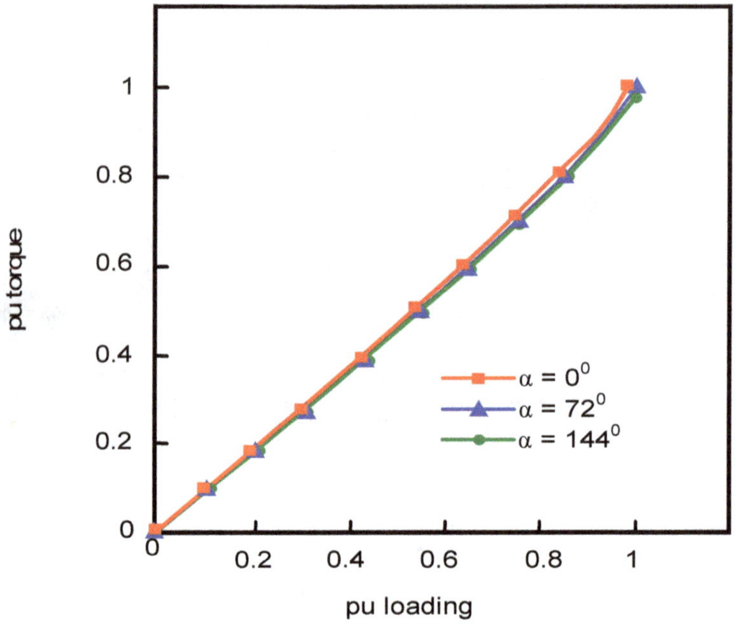

Figure 21. A p.u plot of torque versus p.u. loading for various angles

Since the effective capacitance varies with load when there is an application with varying load, the machine might not always operate at optimum power factor. A possible solution to this is to implement a thyristor controlled static switch or a PWM controller between the capacitors and the auxiliary winding; this will make the capacitance and therefore power factor controllable. Finally, further experiment is carried out on the active power per phase, the power factor versus firing angle and the per unit torque versus per unit loading where the switched series capacitor is connected to the auxiliary winding. The static switching is such that only the required level of reactive compensation is allowed. Figures 20 and 21 are the experimental results obtained based on this analysis.

7. Conclusion

A study has been conducted on a 0.75 KW machine with 380V, 50Hz supply for an effective power factor correction. This has been achieved by connecting the main winding to the three phase supply and the auxiliary winding to the fixed capacitors for reactive power injection. The modified machine with reactive power injection has potential compared to the conventional three-phase machines. It is seen from the waveform analysis that the machine has capability of reducing the starting current. Simulation results have shown a good improvement on both power factor and efficiency when introducing the reactive power injection with increase of capacitor value. Both steady-state and dynamic analysis together with experimental set up has shown a great improvement compared to the uncompensated machine. Another very important improvement is the supply current decreasing with increasing capacitance. This is not the case with conventional power factor correction techniques because the reactive power needed is still drawn through the only stator winding set. This advantage of the modified machine may potentially reduce installation costs as smaller supply cables can be used.

Despite its good performance it has certain drawbacks. The machine would be bigger in structure than a conventional machine. More copper is needed for the additional winding and more insulating material is needed. This would make the machine much more expensive than the conventional three phase machine. Another drawback of this concept is that the capacitors have to be sized for a specific load. When there is an application with varying load, the machine might not always operate at optimum power factor. A possible solution to this is to implement a PWM controller between the capacitors and the auxiliary winding; this will make the capacitance and therefore power factor controllable.

This modified induction machine has a research potential with the recent focus on energy efficiency. Further research needs to be carried out on the performance behaviour of this machine.

Author details

Adisa A. Jimoh, Pierre-Jac Venter and Edward K. Appiah
Tshwane University of Technology, Pretoria, South Africa

8. References

el-Sharkawi, M. A., Venkata, S. S., Williams, T. J. & Butler, N. G. (1985) An Adaptive Power Factor Controller for Three-Phase Induction Generators. *Power Apparatus and Systems, IEEE Transactions on,* PAS-104, 1825-1831.

Jimoh, A. A. & Nicolae, D. V. (2007) Controlled Capacitance Injection into a Three-Phase Induction Motor through a Single-Phase Auxiliary Stator Winding. *Electric Machines & Drives Conference, 2007. IEMDC '07. IEEE International.*

Krause, P. C. (1986) *Analysis of Electric Machinery,* New York, Mcgraw-Hill.

Lipo, T. A. & Novotny, D. W. (1996) Vector Control and Dynamics of AC Drives. In Hammond, P., Miller, T. J. E. & Kenjo, T. (Eds.). New York, Oxford Science Publications.

Muljadi, E., Lipo, T. A. & Novotny, D. W. (1989) Power factor enhancement of induction machines by means of solid-state excitation. *Power Electronics, IEEE Transactions on,* 4, 409-418.

Park, R. H. (1929) Two-Reaction Theory of Synchronous Machines Generalized Method of Analysis-Part I. *American Institute of Electrical Engineers, Transactions of the,* 48, 716-727.

Stanley, H. C. (1938) An Analysis of the Induction Machine. *American Institute of Electrical Engineers, Transactions of the*, 57, 751-7.

Chapter 4

The Dynamics of Induction Motor Fed Directly from the Isolated Electrical Grid

Marija Mirošević

Additional information is available at the end of the chapter

1. Introduction

An induction motor is the most common machine used for industrial drives. It is used in variety of drives due to its robust construction, relatively low cost and reliability. In isolated electrical network, such as marine and offshore power systems and emergency generation plant, induction motors are the most power consuming loads and are used for winches, water pumps, compressors, fans and for other on-board applications, in continuous mode or intermittently.

In many applications induction motors are direct-on-line switch-started. Their dynamic characteristics have an obvious influence on the transient process of power system; however, they cause a significant disturbance in transients (significant impacts loads) that can produce disturbances in the isolated electrical network, which in turn affects the quality of electric power system and thus, on the dynamic behavior of induction motors.

Direct-on-line starting represents the simplest and the most economical system to start squirrel-cage induction motor. During starting induction motors draw high starting currents which are several times the normal full load current of the motor. This current causes a significant voltage dip on the isolated electrical grid until the induction motors reach nearly full speed. This voltage drop will cause disturbances in the torque of any other motor running on the isolated electrical grid. Significant disturbances in transients are caused by direct-on-line switch-started induction motors, especially if the load torque on the motor shaft is increased and beside that, also by the sudden change load, such as the impact load on the motor shaft (McElveen, et al., 2001; Cohen, 1995). This situation is particularly difficult because of relatively strong electrical coupling between electrical grid and loads. Besides the voltage dips, an interruption can also appear, which further affects on the fatigue of induction motors connected to the grid. When an interruption of the supply lasts longer than one voltage period, many AC contactors will switch off the motor. In some cases, faulty contactor may produce multiple switching on and off. However, these interruptions will affect the dynamics of both electrical and mechanical variables. Therefore, it is interesting to analyze dynamics of the induction motor in case when it comes to short-term interruptions in the motor power supply.

That's why the dynamic behavior of induction motors fed directly from isolated electrical grid, as well as dynamics of aggregate is in focus of researchers. Presently, advanced modeling and digital simulation techniques can be used to analyse the dynamics behavior of electrical as well as mechanical systems.

2. Mathematical models used in isolated electrical grid

The aim is to analyze the dynamics of the induction motors fed directly from the isolated electrical grid. For this purposes the mathematical model of isolated electrical grid has been develop consisting diesel electrical aggregate and unregulated induction motors.

Diesel generators are used as the main sources of electricity in cases of isolated systems. In many applications the diesel generator can suffer significant impacts loads that can produce disturbances in the isolated grid. However, the autonomous operation of the synchronous generator is characterized by a change in steady state which causes a change in voltage and frequency, which in turn affects the quality of electric power systems.

The model of diesel electrical aggregate considered in this study consists of: a diesel engine and a speed controller, a synchronous generator and a voltage controller, a mechanical connection between engine and electrical machine shaft.

The synchronous generator is presented as a machine with three armature windings, a magnetizing winding on the rotor and damper windings. One damper winding is located along direct-axis (D), and one along the quadrature-axis (Q). The basis of the mathematical model is a set of differential equations of the synchronous generator in the standard *dq*-axis form (Kundur, 1994). The voltage equations are written in generator (source) convention system, in which synchronous machines are usually represented:

$$-u_d = r \cdot i_d + \frac{d\psi_d}{dt} - \omega \cdot \psi_q \tag{1}$$

$$-u_q = r \cdot i_q + \frac{d\psi_q}{dt} + \omega \cdot \psi_d \tag{2}$$

$$E_q = e_q + \frac{x_{1d}}{r_1} \cdot \frac{d\psi_1}{dt} \tag{3}$$

$$0 = r_D \cdot i_D + \frac{d\psi_D}{dt} \tag{4}$$

$$0 = r_Q \cdot i_Q + \frac{d\psi_Q}{dt} \tag{5}$$

where u, i, r, and ψ denote voltage, current, resistance and flux respectively.

The model of a synchronous generator is given in rotor reference frame (ω is rotor electrical speed). The equations of excitation in motor (load) convention system are written. The voltage controller, modeled as PI, is implemented in the model of the synchronous generator.

The model of the prime mover - the diesel engine assumes that the engine torque is directly proportional to the fuel consumption. In order to describe the dynamic behavior of the diesel engine it is necessary to set up a system of differential equations which includes an equation of the engine, the turbocharger, the air collector, the exhaust system, and the speed controller. Taking into account these equations requires the knowledge of characteristics of diesel engines that require complex experimental measurements, according to (Krutov, 1978; Tolšin 1977). The studies carried out in (Erceg & et al. 1996) showed that the mentioned omissions do not affect significantly the results and that the proportionality of torque to the amount of injected fuel can be assumed. This simplification is allowed when it is of interest to observe dynamics of a synchronous generator as well as induction motors. The speed controller is modeled as PI and implemented in the model.

Sudden impact load on the diesel electrical aggregate is the most difficult transition regime for units due to electricity loads and also due to torsional strains in the shaft lines. A more significant disturbance, which is at the same time very common in practice, is the direct-on-line starting of induction motors to such a grid. Starting of induction motors will cause voltage dips and will reduce engine speed depending on the time of the starting of each motor. This will also cause torsional stresses in the shaft. Thus, the mechanical coupling of a diesel engine and a synchronous generator is considered to be a rotating system with two concentrated masses. Masses are connected by flexible coupling. The flexible coupling allows these masses to rotate at a different speed in transients.

The variable angle of rotation between these masses occurs during the transient, in period when mechanical balance between diesel engine and electric generator is disturbed., The torque which appears at coupling zone between two concentrated rotational masses allows thus the analysis of the torsional dynamics in the coupling.

Induction motor as an active consumer and its parameters were analyzed in (Maljkovic, 2001; Amezquita-Brook et al., 2009). According to (Jones, 1967; Kraus 1986) three phase squirrel-cage induction motors are represented with stators and rotors voltage equations:

$$u_{dIM} = R_{sIM} \cdot i_{dIM} + \frac{d\psi_{dIM}}{dt} - \omega \cdot \psi_{qIM} \tag{6}$$

$$u_{qIM} = R_s \cdot i_{qIM} + \frac{d\psi_{qIM}}{dt} + \omega \cdot \psi_{dIM} \tag{7}$$

$$0 = R_r \cdot i_{DIM} + \frac{d\psi_{DIM}}{dt} - \left(\omega - \omega_{IM}\right) \cdot \psi_{QIM} \tag{8}$$

$$0 = R_r \cdot i_{QIM} + \frac{d\psi_{QIM}}{dt} + (\omega - \omega_{IM}) \cdot \psi_{DIM} \tag{9}$$

where: *u*, *i*, R, and ψ denote voltage, current, resistance and flux respectively of an induction motor.

All winding currents, in the transient *dq* axis model of induction motors as well as in a synchronous generator model, are selected as state variables. The model is completed with an equation of the rotational mass motion (Vas, 1996). All variables and parameters are in per unit (p.u.). The motor's equation of motion involves electrical torque (T_{eIM}), whereas (T_{lIM}) represents load torque on the motor's shaft.

When the induction motor starts unloaded, then the torque T_{lIMn} equals zero. Also, for this analysis the loading with a constant load was selected.

Loads, induction motors (index IM), are connected directly to a synchronous generator (index SG), what means that they are on the same voltage as the generator terminals: $-u_d = u_{dIM1} = u_{dIM2}$, $-u_q = u_{qIM1} = u_{qIM2}$. According to the Kirchhoff's law, the current relationship between supplying and receiving elements are: $i_d = i_{dIM1} + i_{dIM2}$, $i_q = i_{qIM1} + i_{qIM2}$.

The validity of the mathematical model of the generator-unit at impact load, direct-on-line starting of non-loaded induction motor, was checked in the previous work (Mirosevic, et al. 2002a, 2011b) by comparing the results of the simulation and the measurement on the generator-unit with a diesel engine of 46.4 kW, 1500 r/min and a synchronous generator of 40 kVA (3x400/231 V, cos φ=0.8; 57.7 A; 1500 r/min; 50 Hz), to which a motor drive of 7.5 kW (Δ 380 V, 14.7 A, 2905 r/min, cos φ=0.9) was connected. The results obtained by numerical calculation indicate that, the set mathematical model can be applied with sufficient certainty

The analysis of the dynamics of induction motors fed directly from the isolated electrical grid was performed by the application of program package "Matlab/Simulink". The block diagram of integral motor drives is presented in Figure 1 involves: a diesel engine (DM, SC), a three phase synchronous generator and voltage controller, their mechanical coupling and induction motors fed directly from the synchronous generator terminals.

Block diagram of Diesel engine and speed controller is presented in Figure 2 and represents subsystem of block naimed as DM SC in Fig. 1.

The Simulink is used to obtain a model of a diesel generator unit, as well as induction motors by means of basic function blocks that can be linked and edited to subsystem such as subsystems IM 1 and IM 2 in Figure 1 which representthe first and the second induction motors respectively. As one can see in Figure 3 components of the subsystem IM 1 that are used in the calculation of variables are presented.

Induction motors are connected to the network using power supply subsystem, while the load on the motor shaft is represented with subsystems: load IM1 and load IM2 in Figure 1.

Moment of switching/disconnecting on the network is controlled by means of the subsystem ON/OFF, also, the part of this subsystem is used to set the time of switching/disconnecting the load on the motor shaft.

All calculations were carried out by means of the "Variable-Step Kutta-Merson" method – an explicit method of the fourth order for solving the systems of differential equations, with variable integration increment.

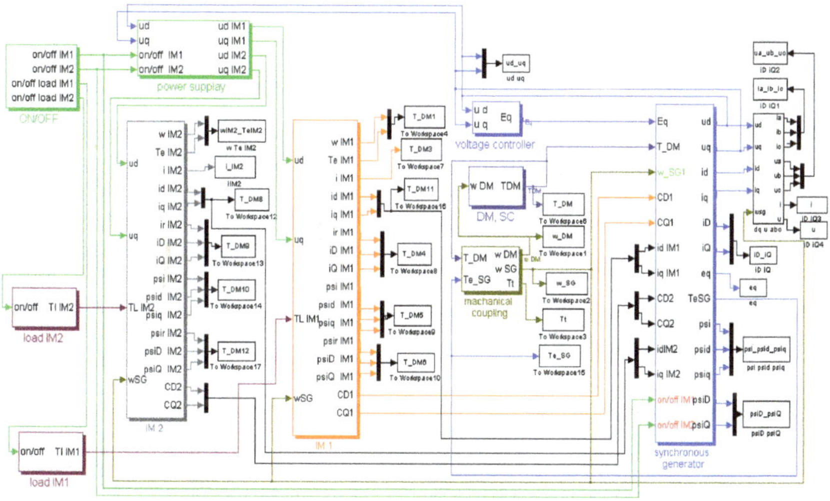

Figure 1. Simulation of the applied mathematical model

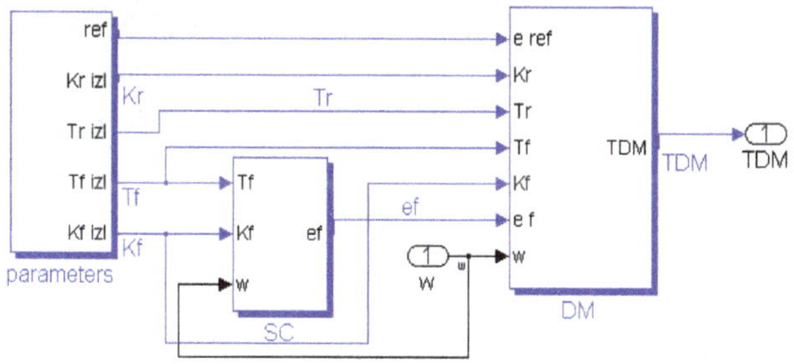

Figure 2. Diesel engine and speed controller model

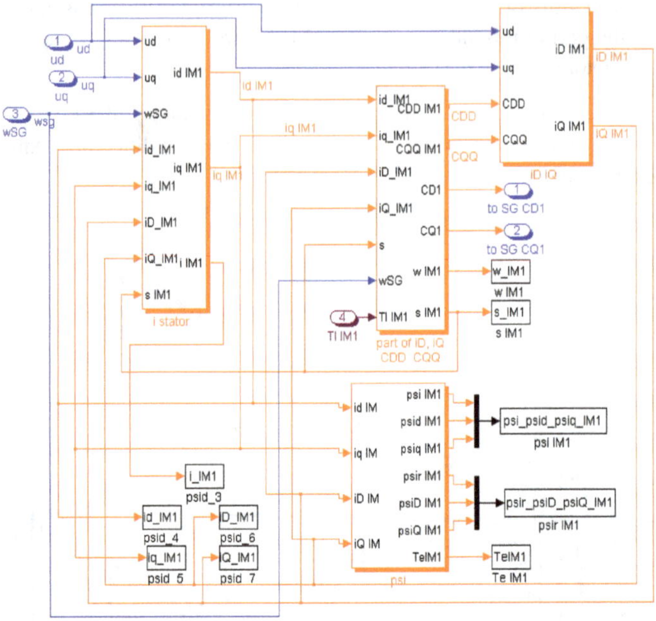

Figure 3. Induction motor model

3. Induction motor starting

At the beginning, two induction motors are connected directly to an isolated electrical grid. The first induction motor (IM1) is connected to the terminals of the aggregate and later, when the first motor has run-up successfully, the second induction motor (IM2) is connected to the grid.

The synchronous generator is initially in a steady state unloaded condition. However, in this condition stator current is zero, rated voltage is on its terminals, while rotation speed of the diesel engine (ω_{DM}) is equal to the speed of the generator (ω_{SG}) and is equal to 1 p.u.

The first observed dynamics is for the case of the starting of unloaded induction motors and in the second case dynamics of loaded induction motors are considered. Transients of: air-gap torque and speed transient of induction motors, terminal voltage, speed transient of synchronous generator and diesel engine, for both cases are presented in Figure 4.

Initially, the first induction motor is starting from rest, the rated voltage is applied on its terminals and there is no mechanical load on the motor shaft. When the induction motor is connected, the load on the aggregate is instantaneously increased, defined in the initial (sub-transient) phase of the transitional phenomenon by locked-rotor torque of the induction motor. As the motor accelerates, its torque grows and the generator load rises. When maximum torque is achieved, the load of the synchronous generator reaches its maximum and then decreases rapidly.

At the instant of starting, as one can see in Fig. 4a, the air-gap torque is momentarily increased; reaches maximum value of 0.86 p.u. and change in it can be noticed during the whole start-up period of the first induction motor. The instantaneous torque oscillates about positive average value.

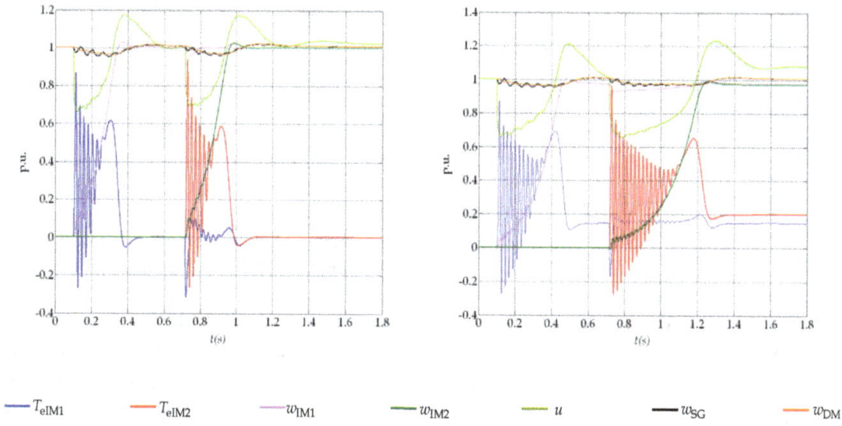

(a) (b)

Figure 4. Transients of: air-gap torque of induction motors (T_{eIM1}, T_{eIM2}) and speed transient of induction motors (ω_{IM1}, ω_{IM2}); terminal voltage (u), speed transient of synchronous generator (ω_{SG}) and diesel engine (ω_{DM}), during direct-on-line starting of: a) unloaded b) loaded induction motors

The oscillations in the air-gap torque are caused by the interactions between the stator and rotor flux linkage. The negative oscillations in the electromagnetic torque of the induction motor are presented at the beginning of the start-up period. These are periods of momentary deceleration that occur during regeneration when the electromagnetic torque becomes negative. The rotor speed only increases when the torque is positive. The oscillations that are present in transient of air-gap torque of the first induction motor are damped at the end of start up period and finally the steady state condition is attained without oscillations.

The response of the air-gap torque is in accordance with the response of the motor currents. Transients of stator currents of induction motors and their components, for both cases, are presented in Figure 5.

Under this condition the starting current is large. The starting current of an induction motor is several times larger than the rated current since the back emf induced by Faraday's law grows smaller as the rotor speed increases. However, a large starting current tend to cause the supply voltage to dip during start-up and can cause problems for the other equipment that is connected to the same grid.

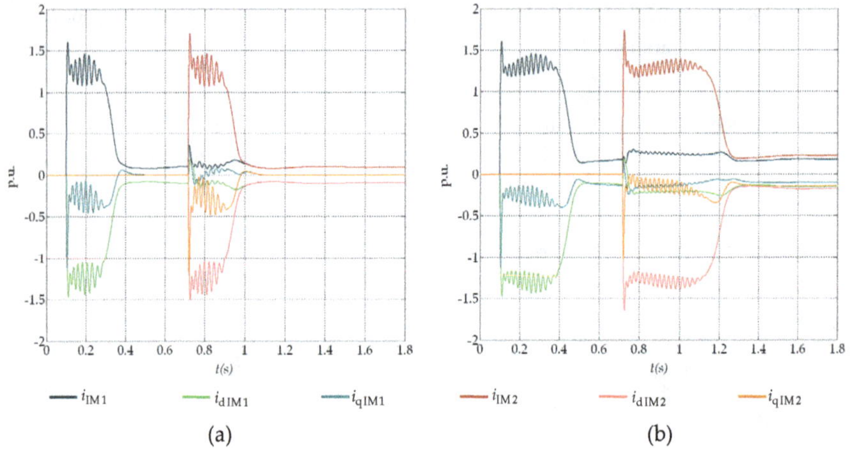

Figure 5. Transients of stator currents of induction motors (i_{IM1}, i_{IM2}) and their components in d (i_{dIM1}, i_{dIM2}) and q (i_{qIM1}, i_{qIM2}) axis during direct-on-line starting of: a) unloaded b) loaded induction motors

At the instant of starting, when the supply has just been switched on the induction motor, the first magnitude of starting current momentarily reaches maximum value of 1.61 p.u. as it is presented in Fig. 5a. The damped oscillations, that are present in stator current transients, disappear at the end of the starting period of the induction motors. When the first motor has run-up successfully, the second induction motor (IM2) is connected to the loaded synchronous generator. Involvement of the second induction motor to the isolated electrical grid the network load instantaneously increased and voltage drop occurs.

The terminal voltage is momentary decreased (Fig. 4a) and after few damped oscillations reached minimal value. However, the high starting currents are appeared. High inrush current, in the first moments, as one can see in Figure 5, reaches the magnitude of the first oscillation of 1.73 p.u. The air-gap torque of the second induction motor T_{eIM2} momentarily achieves 0.94 p.u.

At the moment of switching on to the grid, the second induction motor begins to accelerate and oscillations are present in its transients of air-gap torque during the acceleration period. At the time of the starting of second induction motor (IM2), the reverse torque impulse of 0.31 p.u. in air-gap torque of the first one (IM1) is appeared but decayed rapidly. As the torque of the first induction motor becomes negative motor speed slows down. Thereafter, damped oscillations that are present in the response of electromagnetic torque of the first motor as well as oscillations in electromagnetic torque of the second one are stifled at the end of the run-up period of the second induction motor.

At the beginning of the start-up period of the second motor, the speed of the first one decreases and afterwards recovers. Overshoot in the speed transient occurs at the end of start-up period of the second induction motor.

The starting of loaded induction motor is more difficult transition regime for the aggregate, therefore, the transients of aggregate are slower.

Thus, in the second case, the dynamics of the starting of loaded induction motors is analyzed, however, the load on the first one is T_{lIM1}=0.15 p.u. while T_{lIM2}=0.2 p.u. is applied on the second one. In this case the acceleration time is longer than in the previous case when motors are started unloaded (Fig. 4a), the voltage of synchronous generator is recovering slower, and will be lower then 80%U_n during greater part of start up period, as presented in Fig. 4b.

The initial part of the transients of electromagnetic torque is equal in both cases (load and non-load condition). At the time of the starting of the first induction motor the torque T_{eIM1} reaches a value of 0.86 (p.u.).

At the instant of connection to power supply the instantaneous torque is independent of the balanced source voltages because the machine is symmetrical, even air-gap torque depends upon the values of source voltages though the stator currents. In addition, the air-gap torque oscillates with higher magnitude, about lower average then in case of unloaded motor. These oscillations are damped at the end of the start-up period of the induction motor as presented in Figure 4b.

The component of air-gap torque, which appears because of mutual acting of free currents in stator as well as in rotor windings acts as counter torque on motors shaft and disappears before the end of the run-up. As one can see in Fig 4 the duration of the start-up of both induction motors is longer than in the previous case, in which the motors run-up unloaded. Thus, this acceleration period of second induction motor is 500 ms, while in the previous case lasted 230 ms. In Fig. 5b stator currents, in *dq* axis, and their components, during start up of loaded induction motor, are presented. As the induction motor is directly connected to the terminals of unloaded synchronous generator that means that stator current of induction motor is at the same time the armature current of synchronous generator.

At the beginning of the transient phenomena inrush current which appears during the first half period is dominating but disappears quickly. After initial damping, oscillations of free currents will continue with slightly greater magnitude than at the beginning of transients. These currents, which also can be seen during the start up period of unloaded induction motor, disappear at the end of start up. Corresponding stator flux linkage, during direct-on-line starting of unloaded and loaded induction motor are presented in Fig. 6.

The transients of the first induction motor current in *abc* coordinate system (i_{abcIM1}), in both cases, are presented in Fig. 7. The current of synchronous generator (i_{abcSG}), in both cases, is presented in Fig. 8. This is the total current that motors draw from the electrical grid.

The phenomena of voltage and frequency deviation are typical for isolated electrical grid which in turn affects the quality of electric power systems. The short-term frequency deviation, during direct-on-line starting unloaded and loaded induction motors are presented in Figure 9. There is relatively strong electrical coupling between synchronous generator and loads as well as torsional strains in the shaft line. However, the torque in the coupling for both cases is presented in Figure 10. Oscillations in transients of torsion torque are longer present during direct-on-line starting loaded induction motor and damped at the end of start-up period.

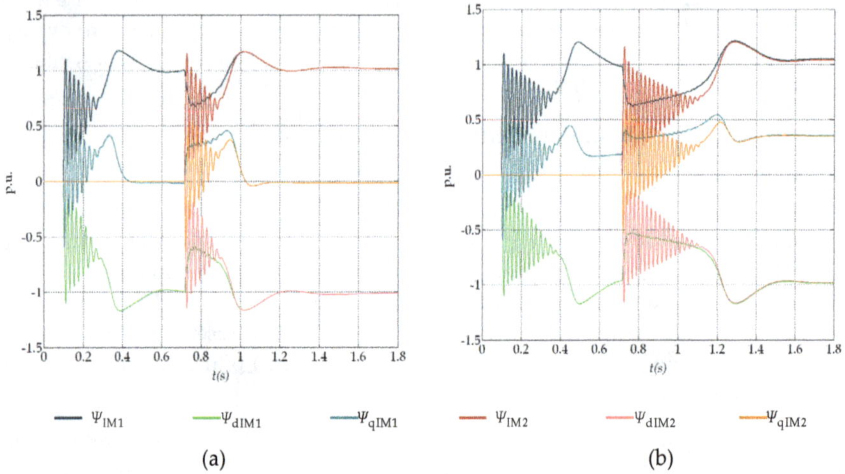

Figure 6. Transients of stator flux linkages of induction motors (Ψ_{IM1}, Ψ_{IM2}) and their components in d (Ψ_{dIM1}, Ψ_{dIM2}) and q (Ψ_{qIM1}, Ψ_{qIM2}) axis during direct-on-line starting of: a) unloaded b) loaded induction motors

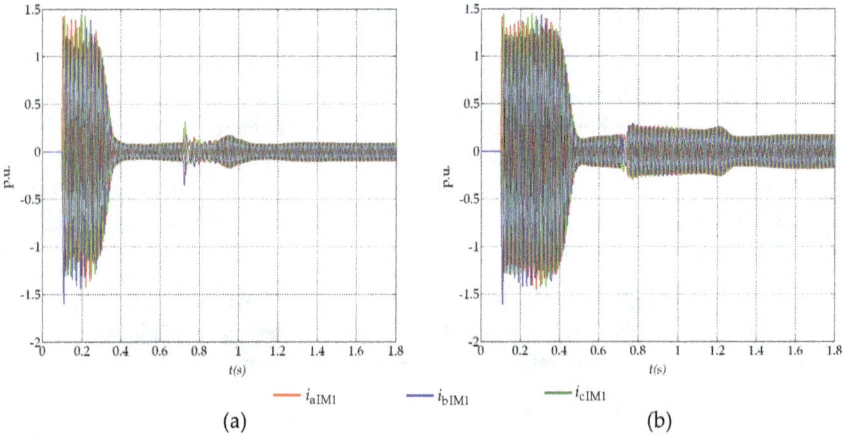

Figure 7. Stator current (i_{abcIM1}) of first induction motor, during direct-on-line starting of: a) unloaded b) loaded induction motors

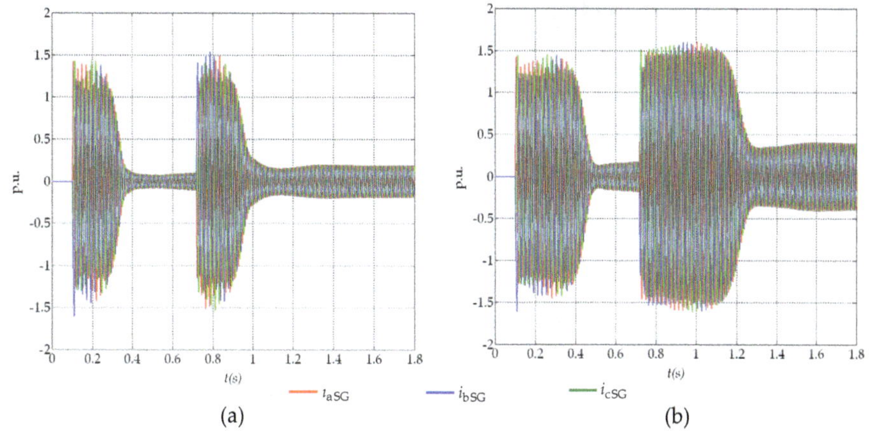

Figure 8. Stator current (i_{abcSG}) of the grid (synchronous generator), during direct-on-line starting of: a) unloaded b) loaded induction motors

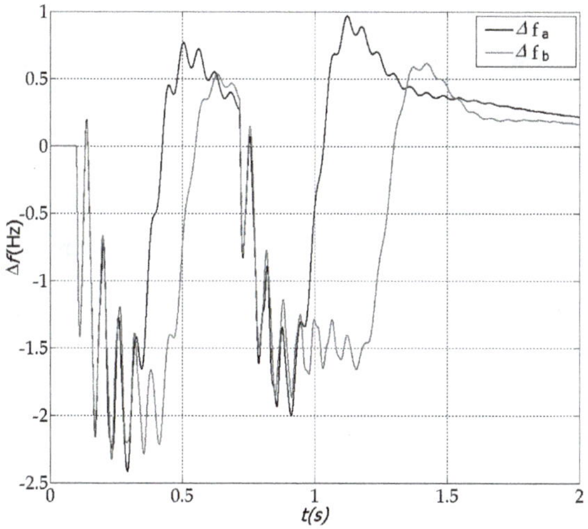

Figure 9. Frequency variations during direct-on-line starting: Δf_a-motors are unloaded, Δf_b -motors are loaded

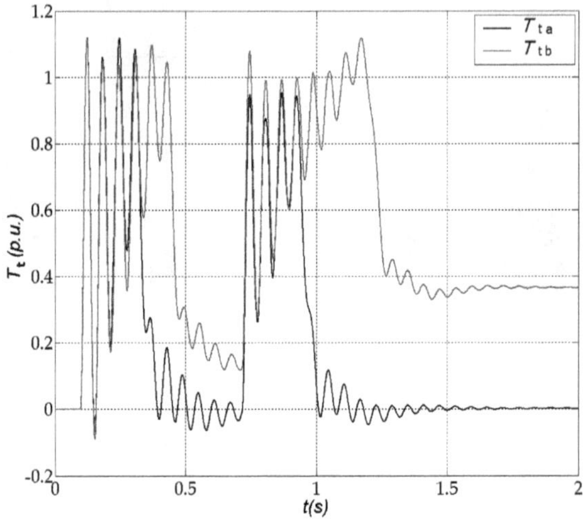

Figure 10. Torsional torque during direct-on-line starting: T_{ta}-motors are unloaded, T_{tb}-motors are loaded

4. Induction motor under sudden change load

The dynamics of sudden change load on the motor shaft is considered. Induction motors are connected directly to the grid, as in previous case, and at chosen moment the first induction motor is starting up, loaded of T_{IIM1} =0.05 p.u. When the first motor has run-up successfully, the second induction motor (IM2) is connected and it starts with T_{IIM2} =0.05 p.u. load on its shaft. Before the second induction motor is connected to the grid the aggregate was led to the steady operation conditions. During start-up of the second induction motor, sudden load change at the first induction motor shaft appeared.

Two cases are considered. In the first case, the impact load of additional 0.1 p.u. on the first induction motor IM1, is applied. While in the second case, the motor is suddenly unloaded, 0.05 p.u. is disconnected from its shaft. Thus, after disconnecting the load the first induction motor run at idle. Transients of air-gap torque and speed transient of induction motors for both cases are presented in Figure 11 and 12.

Direct on line starting of induction motors induces high strain on the power system. This strain arises when the second motor is connected, and additionally is growing up at the instant of impact 0.1 p.u. load on the first motor shaft. When the second induction motor is connected to the grid, it begins to accelerate and the torque of the first induction motor is changed. The electromagnetic torque of the first induction motor becomes negative and motor speed slows down, achieving about 0.95 p.u. Later on, the speed of the first induction motor starts recovery. At the instant of impact additional load on the motor shaft the speed is decreasing again. The speed of the first induction motor is recovering with strongly damped oscillations at the end of start-up period of the second induction motor (Fig. 11a and Fig. 12a).

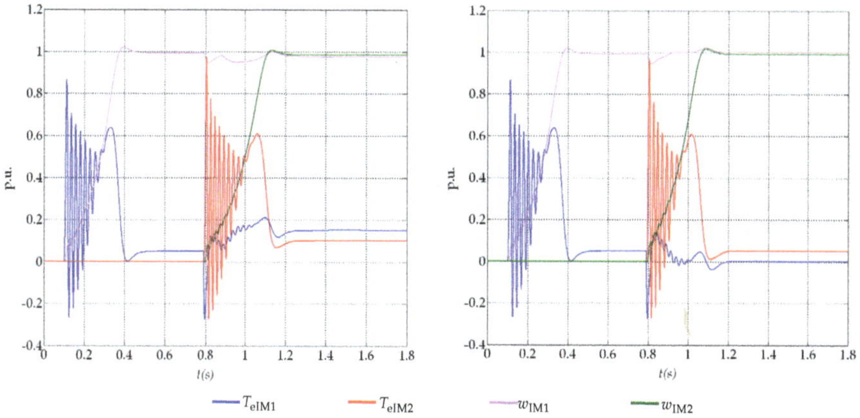

(a) (b)

Figure 11. Transients of: air-gap torque (T_{eIM1}, T_{eIM2}) and speed transient of induction motors (ω_{IM1}, ω_{IM2}); during sudden change load: a) impact load b) load disconnected

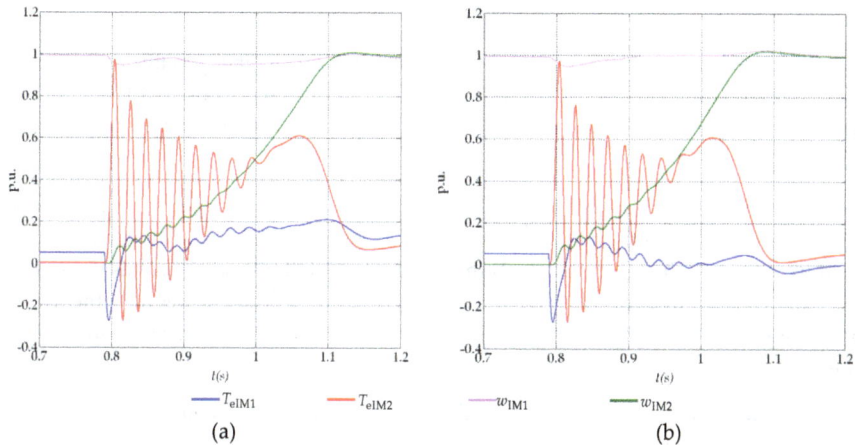

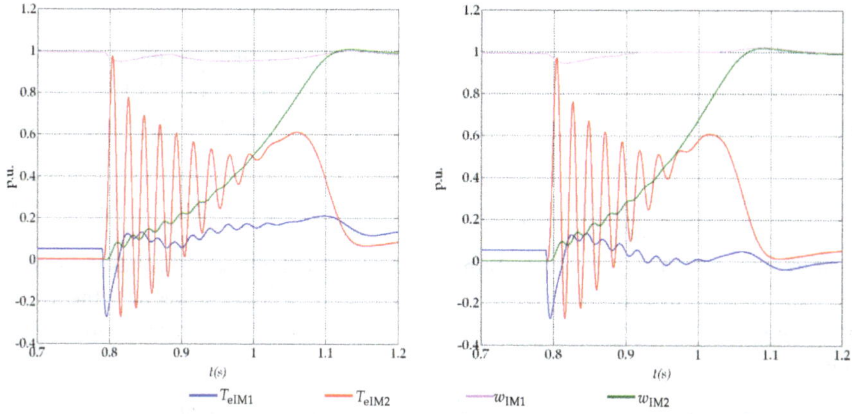

(a) (b)

Figure 12. Transients of: air-gap torque (T_{eIM1}, T_{eIM2}) and speed transient (ω_{IM1}, ω_{IM2}); during sudden change load: a) impact load b) load disconnected, detail of Fig. 11.

Transients of stator currents of induction motors and their components, for both cases, are presented in Figure 13. The current of the first motor, at the moment of connection IM2, momentarily increases to 0.31 p.u. and then, with damped oscillation tend to decrease. At the moment of impact additional load the current of the first induction motor is growing up (Fig. 13a). Air-gap torque of the first induction motor, after short term negative value of 0.27 p.u., oscillate about the positive average and tend to decrease.

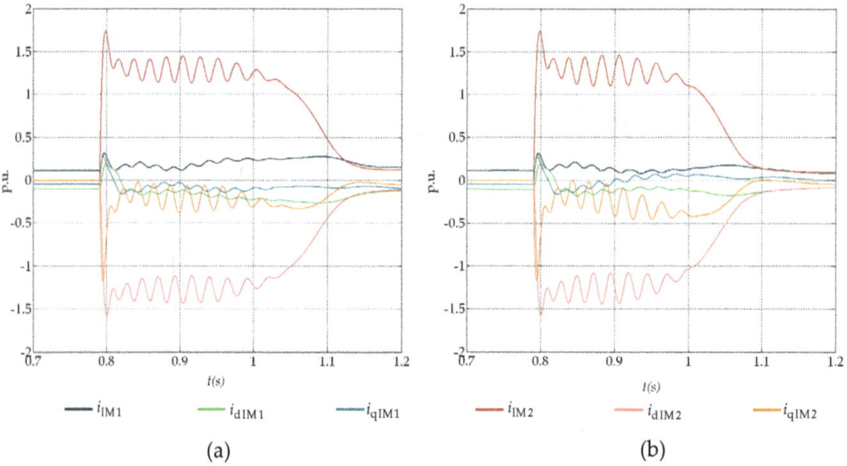

Figure 13. Transients of stator currents of induction motors (i_{IM1}, i_{IM2}) and their components in d (i_{dIM1}, i_{dIM2}) and q (i_{qIM1}, i_{qIM2}) axis during sudden change load: a) impact load b) load disconnected

Thereafter, the air-gap torque is increased, oscillates about the higher positive average then before impact additional load. The electromagnetic torque, as one can see in Fig. 12a, with damped oscillation reaches maximal value of 0.2 p.u. at the end of the start up period of the second induction motor.

The inrush current that is appeared at the beginning of the start-up period of the second induction motor reaches the magnitude of the first oscillation of 1.73 p.u. This current causes the voltage drop at the motor terminals that are connected on the same grid as the second one. The voltage drop results that the motor speed slows down. Because the motor rotor slows down the higher current is appeared on the grid and voltage are reduced even more.

As the additional load is connected the voltage is slowly recovered and start-up period of the second induction motor takes 300 ms.

In the second case, the motor is suddenly unloaded, 0.05 p.u. is disconnected from its shaft at the beginning of the start-up period of the second induction motor. After starting, the second motor begins to accelerate and the torque of the first one is changed, as mentioned before it becomes negative and motor speed slows down. Thereafter, the speed of the first induction motor continues recovering the whole start-up period of the second induction motor (Fig. 11b). At the moment of sudden unload, the current of the first motor continues decreasing and with damped oscillation reaches steady state (Fig. 13b). Air-gap torque of the first induction motor continues decreasing reaches steady state faster than in the first case, in case of sudden impact load (Fig. 12b).

The inrush current that is appeared at the beginning of the start-up period of the second induction tends to reducing, and after the first induction motor is suddenly unloaded continues decreasing (Fig. 13b). Oscillations in transients of air-gap torque are shorter present and are damped at the end of start up period of the second induction motor, as one can see in figure (Fig. 12b). Corresponding transients of stator flux linkage, in both cases, are presented in Fig. 14. As the grid is unloaded the voltage is recovered, and start-up period of the second induction motor is now shorter, it takes about 260 ms.

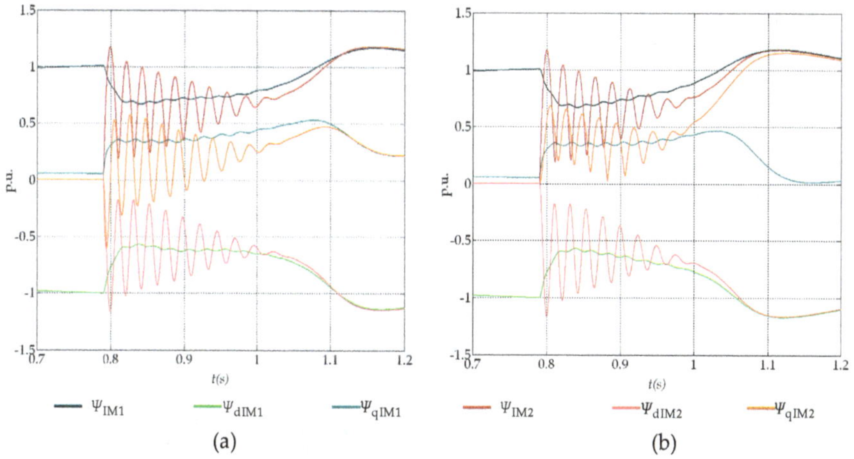

Figure 14. Transients of stator flux linkages of induction motors (Ψ_{IM1}, Ψ_{IM2}) and their components in d (Ψ_{dIM1}, Ψ_{dIM2}) and q (Ψ_{qIM1}, Ψ_{qIM2}) axis during sudden change load:
a) impact load b) load disconnected

The transients of the first induction motor current in *abc* coordinate system (i_{abcIM1}), in both cases, are presented in Fig. 15, while the current of synchronous generator (i_{abcSG}) is presented in Fig. 16.

Frequency variation during sudden change load is presented in Figure 17 and, as one can see, that impact load on the motor shaft results in short-term frequency decreasing. When load is switched off, short-term increase of frequency is appeared. Sudden change load affects on speed of aggregate and torque in the coupling, for both cases, is presented in Figure 18.

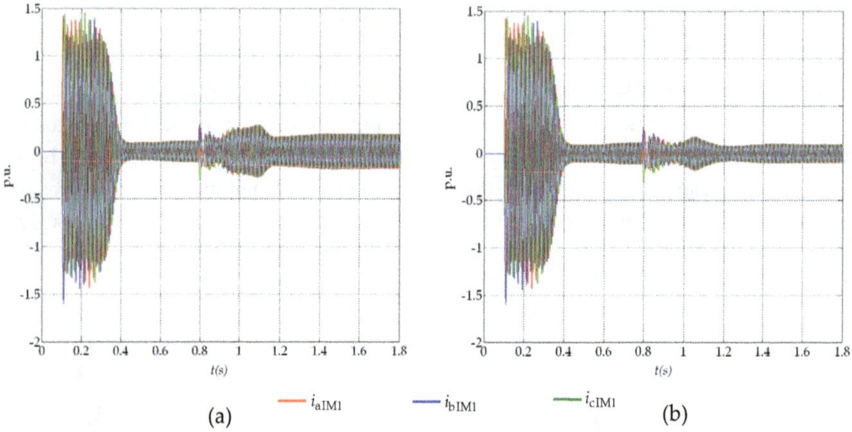

Figure 15. Stator current (i_{abcIM1}) of first induction motor, during sudden change load: a) impact load b) load disconnected

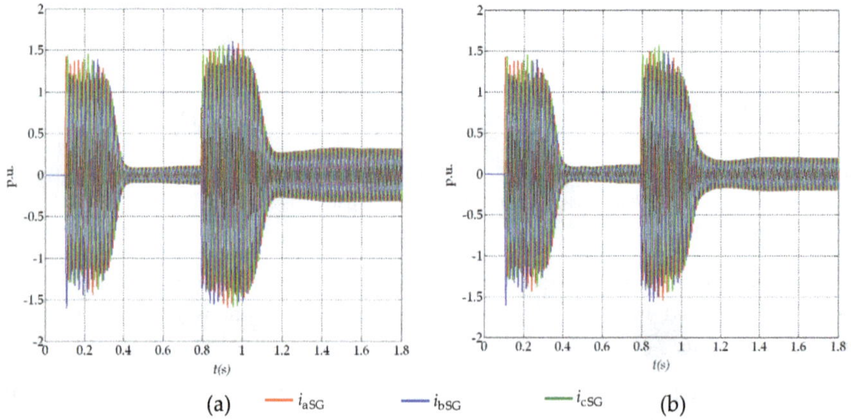

Figure 16. Stator current (i_{abcSG}) of the grid (synchronous generator), during sudden change load: a) impact load b) load disconnected

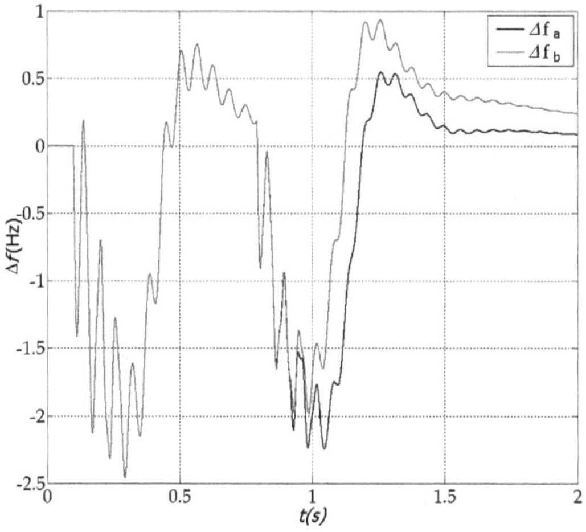

Figure 17. Frequency variations during sudden change load: Δf_a–impact load, Δf_b-load disconnected

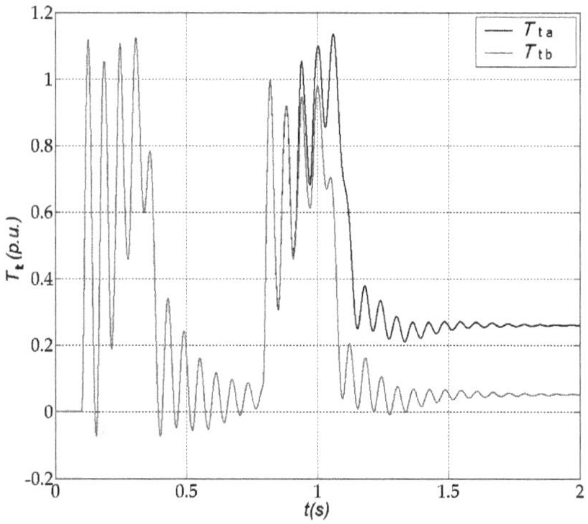

Figure 18. Torsional torque during sudden change load: T_{ta}-impact load, T_{tb}-load disconnected

5. Induction motor under short term voltage interruption

With the aim to get a better insight into the dynamics of induction motor fed directly from the isolated electrical grid the short-term interruptions in the motor power supply are considered.

Direct starting of induction motors on an isolated electrical grid produce disturbance for supply network and local consumers. This disturbance, such as voltage dips and reduction of aggregate speed, will cause decrease of network power quality. The significant voltage dips appear due to faults in power supply, and also, due to certain faults on loads connected to the isolated electrical grid. These voltage dips cause changes in transients of induction motors, as well as in transients of diesel generator unit. Besides the voltage dips, voltage interruption can also appear, which further effects the operation of the induction motor.

Two cases are considered. In the first case, a short term interruption appeared after both motors have run-up successfully. In that moment motors are in steady state condition. In the second case, interruption of power supply on the first induction motor appeared at the beginning of the start up period of the second one.

Motors are started loaded and load on the first induction motor shaft is 0.1 p.u., while the load of 0.05 p.u. is applied on the second induction motor shaft. At the beginning, the first induction motor (IM1) starts and after it has run-up successfully, the second one (IM2) is connected to the grid. At the chosen moment, as motors are in steady state condition, the first induction motor is shortly disconnected from the network, and then, after 100 ms reconnected to the power supply (Fig. 19 and 20).

At the time of voltage interruption, a large negative impulse of the torque of the first induction motor occurs. This reverse torque impulse rapidly decays and the air-gap torque (T_{eIM1}) becomes equal to zero. Changes in the electromagnetic torque of the second induction motor (T_{eIM2}) occurs at the time of disconnection of the first one.

Transients of air-gap torque and speed transient of induction motors; during the short-term interruptions that is appeared on the first induction motor, in case of steady state condition is presented in Figure 19a. However, in Figure 19b, the transients obtain in case of interruption in the first induction motor power supply that appeared at the beginning of the start up period of the second induction motor is presented. Transients of stator currents of induction motors and their components in d and q axis during the short-term interruptions, for both cases are presented in figure 20.

As one can see in Fig. 20a, at the time of voltage interruption, current of the first induction motor momentarily reaches approximately 2 p.u., while current of the second one reaches approximately 0.6 p.u. Thus, by restoring the supply, at the beginning of the transients the current of the first induction motor momentarily reaches 2 p.u., at the same time as the current of the second induction motor reaches maximal value of 0.35 p.u. A negative impulse of the torques T_{eIM1} and T_{eIM2} instantaneously appears. Thus, the speed of the first induction motor at the beginning of the transients additionally is shortly decreased, and afterwards continues recovering.

At the instant of the supply recovery air-gap torque of second induction motor suddenly reaches negative value and, as one can see in Fig. 21a, the speed of second induction motor is decreased and than rapidly recovers.

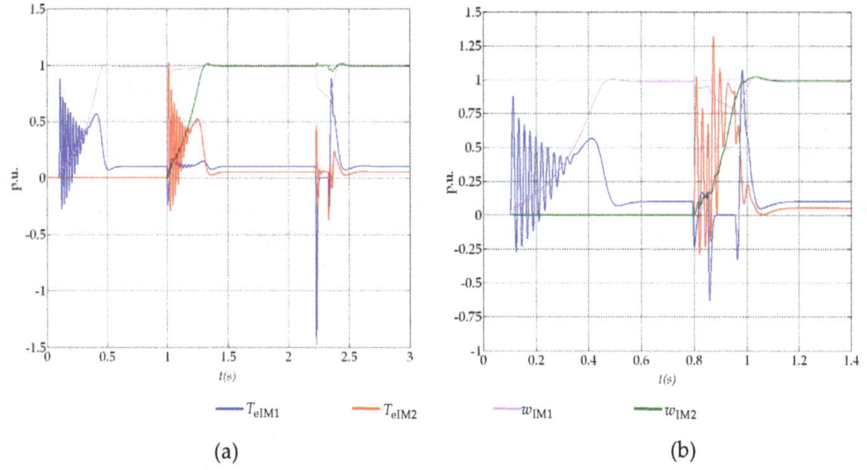

Figure 19. Transients of: air-gap torque (T_{eIM1}, T_{eIM2}) and speed transient of induction motors (ω_{IM1}, ω_{IM2}) during the short-term interruptions that is appeared on the first induction

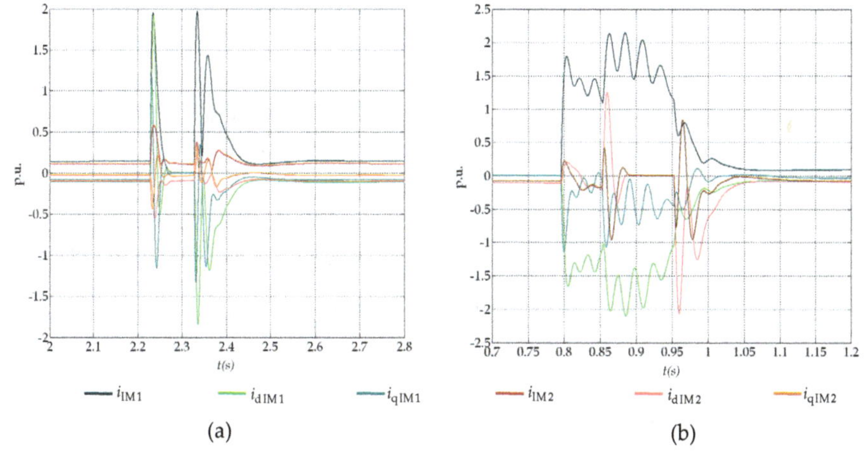

Figure 20. Transients of stator currents of induction motors (i_{IM1}, i_{IM2}) and their components in d (i_{dIM1}, i_{dIM2}) and q (i_{qIM1}, i_{qIM2}) axis during the short-term interruptions: a) motors are in steady state condition b) at the beginning of the start up period of the second one

In the second case, interruption in the first induction motor power supply appeared at the beginning of the start up period of the second induction motor.

At the time of connecting the second induction motor, the current is momentarily increased to 1.7 p.u. and has a decreasing tendency. By the time of short term power interruption on the first induction motor, the current of the second one suddenly increases to 2.2 p.u. Further, the current oscillates around a higher average value than before and the damped oscillations are rapidly decreasing (Fig. 20b).

The transient of air-gap torque of induction motors are presented in Fig. 21. However, the air-gap torque of the first induction motor, at the time of voltage interruption occurred, momentarily reaches negative impulse of approximately 0.6 p.u. (Fig. 21b). The speed of the first induction motor (IM1) decreases at the beginning of the start up period of the second induction motor and tend to increase when fault occurs. As voltage interruption occurred the speed continuous reducing. In the speed transient of the second induction motor (IM2) one can see that is short term decreased appeared at the beginning of interruption and after that the motor continuous accelerates.

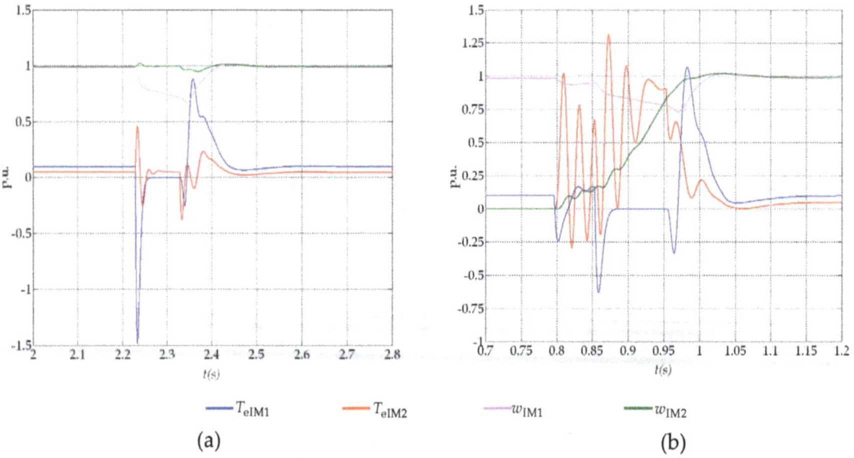

Figure 21. Transient of electromagnetic torque (T_{eIM1}, T_{eIM2}) and speed transient of induction motors (ω_{IM1}, ω_{IM2}); during the short-term interruptions: a) motors are in steady state condition b) at the beginning of the start up period of the second one detail of Fig. 20.

Terminal voltage (u_{abcIM1}) of the first induction motor, during the short-term interruptions, for the both cases is presented in Figure 22, while corresponding stator currents (i_{abcIM1}) of first induction motor are shown in Figure 23.

The voltage dips as well as voltage interruption, which further effects the operation of the induction motor, causes changes in transients of diesel generator unit. A short-term power interruption will result, as shown, in significant changes of electrical and mechanical variables and will also cause torsional strain in the shaft line. The mechanical coupling of a diesel engine and a synchronous generator is considered to be a rotating system with two concentrated masses that are connected by flexible coupling. The flexible coupling allows these masses to rotate at a different speed in transients. Thus, in Fig. 24 the speed transients of diesel engine and synchronous generator are presented, while torque at coupling zone is presented in Figure 26.

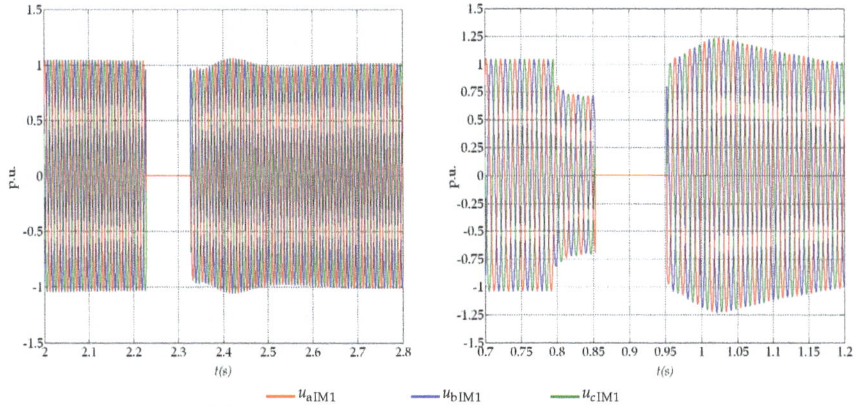

(a) (b)

Figure 22. Voltage (u_{abcIM1}), during the short-term interruptions: a) motors are in steady state condition b) at the beginning of the start up period of the second one

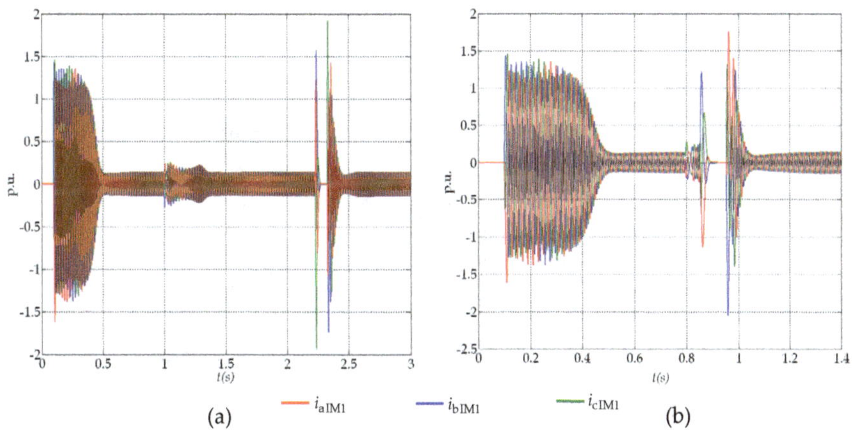

(a) (b)

Figure 23. Stator current (i_{abcIM1}) of first induction motor, during the short-term interruptions: a) motors are in steady state condition b) at the beginning of the start up period of the second one

Sudden impact load on the isolated electrical grid induces a large strain on the diesel generator unit shaft. After the first induction motor is started, the diesel engine accepts the load and later reaches the steady speed. As a result of the starting of the loaded IM2, the speed of the diesel engine decreased and reached a minimum value of 0.97 p.u., while the speed of the synchronous generator reached a value of about 0.96 p.u. (Fig. 24a).

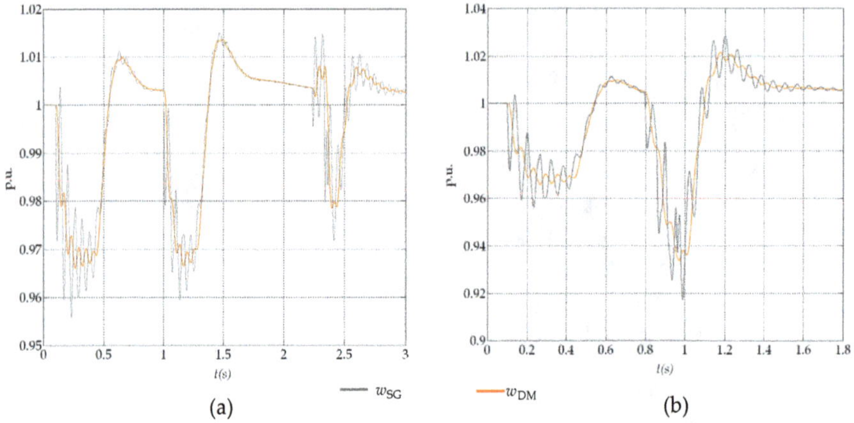

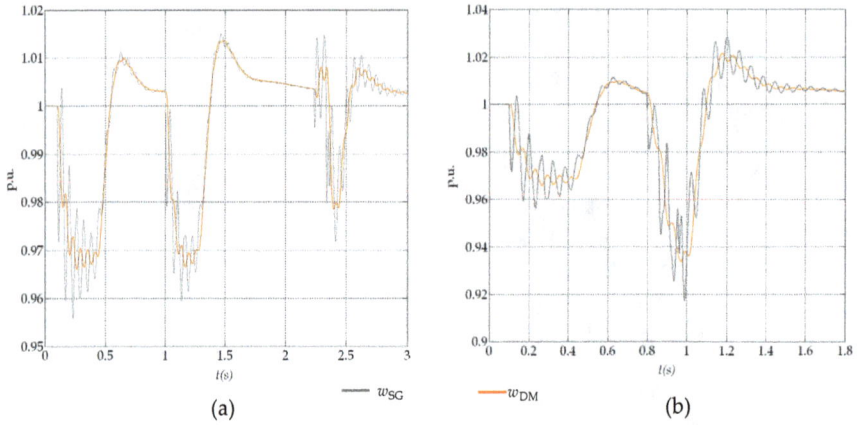

Figure 24. Speed transient of synchronous generator (ω_{SG}) and diesel engine (ω_{DM}) during the short-term interruptions: a) motors are in steady state condition b) at the beginning of the start up period of the second one

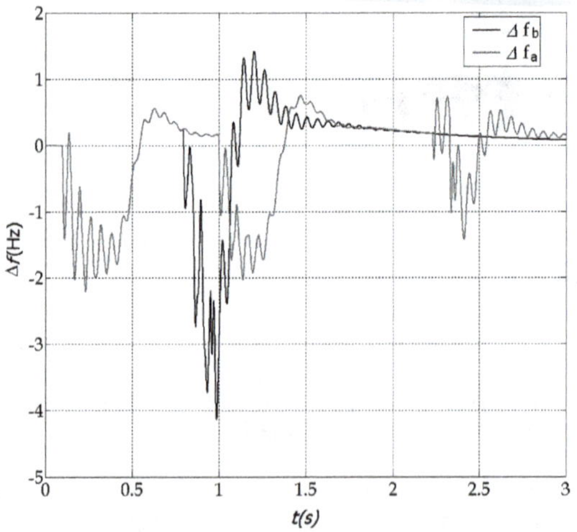

Figure 25. Frequency variations during short-term interruptions: Δf_a- motors are in steady state condition, Δf_b- at the beginning of the start up period of the second one

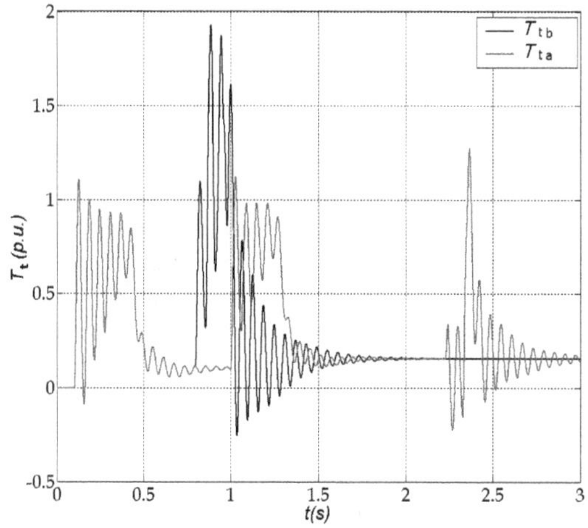

Figure 26. Transient of torsional torque during short-term interruptions:
T_{ta}-motors are in steady state condition, T_{tb}- at the beginning of the start up period of the second one

Occurrence of short-term power loss, at the time that motors are in steady state condition, speed of both machines increased. After, reconnected to the power supply the speed of synchronous generator as well as induction motor is reduced. Damped oscillations are presented during transients. In the second case, when the short-term interruption occurs at the beginning of the start up period of the second induction motor, speed of both machines continue decrease. In this case, the speed of synchronous generator is reduced to less of 0.92 p.u., while the speed of the diesel engine is reduced to less of 0.94 p.u. The oscillations appear with greater magnitude then in previous case. Speed deviation affects on frequency of grid and frequency variations during short-term interruptions, for both cases are presented in Figure 25. Significant disturbances appear in transients of torsional torque especially in case when short-term interruptions is appear at the beginning of the start up period of the second induction motor, as one can see in Figure 26.

6. Discussion

The dynamics of induction motors fed directly from the isolated electrical grid is considered in the cases of: direct-on-line starting of induction motors, sudden change load on the motor shaft and during short-term voltage interruptions. Direct on line starting of induction motors on the isolated power system is the most difficult transition regime for units due to electricity loads and also due to torsional loads on the shaft line. The autonomous operation of the synchronous generator is characterized by a change in steady state which causes a change in voltage and frequency, which in turn affects the quality of electric power systems. During starting induction motors draw high starting current, known as locked rotor current, which are several times the normal full load current of the motor. This current causes a significant voltage dip on the isolated electrical grid, the terminal voltage is momentary decreased and after few damped oscillations reached minimal value. During this short period the voltage regulator does not affect jet. The voltage drop increases in case of starting loaded induction motor, the current momentarily reaches higher value of first magnitude and oscillates about higher average value than in previous case.

Disturbances in the system are present when the second induction motor is connected to the grid and additionally is growing up at the instant of impact additional load on the previous induction motor shaft. As load on the motor shaft grows up, the current of the motor is higher; oscillations are more damped and longer present in comparison with lightly load or no-loaded induction motor.

This current causes the voltage drop at the motor terminals that are connected on the same grid. The mutually effect between source and loads exists. The voltage drop results that the motor speed slows down. Because the motor rotor slows down the higher current is appeared on the grid and voltage are reduced even more.

At the instant of starting of induction motors the locked rotor torque appeared and as the rotor starts to rotate the air-gap torque oscillates about positive average value. Oscillations in transient of electromagnetic torque are damped at the end of start up period and finally the steady state condition is attained without oscillations. The oscillations are longer presented in case of bigger load torque on the motor shaft. Sudden change load on the motor shaft causes changes in air-gap torque. Damped oscillations are longer present in a case of impact additional load than during load disconnected. Both cases influence on the electrical grid and changes in transients of both induction motors as well as in synchronous generator are present.

The start-up period of the induction motor is longer when the bigger load torque on the motor shaft is applied. Also, decreasing the terminal voltage causes longer duration of speed transient in dynamics.

Sudden change load on a motor shaft, which occurs during operation period, results in speed change of the other motors that fed from the same grid.

Direct starting of induction motors on isolated electrical grid, as well as sudden change load, caused voltage dips and also reduces speed of aggregate. The significant voltage dips appear due to faults in power supply, as well as certain faults on loads connected to isolated electrical network. Besides the voltage dips, voltage interruption can also appear. This causes significant disturbances on the grid and affects the operation of other induction motor. If interruption of the supply lasts longer than one voltage period, many AC contactors will switch off the motor. However, in some cases, faulty contactor may produce multiple on and off switching. These interruptions affect the dynamics of both electrical and mechanical variables, which will also cause torsional stresses in the shaft line. The consequences of voltage interruption on the induction motor behavior are current and air-gap torque peaks that appear at instant of fault and recovery voltage.

Sudden changes of active power have impact on both voltage and frequency but a start-up of electric motor influences disproportionately the voltage value due to relatively low power factor during the process.

7. Conclusion

The dynamics of induction motors fed directly from the isolated electrical grid is analyzed. In isolated electrical grid, such as for example ship's electrical grid, the main source is a diesel generator and induction motors are the most common loads. Induction machine plays a very important role in that application and a significant number of induction motors are used at critical points of on board processes. The connection of large induction motors (large relative to the generator capacity) to that grid is difficult transient regime for units due to electrical loads and also due to torsional loads on the shaft line. Direct on line starting of induction motors induces high strain on the power system and this strain arises when the next induction motor is connected to the grid. Sudden impact load on the induction motors shaft is an additional strain on the network, especially when impact load on a motor shaft occurs during start up of another one. This in turn affects the quality of electric power system and thus, the dynamic behavior of induction motors. The significant voltage dips appear due to faults in power supply, and also, due to certain faults on loads connected to the isolated electrical grid. These voltage dips cause changes in transients of induction motors, as well as in transients of diesel generator unit. Besides the voltage dips, voltage interruption can also appear, which further effects the operation of the induction motor. Factory production tests demonstrate the capability of the unit to supply defined loads applied in a defined sequence. In order to make changes to the loading of an existing isolated electrical grid, it is necessary to analyze and document the effect of the additional loads on its normal and transient performance. An induction motor starting study may be of use in analyzing the performance of small power systems. Such systems are usually served by limited capacity sources that are subject to severe voltage drop problems on motor starting, especially when large motors are involved. In some cases, specific loads must be accelerated in specially controlled conditions, keeping torque values in defined limits. The results obtained by this analysis can be used as a guideline in choosing as well as setting parameters of the protection devices.

Author details

Marija Mirošević
University of Dubrovnik, Department of Electrical Engineering and Computing, Croatia

Acknowledgement

The presented results are carried out through the researches within scientific projects „New structures for hydro generating unit dynamic stability improvement" and "Revitalization and operating of hydro generator" supported by Ministry of Science, Education and Sports in the Republic of Croatia.

8. References

Ameziquita-Brook, L., Liceaga-Castro, J., Liceaga-Castro, E., (2009). Induction Motor Identification for High Performance Control Design, *International Review of Electrical Engineering*, Vol. 4, No. 5; (October 2009), pp. (825-836) ISSN 1827-6660.

Cohen, V. (1995). Induction motors-protection and starting. *Elektron Journal-South African Institute of Electrical Engineers.12*, pp. (5-10) Citeseer.

Erceg, G., Tesnjak, S. & Erceg, R., (1996). Modelling and Simulation of Diesel Electrical Aggregate Voltage Controler with Current Sink, *Proceedings of the IEEE International conference on industrial technology*, 1996

Jones, C. V. (1967). *The Unified Theory of Electrical Machines*, Butterworths, London.

Krause, P. C. (1986). *Analysis of Electric Machinery*, McGrawHill, Inc. New York, N.Y.

Krutov, V., Dvigatelj vnutrennego sgorania kak reguliruemij objekt. (Mašinostroenie, 1978).

Kundur, P. (1994). *Power System Stability and Control*, McGraw-Hill

Maljković, Z., Cettolo, M., Pavlica, M. (2001). The Impact of the Induction Motor on Short-Circuit Current, *IEEE Industry Applications Magazine*. Vol. 7, No. 4; pp. (11-17).

McElveen, R., Toney, M., Autom, R., & Mountain, K. (2001). Starting high-inertia loads, *IEEE Transactions on Industry Applications, 37* (1), pp. (137-144).

Mirošević, M., Maljković, Z., Milković, M., (2002), Torsional Dynamics of Generator-Units for Feeding Induction Motor Drives, *Proceedings of EPE-PEMC 9th International Conference on Power Electronics and Motion Control*, Cavtat, Dubrovnik, Croatia, Sept. 2002., T8-069.

Mirošević, M., Sumina, D., & Bulić, N. (2011) Impact of induction motor starting on ship power network, *International Review of Electrical Engineering*, Vol. 6, No. 1; (February 2011), pp. (186-197) ISSN 1827-6660.

Tolšin,V. & Kovalevskij, E., Prethodnie procesi u dizel generatorov. (Mašinostroenie, 1977).

Vas, P. (1996). *Electrical Machines and Drives A Space Vector Theory Approach*, Oxford Science Publications

Chapter 5

Effects of Voltage Quality on Induction Motors' Efficient Energy Usage

Miloje Kostic

Additional information is available at the end of the chapter

1. Introduction

Today, about 50% of electrical energy produced is used in electric drives. Electrical motors consume around 40% of total consumed electrical energy (Almeida et al., 2007) and of that thereof, induction motors account for 96% of energy consumption. Around 67% of this energy is used in induction motors with a rating below 75 kW and it can be shown that 85% of the energy losses are dissipated in these rating motors. Efficiency improvements of constant-speed drives, both constant-torque and variable-torque drives, is very important. It is usual that techniques for efficiency improvements of variable-torque drives are different from those of constant-speed and constant-torque applications. The latter is dealt with through optimization; it is very difficult to design and build a motor with high rating efficiency and rating power factor - it has been shown (Fei et al., 1989) that higher efficiencies are associated with lower power factor. It is especially difficult to design and build a drive operating at high efficiency and power factor over an entire range of loads, say from 25 - 100% of rated load (P_N), i.e. at partial load.

Electrical energy savings in the drive could be realized by improvements of power quality in the consumer network. Term power quality (Linders, 1972; Bonnett, 2000) mostly means quality of supply voltage that should meet the following requirements:

- voltage value (permissive variations are in the range of $U_N \pm 5\%$ of nominal voltage),
- permissible voltage asymmetry is 2% and has greater influence on accurate and economical motor operation,
- permissive total harmonics distortion of voltage is $THD_U \leq 3\text{-}8\%$, where higher values correspond to lower voltage networks.

Power losses and reactive loads depend from on voltage magnitude and they are further increased due to unbalanced voltage and (or) the presence of harmonics in supply voltage.

Unbalance voltage can occur due to the presence of larger single-phase consumers or asymmetrical capacitor banks with damage or capacitors that are switched off due to the fuse burning only in one phase. Nowadays, the presence of higher harmonics in the supply voltage is ever more frequent due to the growth of consumers who are supplied through the rectifiers and inverters: regulated actuators, electrothermical consumers and consumers alike.

The effect of a variation in supply voltage, wave-form or frequency on the motor's efficiency and power factor characteristics depends on the individual motor design. Typical variations of current, speed, power factor and efficiency with voltage for constant output power are given in Fink (1983). The usual characteristics of induction motors within the ±10% voltage band (U_n±10%) are well known. These are included in corresponding table for typical 30-100 kW, 1500 or 1800 1/min motors (Linders, 1972; Fink, 1983), but the effect of saturation has been largely neglected in these tables. However, it is the author's intent to show a correlation between motor characteristics and voltage level.

This proposed has three parts:

1. Study of the effect of voltage magnitude on motor losses and motor reactive loads,
2. analysis of the effect of unbalanced voltages, and
3. analysis of the effect of non sinusoidal voltages to the efficient energy use.

2. Effect of voltage magnitude on motor power losses and motor reactive loads

Voltage magnitude has a significant and different influence on motor loads and electric energy consumption, depending on types, nominal powers and load level ($p=P/P_N$) of motors. Data found in classic references (Fink, 1983) are almost identical to data from older references (Linders, 1972;), although they are not accurate enough for motors made after 1970. The main reason for disagreement is higher values of no-load current and more significant dependency of core losses on voltage magnitude for newer motors ("U" or "T" shape of magnetization curve). Data for loads less than 50% are not available in Fink (1983) and Hamer et al. (1997), although more than half of motors operate in these load regimes.

The influence of voltage magnitude on energy characteristics of standard induction motors (made after 1970.) is significant, as was affirmed by the author's research and verification (Kostic, 1998). It was also ascertained that the changing of energy characteristics is more significant for smaller motors. According to newer literature, efficiency (η) and power factor ($cos\varphi$) dependency on voltage value is more significant than was shown in standard literature.

The paper by Hamer et al. (1997) analyses the effects of voltage magnitude only on two motors (10HP and 100HP) and loads from 50-100%, and the results, and results are illustrated in Fig. 1. Results for standard efficiency motors, given in Fig. 1, are almost equal to the author's results (Kostic, 1998, 2010). A brief theoretical approach for determining dependency of power losses and reactive loads on voltage value will be presented, as well as proceedings for calculation and analysis of power losses and reactive loads on voltage value.

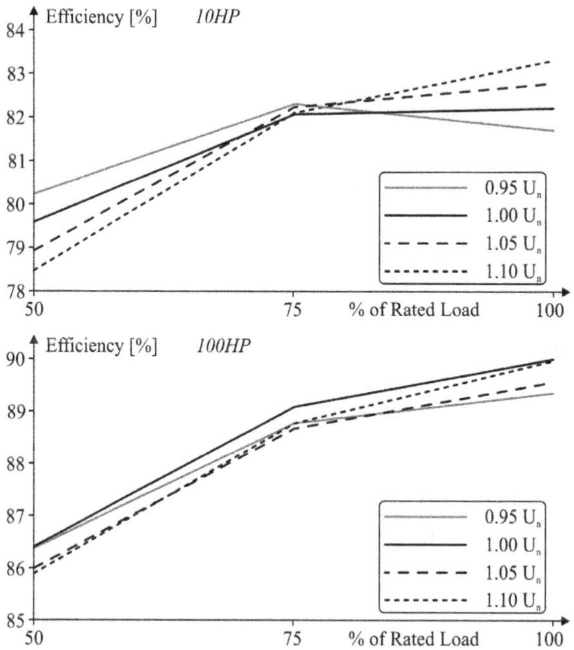

Figure 1. Efficiency versus load for various applied voltages in percent of its 460 V rating for a standard efficiency motor (10 HP and 50 HP)

2.1. Dependency of power losses and reactive loads on voltage value

In order to determine total dependence of power losses on voltage, for the load range from no-load to full load, it is necessary to determine no-load power - voltage dependency, $P_0(u)$:

$$P_0(u) = P_{Cu0}(u) + P_{Fe}(u) + P_{fw} \tag{1}$$

Where are

P_{Cu0} copper losses in no-load,

P_{Fe} core losses in no-load,

P_{fW} friction and windage losses in no-load.

Load losses component (P_{LL}) depend on relative load ($p_L = P_L/P_N$) and relative voltage values ($u = U/U_N$):

$$P_{\gamma P} = P_{LL,N} \cdot p^2 / u^2 \tag{2}$$

where are $P_{LL,N} = P_{Cu,S} + P_{Cu,R} + P_{\gamma ad}$ - a load losses in a nominal regime (P_N, U_N), and $P_{\gamma ad}$ are additional load losses. Load losses, $P_{LL,N}$, can also be calculated as a difference of full load power losses ($P_{\gamma N}$) and no-load power (P_{0N}):

$$P_{LL,N} = P_{\gamma N} - P_{0N} \tag{3}$$

or in per unit ($p_\gamma = P_\gamma / P_N$, $p_0 = P_0 / P$, and $p_{LL,N} = P_{LL,N} / P_N$) as:

$$p_{LL,N} = p_{\gamma N} - p_{0N} \tag{4}$$

Total load losses can be calculated in absolute values as:

$$P_\gamma(p,u) = P_0(u) + P_{LL,N} \cdot p^2 / u^2 \tag{5}$$

or in per unit:

$$p_\gamma(p,u) = p_0(u) + p_{LL,N} \cdot p^2 / u^2 \tag{6}$$

In order to ascertain reactive loads $Q(u)$ dependency, it is necessary to determine no-load reactive power versus voltage, in absolute values ($Q_0(u)$):

$$Q_0(u) = \sqrt{3} \cdot U_0 I_0(u) \cdot \sin\phi_0 \approx \sqrt{3} U_0 I_0(u) \tag{7}$$

Or in per unit values ($q_0(u) = Q_0/P_N$), for the load range from no-load to full load

$$q_0(u) = \frac{\sqrt{3} \cdot u_0 i_0(u)}{\eta_N \cos\phi_N} \tag{8}$$

In the rated regime are: efficiency, $\eta_N = P_N/P_{1N}$, and power factor, $\cos\phi_N = P_N/(\sqrt{3} \cdot U_N I_N)$. Values of reactive power in the load branch, Q_{LN} and q_{LN} are calculated from the quotient of maximum and nominal torque T_m / T_N, as explained in Appendix, (Kostic, 1998, 2001):

$$Q_{LN} = 0.5 \cdot P_N / (T_m / T_N) \tag{9}$$

or in per unit as:

$$q_{LN} = 0.5 / (T_m / T_N) \tag{10}$$

Equations (9) and (10) are obtained by the procedure given in the Appendix, gained from the equivalent Γ-circuit of the induction machine (Kostic, 2001, 2010). Difference of nominal reactive power and no-load reactive power is a little bit less from calculated value of Q_{LN} and q_{LN} because of reactive power reduction on magnetization branch ($\Delta q_{\mu N} = (0.01\text{-}0.10) q_{0N}$). Total reactive load is calculated in absolute values as:

$$Q_1(u) = Q_0(u) + Q_{LN} \cdot p^2 / u^2. \tag{11}$$

or in per unit

$$q_1(u) = q_0(u) + q_{2n} \cdot p^2 / u^2 \tag{12}$$

For motors of nominal power ≤ 3kW, value of nominal reactive power is almost equal to no-load power ($Q_{1N} \approx Q_0$), because $Q_{LN} \approx \Delta Q_{\mu N}$, so $Q_1(u) \approx Q_0(u)$ and $q_1(u) \approx q_0(u)$, (Kostic, 1998, 2010). Expressions (11) and (12) are commonly in use. Instead of Q_{LN} and q_{LN}, ΔQ_N and Δq_N can be used if they are known or if they can be calculated ($\Delta Q_N = Q_{1N} - Q_{0N}$). For calculating the dependency $P_\gamma(u)$ and $Q_1(u)$, according to expressions (5) and (11), it is necessary to know:

- no-load characteristic $I_0(u)$, $Q_0(u)$, $P_0(u)$ and $P_{Fe}(u)$, for the analyzed voltage range,
- motor catalogue data: nominal power (P_N), nominal current (I_N), efficiency (η), power factor ($cos\varphi$), slip (s_N) and the quotient of maximum and nominal torque (T_m/T_N), and
- $P_{\gamma N}$ and Q_{LN} are calculated.

2.2. Dependency of motor input power and reactive loads on voltage values

Dependencies of input power (P_1/P_{1N}) and reactive loads (Q_1/P_{1N}) versus relative voltage value (U/U_N), for P_{LL}/P_N = 25%, 50%, 75% and 100%, for motors of nominal powers 1 kW, 10 kW and 100 kW, have been determined by the procedure described in chapter A; results are illustrated in the Fig. 2, (Kostic, 1998, 2010). Influence of voltage on reactive loads and power loss is notable, especially for small motors and for lower loads P_{LL}/P_N.

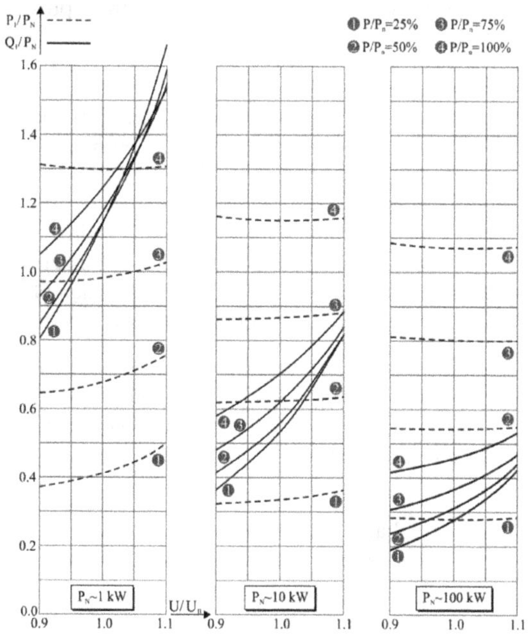

Figure 2. Dependencies of motor input power and reactive loads on supply voltage

Results of the author's research (Kostic, 1998, 2010) confirmed that there are significant possibilities for energy savings by setting voltage values within the voltage band (U_n±5%), because more then 80% of induction motors, especially small and medium power (1 - 30 kW), operate at partial load (≤70%). Dependencies of power loss $P_γ(u)$ and reactive loads $Q_1(u)$ for motor of nominal power 2 MW, for P_1 / P_N = 0% (no-load), 25%, 50%, 75% and 100% are given in Fig. 3, (Kostic et al., 2006 and Kostic, 2010).

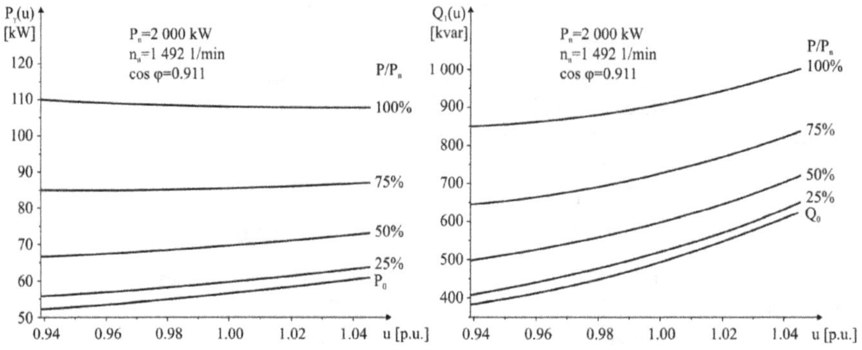

Figure 3. Dependencies of power losses and reactive loads on supply voltage

2.3. Basic reduction of electric energy own consumption of power plants by setting voltage within U_n±5%

Subject of concrete project (Kostic et al., 2006) is reduction of electric energy own consumption of thermal power plant "Nikola Tesla" B, Obrenovac (Serbia), with 2 blocks.. The own consumption of the electric energy, with motors on medium voltage (6.6 kV), is about 90% and with motors on voltage 0.4 kV is about 10%, supplied by a transformer whose primary is directly connected to the generator bus. Nominal powers of the transformers (21 kV/6.6 kV) are approximately equal to total nominal powers of all installed motors which are about 140 MW. Active and reactive loads are about 70 MW and 60 Mvar. As the load of most motors is about 35–70% of full load, reduction of electric energy own consumption could be achieved by setting the voltage magnitude in the range U_n±5%.

Application of this procedure results in reduction: of core losses (P_{Fe}), reactive loads (Q) and active power losses component $P_{CuQ}=RI_Q^2$. Thereby, both active and reactive energy consumptions are decreased. Optimal voltage values are being determined according to appropriate calculations and analysis is based on motor catalogue data and its experimental verification at actual load regimes.

According to the effects of voltage magnitude on motor power losses ($P_γ$) and motor reactive loads (Q_1), for overall own consumption, the following dependencies can be determined: restrictions:

- motor power losses change $ΔP_γ = P_γ(U_i) - P_γ(U_N) = ΔP_1 = P_1(U_i) - P_1(U_N)$, i.e. active loads change, and

- reactive loads change $\Delta Q_1(u) = Q_1(U) - Q_1(U_N)$,

for the voltage range $U/U_N = 0.955\text{-}1.045$, i.e. for $U = 6300V\text{-}6900V$ ($U_N = 6\,600V$), Fig. 4.

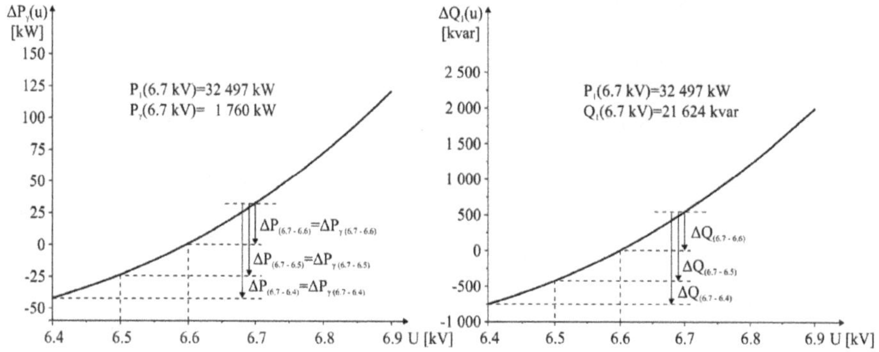

Figure 4. Dependency of total power losses and reactive loads for own consumption

As was shown in Kostic et al. (2006), application of this procedure causes reduction of the electric energy own consumption. Changing voltage value (regulation at own consumption transformers 1BT and 2BT, and at common consumption transformer OBT) from 6.8 kV (1.03 U_N) to 6.6 kV (U_N) or 6.5 kV (0.985 U_N) causes reduction of:

- Active power losses for 161 kW and 213 kW, respectively,
- Reactive loads for 3 544 knar and 4 559 knar.

Power losses addition reduction in the own consumption network for 42.4 kW and 54.7 kW, respectively, due to of the above mentioned of reactive loads' reduction in the own consumption network.

According to values of reduced power losses, reactive loads and assumed operational plan of the thermal plant (estimated 6 000 h/years), savings in active and reactive energy have been determined and given in Table I, (Kostic et al., 2006, and Kostic, 2010).

Consumption objects	Load Reduction		Consumption Energy Reduction	
	Active [kW]	Reactive [kvar]	Active Energy [kWh/year]	Reactive Energy [kvarh/year]
Block 1 Motors	161	3 544	966 000	21 264 000
Block 1 Network	42	-	252 000	-
Block 2 Motors	213	4 559	1 278 000	27 354 000
Block 2 Network	55	-	330 000	-
Total	**471**	**8 103**	**2 826 000**	**48 618 000**

Table 1. Reduction of the active and reactive energy own consumption of Power plant by voltage change, Kostic et al. (2006).

3. Motor voltage asymmetry and its influence on inefficient energy usage

Analysis of the effect of unbalanced voltages on the three-phase induction motor is presented in the paper by Bonnett (2000). Since the unbalanced voltage of 2%, 3.5% and 5% causes an increase in losses, in the same order, for 8%, 25% and 50% of nominal power losses in the motors, it is a reasonable requirement to permit voltage asymmetry ≤2%, so this is the upper limit in most national and international standards. The truth is that with less load, the motor could safely work also at higher values of unbalanced voltage. The literature (Linders, 1972) states that the information given previously is determined from measurements and that they are higher than calculated values. However, it is explained here by the fact that the rotor inverse resistance is higher by 1.41 times compared to the rotor resistance in short-circuit mode (Kostic & Nikolic, 2010), since current frequency of the negative sequence in the rotor winding is twice as high ($f_{r,NS} \approx 2f_1 = 2f_{r,SC}$), i.e. it is higher by 1.41 times than corresponding values given in the literature. Thus, it is increasingly convincing that the requirements which are given in the most appropriate standards are justified. Performed analysis shows that there are some considerations that should be included in current standards. Motor operation is not generally allowed when voltage asymmetry (U_{NS}/U_N) is higher than 5%, because in the (rare) case that the direct and inverse component of the stator currents in one phase are collinear, increase of the current in that phase would be ≥ 1.38 times, and the increase of the losses in the windings of that phase would be ≥ 90% (Linders, 1972; Kostic & Nikolic, 2010).

The effect of voltage asymmetry on the three-phase induction motor is equivalent to the appearance of negative sequence voltage system that creates a rotating field which rotates contrary to the rotation of the positive sequence field and motor rotating direction. The consequence is that small values of negative sequence voltage produce relatively high values of negative sequence currents. By definition, the coefficient of asymmetry (K_{NS}%) is the ratio of negative sequence voltage (U_{NS}) and positive sequence voltage (U_{DS}). For simplification, Standard NEMA use the following definition of voltage unbalance

Voltage unbalance=100· (maximum voltage deviation from average value)/average value

For instance, for measured voltages of 396V, 399V and 405V, average value is 400V. Then, the highest variation from average voltage is determined (405V – 400V = 5V). At the end, the coefficient (percent) of voltage asymmetry is calculated as the quotient of highest variation and average value: $K_{NS}\% = 100\cdot(5/400) = 1.25\%$. Since percentages of negative sequence currents are 6-10 times higher than corresponding voltage asymmetry, negative sequence currents could reach 10%. This causes additional motor heating and the appearance of inverse torque that reduces starting and maximum motor torque, and causes a small increase of motor slip. Because power losses and motor heating are increased, allowed motor loading is decreased. As the percent of asymmetry rises, the derating factor of nominal power decreases, according to NEMA MG1 (Bonnett, 2000), as shown in Fig. 5.

With an increase of voltage asymmetry coefficient, motor efficiency decreases under all load levels. Dependence of motor efficiency is given in Fig. 6 for the voltage unbalance of 0.00%, 2.50%, 5.00% and 7.50% (Bonnett, 2000).

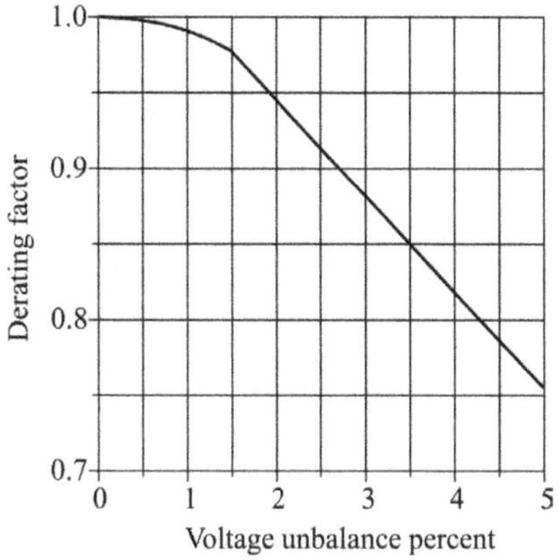

Figure 5. Relation between derating factor and voltage asymmetry

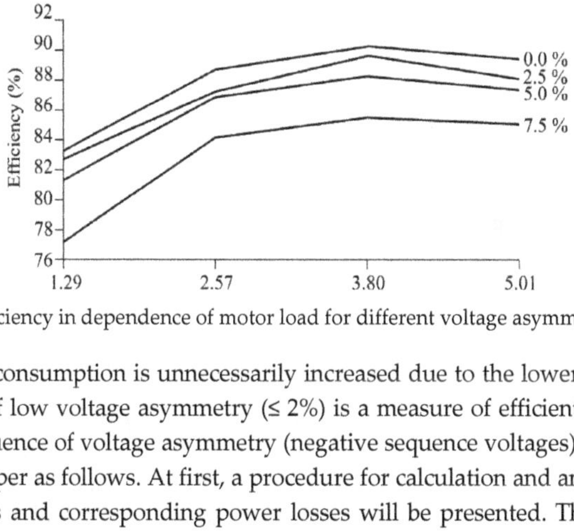

Figure 6. Motor efficiency in dependence of motor load for different voltage asymmetries

Electrical energy consumption is unnecessarily increased due to the lower motor efficiency, so maintenance of low voltage asymmetry (≤ 2%) is a measure of efficient energy usage. In that case, the influence of voltage asymmetry (negative sequence voltages) will be presented in detail in the paper as follows. At first, a procedure for calculation and analysis of negative sequence currents and corresponding power losses will be presented. Then, evaluation of voltage unbalance that could arise in the considered consumer network is presented.

3.1. Equivalent circuit of induction motor for negative sequence

When the induction motor is supplied from the network with asymmetrical voltages, then the three-phase voltage system should be decomposed to positive, negative and zero sequences. Further, using equivalent motor circuits (Boldea & Nasar, 2002; Ivanov-Smolensky, 1982) separately for positive (Fig. 7a) and for negative sequences (Fig. 7b), calculations and analyses of motor energy and operation characteristics (currents, power losses, torques) are performed. At the end, with corresponding superposition of relevance values, their overall values are obtained. Only in that way could real (overall) values of motor power losses be determined.

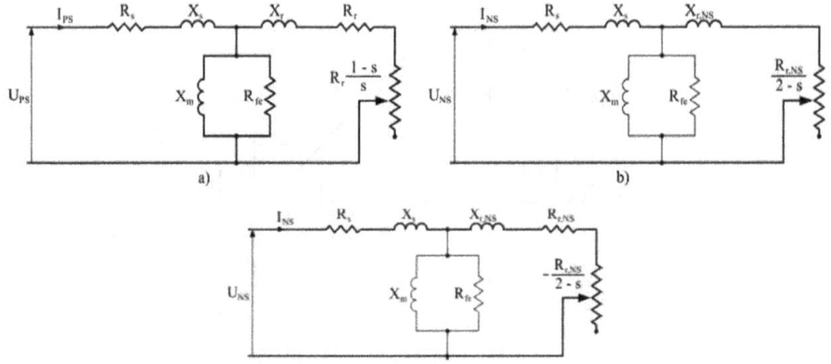

i. (c)

Figure 7. Equivalent circuits of induction motor a) positive and b) negative voltage sequence, and c) completely circuit for negative sequence voltage

3.2. Parameters of equivalent circuit for negative sequence

Stator winding resistance (R_s) and reactance (X_s) are almost the same for positive and negative voltage sequences. The parameters of equivalent circuits that are correlated to the rotor side and negative sequence voltage system (resistance $R_{r,NS}$ and reactance $X_{r,NS}$) are substantially different than those for positive voltage sequences, because the frequencies of the rotor currents in negative sequences are higher for 50÷100 times:

$$f_{NS} = f_1 \cdot (2 - s) \approx 2 f_1 \succ\succ s \cdot f_1 \tag{13}$$

(For example, for load slip s=0.01÷0.05: $f_{r,NS}=(1.90\div1.98)f_1 >> f_{r,PS}=s\cdot f_1=(0.01\div0.05)f_1$).

Values of resistance $R_{r,NS}$ and reactance $X_{r,NS}$ are determined from the corresponding parameters in the short-circuit regime, $R_{r,SC}$ and $X_{r,SC}$. In Boldea & Nasar (2002) and Ivanov-Smolensky (1982) it is noted that values of corresponding resistances and reactances are approximately equal to those in the short-circuit regime: $R_{r,NS} \approx R_{r,SC}$ and $X_{r,NS} \approx X_{r,SC}$.

If we look carefully, we could find that this statement is not correct. Those parameter changes are dependent on the ratio of the rotor conductor height (H_b) and field penetration depth ($\partial_{r,SC} = (2\cdot\rho/(\mu\cdot2\cdot\pi\cdot f_1))^{1/2}$, i.e. the quotient $\zeta = H_b/\partial_{r,SC}$; where: f_1 – frequency of supply voltage first order harmonic, ρ - conductor specific resistance, $\mu = \mu_0$ – magnetic permeability of conductor.

By Fig. 8 (Kostic, 2010), for $\zeta_{SC} = H_b/\partial_{SC} = 1.2\div3$ are obtained: $R_{r,SC} = 1\div3R_r$ and $L_{r,SC} = 1\div0.50L_r$, where R_r and L_r are resistance and inductance in a low slip regime, respectively, for instance in a nominal regime. Since current frequency of the negative sequence in the rotor conductor (bar) is twice as high ($f_{r,NS} \approx 2f_1 = 2f_{r,SC}$), the penetration depth of those currents is lower by √2 times then in a short-circuit regime. Corresponding values of resistance and reactance are higher by √2 times:

$$R_{r,NS} = R_{r,SC} \cdot \sqrt{2} \tag{14}$$

$$X_{r,NS} = X_{r,SC} \cdot \sqrt{2} \tag{15}$$

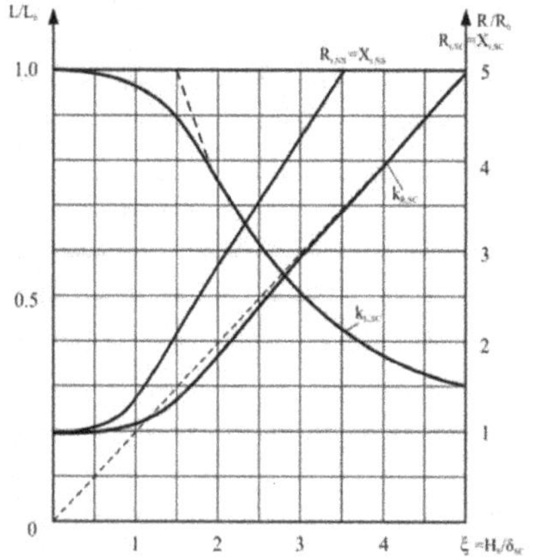

Figure 8. Dependence of rotor resistance and inductance from ratio ζ=Hb/∂,sc

The explanation is the following: for motors with powers higher than 5 kW (or with relative rotor conductor height $\xi_{SC} = H_b/\partial_{SC} \geq 1.2$), already in short-circuit mode, the rotor induced currents not the entire cross section, or rotor conductor (bar) height H_b, (Kostic, 2010). From that it could be concluded that for the negative sequence currents ($I_{r,NS}$ with frequency $f_{r,NS} \approx 2f$) it is used for $\sqrt{2}$ times lower part of the section of the rotor conductor and rotor resistance and rotor reactance are higher by √2 times, Fig. 8.

Since usually values of rotor resistance (R_r) and reactance (X_r) are known in the nominal regime, then it is necessary to know values of the coefficient of rotor resistance increase (k_R>1) and the coefficient of rotor inductance decrease ($k_L < 1$), both in the short-circuit regime. From (14) and (15), the values for $R_{r,NS}$ and $X_{r,NS}$ could be calculated as:

$$R_{r,NS} = k_{R,NS} \cdot R_r = \sqrt{2} k_{R,SC} \cdot R_r \tag{16}$$

$$X_{r,NS} = k_{X,NS} \cdot X_r = \sqrt{2} k_{X,SC} \cdot X_r \tag{17}$$

where values of coefficients $k_{R,NS}$ and $k_{X,NS}$ are determined from Fig.8 for previously established ratio value $\zeta_{NS} = H_b/\partial_{NS}$, where is ∂_{NS} –penetration depth of negative sequence field in rotor conductor. In such a manner approximate values of coefficients k_R and k_X are determined, for different rotor bar height H_b (mm)= 15, 20, 30, 40, 50 and given in Tab. 2.

From the quantitative review of corresponding values for $k_{R,SC}$ and $k_{R,NS}$, and $k_{X,SC}$ and $k_{X,NS}$, it could be seen that valid relations are: $k_{R,NS} \approx 1.41 \cdot k_{R,SC}$ and $k_{X,NS} \approx k_{R,SC}/1.41$, and it could be concluded that relations (14) and (15) are correct. In that way the author's statement that "it

is not correct to believe that values of corresponding resistances and reactances for negative sequence currents are approximately equal to those for motor short-circuit regime" is confirmed, as it is quoted in the literature (Boldea & Nasar, 2002; Ivanov-Smolensky, 1982). To the contrary, it is only correct that those values are in relation by (16) and (17). It is useful to specify common values for rotor conductor (bar) height H_b (mm) and frame sizes (axial height) for standard induction motors, as given in Tab. 3. From these facts more precise calculations and analyses of negative sequence voltage (and negative sequence currents) influencing the motor operation could be performed (Kostic & Nikolic, 2010).

Rotor slot depth H [mm]	15	20	30	40	50
$K_{R,SC}$	1.30	2.00	3.00	4.00	5.00
$K_{R,NS}$	2.00	2.80	4.20	5.60	7.00
$K_{X,SC}$	0.90	0.75	0.50	0.38	0.30
$K_{X,NS}$	0.75	0.54	0.36	0.27	0.22
$\partial_{Al,SC}$ = 10 mm; $\partial_{Al,NS}$ = 7 mm					

Table 2. Coefficients k_R and k_X in short-circuit regime and negative sequence currents

Nominal power, Pn [kW]	1.1 - 2.2	3 - 7.5	11 - 18.5	22 - 45	55 - 160	200 - 355
Axial height [mm]	80 - 90	100 - 112	132 - 160	180 - 200	225 - 280	315 - 400
Rotor slot depth, H [mm]	13 - 17	18 - 22	24 - 34	35 - 44	40 - 50	40 - 50

Table 3. Usual values for rotor bar height H_b, and frame sizes for standard induction motors

3.3. Negative sequence currents and power losses

The negative effect on the motor operation due to the presence of negative sequence voltage is obvious for two reasons:

- it induces negative sequence currents that produces losses in the stator and rotor windings, i.e. on the stator (R_s) and rotor ($R_{r,NS}$) resistance, and
- inverse torque appears which is opposite to the motor operative torque.

It is useful to express the value of negative sequence current ($I_{1,NS}$) in the units of nominal positive sequence current $I_{1N,PS}$:

$$\frac{I_{1,NS}}{I_{1n}} = \frac{U_{1,NS} / Z_{M,NS}}{U_{1N} / Z_{M,n}} \approx \frac{U_{1,NS} / X_{M,NS}}{U_{1,n} / Z_{M,N}} = \frac{U_{1,NS}}{U_{1N}} \cdot \frac{1}{X_{M,NS}} \approx (6 \div 8) \cdot \frac{U_{1,NS}}{U_{1N}} \tag{18}$$

since negative sequence motor impedance $Z_{M,NS} \approx X_{M,NS} \approx$ (0.8-0.9) $X_{M,SC}$ and motor short-circuit reactance $X_{M,SC} \approx$ (0.15÷0.20) $Z_{M,N}$, where $Z_{M,N}$ is motor impedance in the nominal regime. It could be seen from (18) that negative sequence currents in stator and rotor windings are 5÷8 times higher from the values of negative sequence voltage coefficients ($U_{1,NS} / U_n$). Since negative sequence currents are not dependent on motor load and slip, and then we suggest calculating the coefficient of asymmetry in the units of nominal motor voltage. In the following calculations and analyses the value X_{NS} =0.13 (or X_{SC} = 0.16, i.e. I_{SC}/I_N=6.25) is used, so:

$$\frac{I_{1,NS}}{I_{1n}} = 7.7 \cdot \frac{U_{1,NS}}{U_{1N}} \tag{19}$$

The value of increased losses in the phase windings of stator is proportional to the square of negative sequence currents, and then from (19) it could be calculated as:

$$\frac{P_{CuS,NS}}{P_{CuS,n}} = \left(\frac{I_{1,NS}}{I_{1N}}\right)^2 = 60 \cdot \left(\frac{U_{1,NS}}{U_n}\right)^2 \tag{20}$$

The percentage of losses in rotor conductors could be higher by up to 3-6 times (2-5 times due to the higher rotor resistance for negative sequence currents and further up to 1.2 times due to the additional increase of rotor winding temperature under such a high power losses), i.e. $R_{r,NS} \approx R_{r,SC}$=(2÷6) R_r. The equation for losses calculation in the rotor conductors for $R_{r,NS} = 5R_r$, is:

$$\frac{P_{Cur,NS}}{P_{Cur,n}} = 5 \cdot \left(\frac{I_{1,NS}}{I_{1N}}\right)^2 = 300 \cdot \left(\frac{U_{1,NS}}{U_N}\right)^2 \tag{21}$$

Assuming that negative sequence impedance is $Z_{SC,NS} \approx X_{SC,NS} = 0.13$ (or I_{SC}/I_N=7.7) and negative sequence rotor resistance is $R_{r,NS} = 5\, R_r$, the power losses values are calculated for voltage asymmetry of 2.5%, 3.5% and 5%. Such calculated values from (20) and (21), in percent of nominal losses, for particular motor parts (stator, rotor) and whole motor, are given in Tab. 4. Similar data are given in Linders (1972) where it was noted that measured values are 50% higher than calculated values. This difference was explained by an additional temperature increase of the rotor conductor, i.e. an increase of rotor resistance. Although an additional increase of rotor temperature by more than 100°C is not possible. The mentioned difference of measured and calculated values in Linders (1972) can be explained only by the fact that real rotor inverse resistance is 40% higher (based on (16) then its conventional value. Calculated values for permitted motor load are similar to those provided in NEMA standards (Fig. 5).

Negative sequence voltage [%]	0.0	2.0	3.5	5.0
Negative sequence current [%]	0.0	15.0	27.0	38.0
Stator current (RMS) [%]	100.0	101.0	104.0	107.5
Increased stator winding losses [%]	0.0	2.4	7.4	15.0
Increased rotor winding losses [%]	0.0	12.0	37.0	75.0
Increased iron losses [%], Fig. 8	0.0	2.5	7.5	15.0
Increased total motor losses [%]	0.0	5.5	17.0	34.0
Permissible motor load P/P_n [%]	100	97	91	81

Table 4. Influence of negative sequence voltage on permissible motor load ($P_N \geq 100$ kW)

Given the pessimistic assumption, especially for **smaller motors' power (up to 10 kW)** when the rotor resistance is increased only by 2-3 times (up to 1.5-2.5 times due to the higher rotor resistance for negative sequence currents and even up to 1.2 times due to additional increase in temperature of the conductor rotor in such a large loss of power, i.e. $R_{r,NS} \approx 1.4\, R_{r,SC} = (2 \div 3)\, R_r$, calculation of losses in the rotor conductors, for $R_{r,NS} = 3R_r$, was conducted by the expression:

$$\frac{P_{Cur,NS}}{P_{Cur,n}} = 60 \cdot \left(\frac{I_{1,NS}}{I_{1n}} \right)^2 = 180 \cdot \left(\frac{U_{1,NS}}{U_n} \right)^2 \tag{22}$$

since the inverse of impedance $Z_{1,NS} \approx X_{NS} \approx 0.9\, X_{SC} = 0.13$ (i.e. when the motor short-circuit reactance $X_{SC} = 0.143$, or $I_{SC} / I_n = 7$). Thus obtained data are given in Tab. 5 and they are more accurate for motors with power below 10kW. Based on these calculations, it was found that the effect of unbalanced voltage on power losses is smaller for motors with nominal power ≤ 10 kW. Thus, acceptable voltage asymmetry for these motors could be 3%.

Negative sequence voltage [%]	0.0	2.0	3.0	5.0
Negative sequence current [%]	0.0	15.0	22.0	38.0
Stator current (RMS) [%]	100.0	101.0	102.0	107.5
Increased stator winding losses [%]	0.0	2.4	5.4	15.0
Increased rotor winding losses [%]	0.0	7.0	16.0	45.0
Increased iron losses [%], Fig. 8	0.0	2.5	7.5	15.0
Increased total motor losses [%]	0.0	4.0	9.0	25.0
Permissible motor load P/P_n [%]	100	98	95	87

Table 5. Influence of negative sequence voltage on permissible motor load ($P_N \leq 10$ kW)

Based on data given in Tab. 5, it is concluded that the effects of unbalance on increased power loss is less compared to the data specified in the relevant standards for motors of nominal power ≤ 10 kW. Thus, motor total losses increase is 9% (Tab. 5) for the voltage asymmetry of 3%. As it is less then permissible 10% for these motors, for the networks with motors below 10 kW may be accept a voltage asymmetry ($U_{1,NS} / U_N$) ≤ 3%.

Although the unbalanced voltage losses in the rotor conductors are much higher than the corresponding losses in the stator winding, the increase to the losses of one phase of the stator can be the greatest. Specifically, it is unfavorable in the case where the direct and inverse current component could be in phase in one phase. Current in this phase, at the voltages' unbalance of 5%, would be:

- at nominal load $I = I_{PS} + I_{NS} = I_N + 0.38\, I_N = 1.38\, I_N$, while
- at 80% of the motor load: $I = I_{PS} + I_{NS} = 0.8\, I_N + 0.38\, I_N = 1.18\, I_N$.

These corresponding power loss values, respectively, were higher than the nominal 100% $(1.38^2\text{-}1)$ = 90% and 100% $(1.18^2\text{-}1)$ = 39%. Then the current increase in that phase would be 1.38 times at the negative sequence voltage $U_{1,NS} / U_N = 5\%$ and increase of the losses in the windings of that phase could reach 90% = 100% (1.38^2-1). Otherwise, in practice it could rarely be the case when direct and inverse components of the current matching phase angle. For example, it is necessary to stress that the asymmetry is a consequence of only one phase voltage deviation of the values (not phased by the angle) and the phase angles of the direct and inverse impedance are a little different, which is rarely filled because $\tan(\phi_{NS}) = 0.3 \div 0.4$. But it is possible that the phase angle between these components is about 30^0, and a corresponding increase in this phase will be lower:

- current increase would be in the analyzed cases 1.235 I_N and 1.079 I_N, and
- corresponding increase in losses in this phase would be 52.5% and 16.4%.

respectively, for nominal load and 80% load, both for voltages' unbalance of 5%.

For these reasons the motor operation is not allowed when the values of the coefficient of negative sequence voltage is $U_{1NS} / U_N \geq 5\%$.

Experimental measurements (Boldea & Nasar, 2002) showed that the effect of unbalanced voltages on the iron losses increase more if the motor is powered with high voltage, Fig. 9:

- loss increase in iron is 25W (or 15%) and the unbalanced voltage of 5% and nominal voltage, while
- loss increase in iron is 35W (or 25%) for voltage asymmetry of 5% and the 110%voltage.

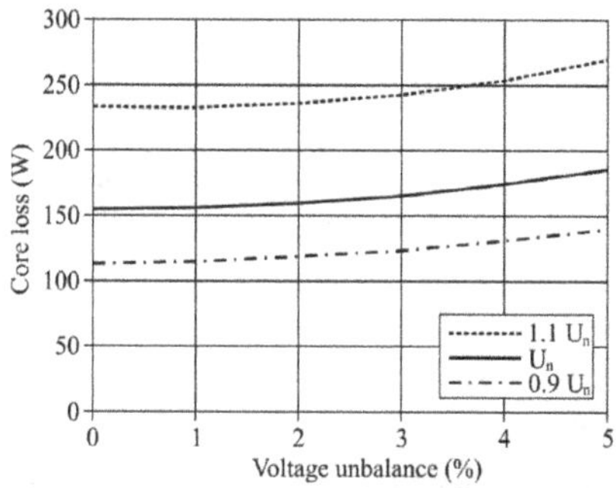

Figure 9. Dependence of iron losses on voltage asymmetry for three values of supply voltage

This additionally has an influence on reducing the coefficient of nominal power (derating factor), as well as reducing motor efficiency and increasing power consumption.

3.4. Causes and evaluation of inverse voltage values

By definition, the coefficient of asymmetry is the relationship between the inverse system voltage (U_{NS}) and direct voltage systems (U_{PS}). Thus, the percentage of unbalanced voltage is calculated using the formula:

$$K_{NS}\% = 100\frac{U_{NS}}{U_{DS}} \tag{23}$$

where the direct and inverse system voltage component are calculated using the formula

$$U_{PS} = \frac{U_{ab} + aU_{bc} + a^2 U_{ca}}{3} \tag{24}$$

$$U_{NS} = \frac{U_{ab} + a^2 U_{bc} + aU_{ca}}{3} \tag{25}$$

Thus, in the case of symmetrical voltages at the motor $\underline{U}_{ab} = a^2\underline{U}_{bc} = a\underline{U}_{ca}$, we get $U_{PS} = U_{ab} = U_{bc} = U_{ca}$ (24) and $U_{NS}=0$ (25).

Asymmetry can arise for several reasons. One is the joining of large consumers to one or two phases. Thus, if a consumer who connected to one phase, e.g. phase "a", creates a voltage drop $\Delta U = 3\%$, then it causes the asymmetry of 1% (U_{NS}= 1%), since the asymmetrical voltage system can be presented as the sum of the symmetric system voltage ($U_{ab} = U_{bc} = U_{ca}$) and the unbalanced system voltages ($U_{ab} = \Delta U = 3\%$, $U_{bc} = 0$, $U_{ca} = 0$). This second voltage system, according to (25), corresponds to unbalanced system voltage of $U_{NS} = 1\%$. If a purely inductive consumer is connected between two phases, so that the voltage levels on each phase of the impedance network is 3%, then a similar analysis leads to the conclusion that this causes asymmetry of 2% (U_{NS} = 2%), at the motor connections. These cases are possible in practice and rarely exceed the specified quantitative values.

Asymmetry may be a consequence of fuse capacitor burning. The fall out of capacitors part between two phases, concerning of the voltage reduction, is equivalent to appear of inductive loads between the two the same phases. As a consequence of that is the appearance of the inverse voltage, which is value equal to 2/3 change of the phase voltages mentioned. The general assessment is that it is almost always the asymmetry coefficient $K_{NS} < 2\%$, except for the interruption of one phase in the network, or interruption in any of the motor phase windings.

3.5. Motor inverse torque

The system of negative sequence voltages leads to the appearance of the inverse torque ($T_{em,NS}$), which is opposed to the torque that drives the motor (the torque that comes from the direct system voltages and currents, $T_{em,PS}$). Resultant driving torque is reduced, T_{em}= $T_{em,PS}$ - $T_{em,NS}$ and the direct torque must be increased to compensate that decrease. Therefore, the slip and direct current systems are increased, and also the corresponding power losses. Direct ($T_{em,PS}$) and inverse ($T_{em,NS}$) electromagnetic torques is:

$$T_{em,PS} = \frac{3U_{1PS}^2 R_r}{s\Omega\left[(R_s + R_r / s)^2 + X_{M,SC}^2\right]} \tag{26}$$

$$T_{em,NS} = \frac{3U_{1NS}^2 R_r / (2-s)}{s\Omega\left[(R_s + R_r / (2-s))^2 + X_{M,SC}^2\right]} \tag{27}$$

Where are values R_r, R_s, $Z_{M,N}$, $X_{M,SC}$, $Z_{M,SC}$ are defined in in 3.3 (page 12).

Assuming that $R_{rNS}=R_{rPS}=R_r$ and $R_r/(2\text{-}s) = R_r/2$ and with less neglect, leads to a relationship:

$$\frac{T_{em,NS}}{T_{em,PS}} \approx (\frac{U_{1NS}}{U_{1PS}})^2 \cdot \frac{Z_{PS}}{2X_{SC}} \tag{28}$$

Sometimes it is convenient to express the inverse torque in units of the nominal torque i.e.:

$$\frac{T_{em,NS}}{T_{em,N}} \approx (\frac{U_{1NS}}{U_{1N}})^2 \cdot \frac{Z_{M,N}}{2X_{SC}} \approx 3 \cdot \left(\frac{U_{1NS}}{U_{1N}}\right)^2 \tag{29}$$

This means that, for the coefficient of unbalanced voltage $U_{1,NS} / U_N = 0.04$, would be $T_{em,NS} / T_{em,N} = 3 \cdot 0.04^2 = 0.0048 = 0.48\%$. Slip and power losses will also be increased for 0.0048 times, or 0.5%. But if we assume that the inverse rotor resistance for 2-5 times higher (up to 2-4 times due to the higher rotor resistance for negative sequence currents and even up to 1.2 times due to additional increase in temperature of the conductor), then the expression for the relative values of the inverse torque is:

$$\frac{T_{em,NS}}{T_{em,N}} \approx (6 \div 15) \cdot \left(\frac{U_{1NS}}{U_{1N}}\right)^2 \tag{30}$$

The lower value of the coefficient related to the motor power up to 10 kW, and the upper value for motor power above 100 kW. When coefficient of unbalanced voltage $U_{1,NS}/ U_N = 0.05$, inverse torque in units of the nominal torque could be at $T_{em,NS}/T_{em,N} = 10 \cdot 0.05^2 = 0.025 = 2.5\%$. Slip and power losses will be increased also for 0.025 times, or 2.5% P_N. As this is a medium power motor, with $\eta_N \approx 90\%$ or with losses of $P_{\gamma N} \approx 10\%$ P_N power losses are increased by $\Delta P_\gamma = 25\%$ $P_{\gamma N}$, which is less than half of the determined value of increasing losses $\Delta P_{\gamma, NS} = 50\%$.

The explanation lies in the fact that in this way (i.e. through the power that is allocated to resistance $R_{r,NS}/2$, Fig.7c) covers only half the power losses in resistance $R_{r,NS}$ while the remaining half of the compensated part of mechanical power (P_m), which is obtained through the axis of direct voltage system, Fig. 10. Thus, another procedure confirmed adequate accuracy of quantitative estimates given in Table 5.

4. Influence of motor non-sinusoidal voltage on efficient energy usage

Analysis of the effect of non-sinusoidal voltages on the three-phase induction motor is presented in this chapter. Power losses and reactive load are increased due to the presence of harmonics in supply voltage. Nowadays, the presence of harmonics in the supply voltage is even more frequent due to the growth of consumers which are supplied through the rectifiers and inverters.

Two interesting cases have been analyzed (Kostic M. & Kostic B., 2011).

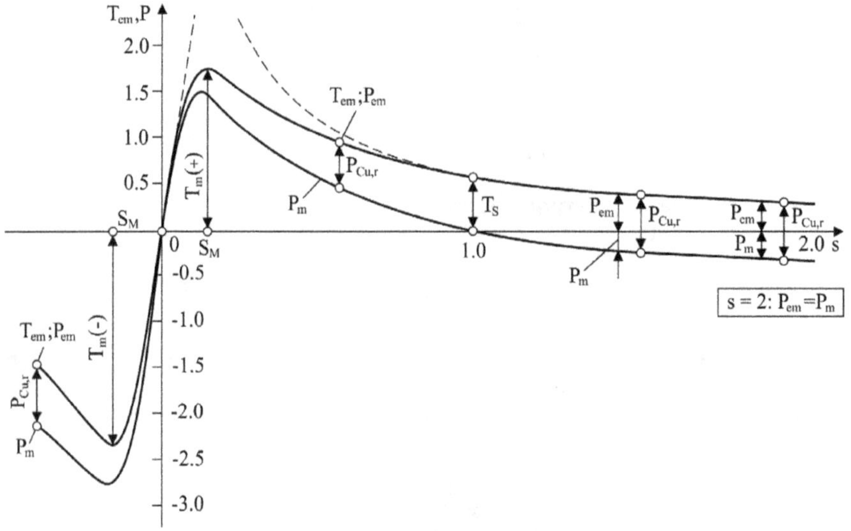

Figure 10. Electromagnetic torque characteristics (T_{em}) and the electromagnetic power (P_{em}) in the following regimes: generator (s < 0), motor (0 ≤ s ≤ 1) and braking (s> 1), as well as at point s = 2 (for the inverse voltage)

1. The case that the voltage, containing harmonics of order $h = 1, 5, 7, 11, 13, 17, ..., 37$, whose amplitudes are equal $U_h = 5\%$.
2. As the induction motors are supplied by a rectangular shape of the voltage inverter with high levels of harmonic voltage ($U_{h,i} = 1/h_i$).

The reason for this (new analysis) lies in the fact that it is (wrongly) believed that the resistance of the smaller motor's rotor does not change for higher harmonic frequencies, i.e. that is identical for all harmonics ($R_{r,h} = R_{r,1} = R_r = Const.$), which brings the difference mentioned above – and error. It is shown in Kostic M. & Kostic B. (2011) that values of the rotor resistance and the rotor reactance in a short-circuit regime could be estimated on the basis of the induction motor's catalogue date. Two important conclusions are established (probably for the first time) in Kostic (2010):

1. that rotor slot resistance and the rotor slot reactance values, $R_{r\text{-}sl,SC}$ and $X_{r\text{-}sl,SC}$, are practically equally in a short-circuit regime, and
2. that they have approximately equally values, per unit, for all motors of the same series.

4.1. Equivalent circuit of induction motors for harmonics

The equivalent circuit for the harmonics is identical to the corresponding equivalent circuit for a short-circuit regime for fundamental frequency f_1. Only, instead of rotor short-circuit resistance (R_{isk}) and rotor short-circuit reactance (R_{isk}), the two rotor resistances, rotor end (R_{ah}) and rotor slot resistance ($R_{r\text{-}sl,h}$), and two rotor reactances, rotor end ($X_{r\text{-}e,h}$) and rotor slot reactance ($X_{r\text{-}sl,h}$), are used for the harmonics order (h), Fig. 11a (Kostic M. & Kostic B., 2011).

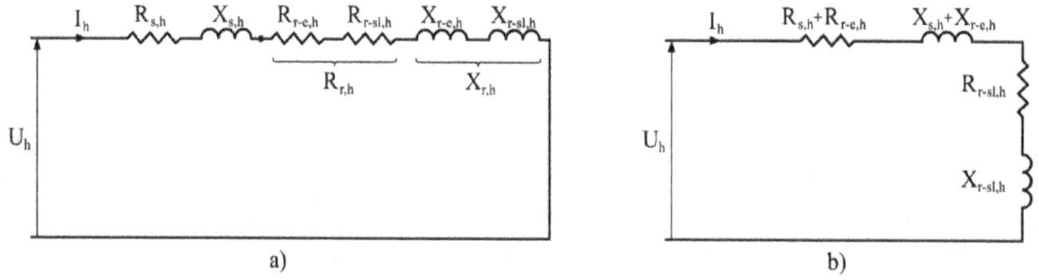

Figure 11. Motor equivalent circuit for harmonics: a) with separate rotor resistances and rotor reactances and b) with grouped motor resistances and motor reactances

Increasing the order of harmonic causes increased frequency of induced currents in rotor conductors, compared to the one in the short–circuit regime. The skin effect is practically present only in the part of the conductor in the slot of the rotor, i.e. it leads only to an increase of rotor slot resistance ($R_{r\text{-}sl,h}$) and a reduction of rotor slot inductance ($L_{r\text{-}sl,h}$). As the depth of penetration is $\partial_{Al,h=5}$ = 4.5 mm, already for the fifth harmonic, $R_{r\text{-}sl,h}$ is always equal to $X_{r\text{-}sl,h}$, Fig. 12. For this reason, similar to the corresponding scheme for short-circuit mode (Kostic, 2010), the rotor reactance ($X_{r,h}$) and resistance ($R_{r,h}$) are separated into two components in the equivalent circuit for the harmonics, Fig. 11a, i.e.:

$$R_{r,h} = R_{r-sl,h} + R_{r-e,h} \tag{31}$$

$$X_{r,h} = X_{r-sl,h} + X_{r-e,h} \tag{32}$$

where $R_{r\text{-}e,h}$ is rotor winding end resistance and $X_{r\text{-}e,h}$ is rotor winding end reactance ("e" in the index comes from the abbreviation of the word "end").

Finally, resistance and reactance of stator windings, and resistance and reactance of rotor conductors outside of slots are grouped (Fig. 11b), for which the influence of skin effects can be neglected from the nominal regime to the short–circuit regime. These are:

- grouped resistance $R_{s,h} + R_{r\text{-}e,h}$ and
- summary reactance, $X_{s,h} + X_{r\text{-}e,h}$.

The remaining resistance $R_{r\text{-}sl,h}$ and reactance $X_{r\text{-}sl,h}$, are separated (Fig 11).

4.2. Parameters of equivalent circuit for harmonics

In the paper by Caustic (2010) it is shown that values of rotor slot resistance ($R_{r\text{-}sl,SC}$) and rotor slot reactance ($X_{r\text{-}sl,SC}$) in the short-circuit mode are approximately the same for motors of all powers in a series and they are approximately equal to each other, i.e.:

$$R_{r-sl,sc} = X_{r,sc} \approx 0.030 \text{ (p.u.)} \tag{33}$$

Their values are within the narrow range of $X_{r\text{-}sl,SC} = R_{r\text{-}sl,SC} \approx 0.027 \div 0.033$ p.u., respectively for the motors of large (> 100 kW), medium (11 - 50 kW) and low power (1 - 7.5 kW).

The explanation is the following: for motors with powers higher than 5 kW (or with relative rotor conductor height $\xi_{SC} = H_b/\partial_{SC} \geq 1.5$), already in short-circuit mode, the rotor induced currents not the entire cross section, or height of rotor bars H, (Kostic, 2010). Current frequencies of individual harmonics in the rotor winding are h times higher ($f_{r,h} \approx h \cdot f_1 = h \cdot f_{r,SC}$), so the actual depth of penetration of individual harmonic currents (∂_h) is $\sqrt{h}$ times lower. Therefore, the corresponding cross section of rotor conductor is $\sqrt{h}$ times lower, so the values of rotor slot resistance are $\sqrt{h}$ times higher and the values of rotor slot inductance are $\sqrt{h}$ times lower as compared to those values for the fundamental harmonic in the short-circuit regime. In general, rotor slot resistance ($R_{r\text{-}sl,h}$), rotor slot inductance ($L_{r\text{-}sl,h}$) and rotor slot reactance ($X_{r\text{-}sl,h}$), in the function of harmonic order $h = f/f_1$, are shown in Fig. 12.

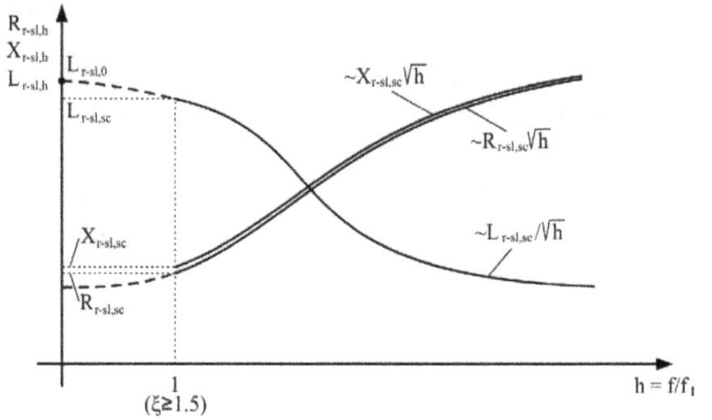

Figure 12. Dependencies of rotor slot resistance, inductance and reactance on harmonic order

If the ratio $\xi = H_b/\delta_{r,SC} \geq 1.5$, then equality $X_{r\text{-}sl,SC} = R_{r\text{-}sl,SC}$ is already true for the fundamental harmonic. As for the harmonics of order $h \geq 5$, the relative rotor conductor height is always equal to $\xi_h = H_b/\delta_h \geq 2$, then for motors of all powers is shown in Fig. 12:

$$R_{r-sl}(hf_1) / R_{r-sl,SC}(f_1) = \sqrt{h} \tag{34}$$

$$L_{r-sl}(hf_1) / L_{r-sl,SC}(f_1) = 1/\sqrt{h} \tag{35}$$

According to this, it is concluded that the following equations can be written:

$$R_{r-sl,h} = R_{r-sl,SC} \cdot \sqrt{h} \tag{36}$$

$$L_{r-sl,h} = L_{r-sl,SC} \Big/ \sqrt{h} \tag{37}$$

$$X_{r-sl,h} = X_{r-sl,SC} \cdot \sqrt{h} \tag{38}$$

This means that, based on given values of rotor slot resistance ($R_{r\text{-}sl,SC}$), rotor slot inductance ($L_{r\text{-}sl,SC}$) and rotor slot reactance ($X_{r\text{-}sl,SC}$) for fundamental frequency f_1 in short-circuit mode, the corresponding parameter values for harmonics $h = f_h/f_1$ can be calculated, i.e. values: resistance ($R_{r\text{-}sl,h}$), inductance ($L_{r\text{-}sl,h}$) and reactance ($X_{r\text{-}sl,h}$).

The values of penetration depth in the stator copper conductors are $\delta_{Cu,} \geq 1.6$ mm for frequencies $f \leq 2000$ Hz. As a rule, since the diameter of the stator winding conductors is $d_{Cu} \leq 2$ mm, it can be assumed that the value of stator windings' resistance keeps almost the same value for all harmonics of order $h \leq 40$ (or 2000 Hz/50 Hz), i.e. the following equation is valid:

$$R_{s,h} \approx R_{s,1} = R_s, \ \ for \ h \leq 40 \tag{39}$$

The same assumption approximately applies to the rotor end resistance ($R_{r\text{-}e,h}$) and to the rotor end inductance ($L_{r\text{-}e,h}$). On this basis, the following equations can be written:

$$L_{r-e,h} \approx L_{r,e} \tag{40}$$

$$X_{s,h} + X_{r-e,h} = h \cdot \left(X_s + X_{r-e} \right) \tag{41}$$

$$R_{s,h} + R_{r-e,h} = R_s + R_{r-e} \tag{42}$$

The total value of motor resistance ($R_{M,h}$), motor reactance ($X_{M,h}$) and motor impedance ($Z_{M,h}$), for current harmonics, are given in the following expressions:

$$R_{M,h} = \left(R_s + R_{r-e} \right) + R_{r-sl,SC} \cdot \sqrt{h} \tag{43}$$

$$X_{M,h} = \left(X_s + X_{r-e} \right) \cdot h + X_{r-sl,SC} \cdot \sqrt{h} \tag{44}$$

$$Z_{M,h} = \sqrt{R_{M,h}^2 + X_{M,h}^2} \tag{45}$$

The summary value of resistance ($R_s + R_{r\text{-}e}$) ≈ *Const* and the summary value of inductance ($L_s + L_{r\text{-}e}$) ≈ *Const*, retain approximately the same value in all modes: operating regime, the short-circuit regime and for regimes with harmonics. For calculation values of $R_{M,h}$ and $X_{M,h}$ by equations (43) and (44), values (R_s+$R_{r\text{-}e}$) and (X_s+$X_{r\text{-}e}$) ought to be determined from a locked rotor test, and assumed value $R_{r\text{-}sl,\ SC}$= $R_{r\text{-}sl,SC}$=*0.030pu*.

Harmonics currents ($I_{M,h}$), due to the existence of the corresponding harmonics voltages ($U_{M,h}$), are calculated using the formula (as a percentage of nominal current, %I_N):

$$I_{M,h} = 100 \cdot U_{M,h} / Z_{M,h} \ \left(\% I_N \right) \tag{46}$$

Harmonic losses in the motor, which are caused by harmonic currents through the windings of stator and rotor, are calculated using the formula (as a percentage of nominal motor power, %P_N):

$$P_{Cu,h} = 100 \cdot \frac{R_{M,h} \cdot I_{M,h}^2}{\eta \cdot \cos\phi} \ \left(\% P_N \right) \tag{47}$$

Commonly, these power losses are calculated as a percentage of nominal power losses in motor windings ($\%P_{CuN}$). Thus, assuming that losses P_{CuN} make up one half of the total power losses in the motor, their value can be determined from the formula:

$$P_{Cu,h} = 100 \cdot \frac{R_{M,h} \cdot I_{M,h}^2}{\eta \cdot \cos\phi} \cdot \frac{2\eta}{1-\eta} \left(\%P_{CuN}\right) \tag{48}$$

4.3. Current harmonics and harmonic losses in motors supplied from the network with voltage harmonics

The harmonic fields induce a current in the rotor and, as a result of interaction, are given corresponding asynchronous torques. The direction of these harmonic torques coincides with the fundamental torque direction, when h=6n+1, and harmonic torque is the opposite to the fundamental torque direction when h=6n-1. In a motor regime with slip equal to $s = 0.01$–0.06, in relation to the rotating fields of harmonics, the slip is approximately equal to 1, $s_h = 1 \pm 1/h \approx 1$.

In Radin et al. (1989), the following assumptions (partly wrong ones) are often listed:

- for motors of lower powers (≤ 20 kW), stator and rotor coil resistances and inductances practically do not depend on frequency, so the only values that become increased are the values of the stator and rotor leakage reactance ($X_{sh} = hX_s$, $X_{rh} = hX_r$),
- for motors of medium power (30-100 kW), rotor resistance is increased according to formula $R_{rh} \approx hR_r$, and
- for motors of high power (≥ 110 kW), both stator and rotor resistance are increased according to formula $R_{sh} \approx hR_s$ and $R_{rh} \approx hR_r$,

while the reactance values are somewhat reduced, $X_{sh} \leq hX_s$, $X_{rh} \leq hX_r$, since the values of corresponding inductances are decreased, $L_r(hf) < L_r$ and $L_s(hf) \leq L_s$.

The truth however is slightly different (Kostic, 2010):

- values of the stator resistance and inductance are practically unchanged in all low-voltage motor powers of 100–300 kW,
- the value of the slot resistance of the rotor in short-circuit mode (in relative units) is slightly changed with motor power, so it could be considered $R_{r\text{-}sl,sc} \approx 0.030$ p.u., as it is shown in Kostic (2010),
- rotor resistance on the part outside the slot is approximately given by the expression $R_{r\text{-}e\text{-}} \approx R_r/3 \approx R_s/3$.

By this and (43), an expression for determining the value of the motor resistance $R_{M,h}$ for harmonic order $h \geq 5$ is obtained:

$$R_{Mh} = \frac{4}{3} R_s + 0.03 \cdot \sqrt{h} \tag{49}$$

Example 1

For motors with power 5 kW - 400 kW, in the same order, the ranges of parameter values are given (Kravcik, 1982) for:

- efficiency factor $\eta = 0.85 - 0.95$ and power factor $cos\phi = 0.85 - 0.92$, i.e. $\eta \cdot cos\phi = 0.72 - 0.875$,
- stator resistance (R_s), for example from $R_s = 0.045Z_N$ to $R_s = 0.015Z_N$,

and the

- the corresponding values of stator harmonic resistances $R_{s,h} = R_s = 0.015Z_N \div 0.050Z_N = Const.$,
- the corresponding value of rotor resistance $R_{r,h} = 0.03 \sqrt{h}$,
- the corresponding values of motor resistance $R_{M,h}$, according to (43)
- the corresponding values of motor reactance $X_{M,h}$, according to (44),
- the corresponding values of motor impedance $Z_{M,h}$, according to (45),
- harmonic currents, according to (46), and
- the value of the harmonic losses, as a percentage of nominal motor power $P_{M,h}$ [$\%P_N$], according to (47), and as a percentage of nominal power losses in the windings $P_{M,h}$ [$\%P_{Cu,N}$] according to (48).

Application of the suggested method is illustrated by Tab. 6. The results show the amounts of increase in power losses due to the presence of harmonics in a given amount ($U_i = 5\%$, $i = 5, 7, .., 37$) in the supply voltage.

$h=f/f_1$	$U_{h,i}$	R_s	$R_{r,h}$	$R_{M,h}$	$X_{M,h}$	$Z_{M,h}$	$I_{M,h}$ [$\%I_n$]	$P_{M,h}$ [$\%P_n$]	$P_{M,h}$ [$\%P_{Cun}$]
1	1.00	0.015 - 0.05	0.030	0.045 - 0.080	0.161	0.167 - 0.180			
5	0.05	0.015 - 0.05	0.067	0.072 - 0.117	0.735	0.739 - 0.744	6.748	0.039 - 0.074	1.482 - 0.839
7	0.05	0.015 - 0.05	0.079	0.094 - 0.129	1.018	1.022 - 1.053	4.822	0.026 - 0.042	0.988 - 0.476
11	0.05	0.015 - 0.05	0.099	0.114 - 0.144	1.579	1.583 - 1.586	3.157	0.014 - 0.020	0.532 - 0.227
13	0.05	0.015 - 0.05	0.108	0.123 - 0.158	2.416	2.419 - 2.421	2.066	0.006 - 0.010	0.228 - 0.113
17	0.05	0.015 - 0.05	0.124	0.129 - 0.174	2.694	2.643 - 2.646	1.891	0.006 - 0.009	0.228 - 0.102
19	0.05	0.015 - 0.05	0.131	0.146 - 0.181	3.249	3.252 - 3.254	1.537	0.004 - 0.006	0.152 - 0.068
23	0.05	0.015 - 0.05	0.144	0.159 - 0.194	3.526	3.534 - 3.536	1.414	0.004 - 0.005	0.152 - 0.057
25	0.05	0.015 - 0.05	0.150	0.165 - 0.200	3.833	3.836 - 3.838	1.303	0.003 - 0.005	0.114 - 0.057
29	0.05	0.015 - 0.05	0.212	0.177 - 0.212	4.080	4.084 - 4.086	1.224	0.003 - 0.005	0.114 - 0.057
31	0.05	0.015 - 0.05	0.167	0.182 - 0.217	4.357	4.361 - 4.362	1.147	0.003 - 0.004	0.114 - 0.045
35	0.05	0.015 - 0.05	0.177	0.192 - 0.227	4.910	4.919 - 4.920	1.016	0.002 - 0.003	0.076 - 0.034
37	0.05	0.015 - 0.05	0.182	0.197 - 0.232	5.187	5.191 - 5.192	0.963	0.002 - 0.003	0.076 - 0.034
Total	THD_u = 7.3%						THD_i=9.87%	$\Sigma P_{M,h}$ 0.112 - 0.186	$\Sigma P_{M,h}$ 4.256 - 2.109

Table 6. Values of harmonic resistances ($R_{M,h}$), reactances ($X_{M,h}$) and impedances ($Z_{M,h}$) and corresponding currents and harmonic losses for motors > 100 kW (left) and < 5 kW (right), for the given values of voltage harmonics $U_{h,i}$ (p.u.) = 5%

The results in Tab. 6 show that, at the maximum permitted content of harmonics in supply voltage ($U_{h,i}$ = 5%, i = 1–37), the percentage of harmonic losses, (in units of the nominal motor power $P_{M,h}$ [%P_N]), is relatively low:

- for motors of lower power (< 5 kW), an increase of losses is by about 0.186%P_N, so a decrease in efficiency is by about 0.2%, which is slightly less compared to 0.25% in Radin et al. (1989),
- for motors of greater power (> 100 kW), an increase of losses is by about 0.112%P_N, so a corresponding decrease in efficiency is by about 0.12%.

Increments of harmonic losses are relatively small, as a percentage of nominal power losses $P_{M,h}$ [%$P_{Cu,N}$]. Apparently:

- increase of losses is about 2.109% $P_{Cu,N}$, for motors of lower power (< 5 kW),
- increase of losses is about 4.256%$P_{Cu,N}$, for motors of greater power (> 100 kW).

By the results in Tab. 6, for $U_{h,i}$ = Const (example $U_{h,i}$ = 5%, h_i = 1–37, as in Tab. 6), the following approximate equations is confirmed:

$$\frac{P_{Cu,h2}}{P_{Cuh1}} \approx \frac{h_1}{h_2}\sqrt{\frac{h_1}{h_2}}, \; for \; U_{h5} = U_7 = U_{hi} = Const, forh_i \leq 40 \tag{50}$$

Equation (50) is derived by the following approximate assumptions: $Z_{Mh} \approx X_{Mh} \approx hX_{M,SC}$ and $R_{Mh} \approx R_{r,h} \approx R_{r,SC} \cdot \sqrt{h}$.

4.4. Harmonic losses when the motor is operating with rectangular shaped voltage

When a motor is supplied by rectangular shaped voltage $U_{h,i}$ (p.u.) = $1/h_i$, h_i = 1 to 37, it is required to calculate the correspondent approximate values of harmonic losses for motors with nominal powers from 5 kW to 400 kW (i.e. for the values of stator resistance from R_s = 0.045Z_N to R_s = 0.015Z_N and correspondent approximate values in a short-circuit regime, $R_{r,SC} \approx$ 0.03, for each motor. For motors within the power range 3 kW-400 kW, for which parameters are given in chapter 4.3, power losses are determined.

The given results in Tab. 7 show that, in the specified harmonic content h = 1, 5, 7, 11, 13, 17, 19 ...35 and 37, the percentage of additional power losses, $P_{M,h}$ [%P_N], is relatively high:

- for motors of greater power (>100 kW), an increase of losses is by about 0.94%P_N, so a decrease in efficiency is by 1%,
- for motors of lower power (3-10 kW), an increase of losses is by about 1.68%P_N, so a decrease in efficiency is by 1.7%.

The literature (Radin et al., 1989) often states that the percentage of increase of power losses in the windings of stator and rotor is due to the harmonics in $P_{M,h}$ [%$P_{Cu,N}$]. Data from Tab. 6, column $P_{M,h}$ [%$P_{Cu,N}$], show that:

- an increase of losses is by 19.05%$P_{Cu,N}$, for motors of power (3–10 kW),
- an increase of losses is by 34.92%$P_{Cu,N}$, for motors of greater power (>100 kW).

This last figure corresponds to the values that are found in the literature, while the value of 19.05%$P_{Cu,N}$, for motors of lower power (3–10 kW), is much higher than the figure which is referred to in the literature (by about 5–10%). The reason for this lies in the fact that it is (wrongly) believed that the resistance of the rotor does not change for harmonic frequencies, i.e. that is identical for all harmonics ($R_{r,h} = R_{r,1} = R_r = Const$), which brings the difference mentioned above - and error. However, things are different because the rotor resistance is variable: $R_{r,h} > R_{r,SC} > R_{r,1}$, (Kostic, 2010; Kostic M. & Kostic B., 2011). To be precise, the values of rotor slot resistance are higher and the values of rotor slot inductance are $\sqrt{h}$ times lower as compared to those values for the fundamental harmonic in short-circuit mode.

Some examples from the literature can be used as proof of the view that the rotor resistance changes for low power motors. Specifically in Vukic (1985), the influence of harmonics on the motor of low power (1.6 kW) was tested. The calculation results, which were carried out assuming that $R_r = Const$, gave an increase in power losses of 12.6%, while the experimental measurements showed that the actual increase in losses was 18.5%. Our calculations give rise to losses of 19%, which slightly differs from the measured values. The accuracy of our calculations has been increased with respect to the fact that slot reactance of the rotor increases $\sqrt{h}$ times, for the harmonics of order h.

$h=f/f_1$	$U_{h,i}$	R_s	$R_{r,h}$	$R_{M,h}$	$X_{M,h}$	$Z_{M,h}$	$I_{M,h}$ [$\%I_n$]	$P_{M,h}$ [$\%P_n$]	$P_{M,h}$ [$\%P_{Cun}$]
1	1.00	0.015-0.05	0.030	0.045-0.080	0.161	0.167-0.180			
5	0.20	0.015-0.05	0.067	0.072-0.117	0.735	0.739-0.744	26.990	0.618-1.184	23.447-13.418
7	0.14	0.015-0.05	0.079	0.094-0.129	1.018	1.022-1.053	13.790	0.213-0.341	7.668-3.864
11	0.11	0.015-0.05	0.099	0.114-0.144	1.579	1.583-1.586	7.010	0.069-0.099	2.504-1.122
13	0.08	0.015-0.05	0.108	0.123-0.158	2.416	2.419-2.421	3.180	0.015-0.022	0.556-0.252
17	0.06	0.015-0.05	0.124	0.129-0.174	2.694	2.643-2.646	2.230	0.007-0.012	0.244-0.136
19	0.05	0.015-0.05	0.131	0.146-0.181	3.249	3.252-3.254	1.600	0.005-0.007	0.180-0.079
23	0.04	0.015-0.05	0.144	0.159-0.194	3.526	3.534-3.536	1.220	0.003-0.004	0.106-0.046
25	0.04	0.015-0.05	0.150	0.165-0.200	3.833	3.836-3.838	1.040	0.002-0.003	0.075-0.036
29	0.03	0.015-0.05	0.162	0.177-0.212	4.080	4.084-4.086	0.830	0.001-0.002	0.036-0.025
31	0.03	0.015-0.05	0.167	0.182-0.217	4.357	4.361-4.362	0.730	0.001-0.002	0.036-0.025
35	0.03	0.015-0.05	0.177	0.192-0.227	4.910	4.919-4.920	0.590	0.001-0.002	0.036-0.025
37	0.03	0.015-0.05	0.182	0.197-0.232	5.187	5.191-5.192	0.520	0.001-0.002	0.036-0.025
Total	THD_u = 30.3%						THD_i=31.5%	$\Sigma P_{M,h}$ 0.94 -1.68	$\Sigma P_{M,h}$ 34.92-19.05

Table 7. Values of harmonic resistances ($R_{M,h}$), reactances ($X_{M,h}$) and impedances ($Z_{M,h}$); as harmonic currents ($I_{M,h}$) and harmonic losses ($P_{M,h}$) for motors with power > 100 kW (left) and lower power, 3–10 kW (right), when the motor is supplied by the rectangular voltage, i.e. by voltage with harmonics $U_{hi} = 1/h_i$.

As R_s=0.050÷0.015, respectively, for motors of power 3÷200 kW, the given results are useful for the evaluation of harmonic currents ($I_{M,h}$) and harmonic losses ($P_{M,h}$) for all mentioned motors, by extrapolation.

5. Summary

The results of the analysis presented in this chapter are summarised in the following.

A) Effect of voltage magnitude on motor power losses and motor reactive loads

The results of the research show dependencies of the input power and reactive load on voltage magnitude, as given in Fig. 2.

1. Decreasing voltage magnitude by 1%, by setting voltages in range of U_n±5%, leads to:
   a. reactive loads decreasing
      - from 1% to 2%, at loads from 100% to 75%, respectively, for motors above100 kW,
      - up to 3% to 4%, at loads from 75% to 25%, respectively, for motors below 10 kW;
   b. power losses (and active input powers) decreasing/increasing (sign „ - ")
      - from (-0.1%) to 0.1%%, respectively at loads from 100% to 50%, for motors above100 kW
      - from 0% to 0.6%, respectively at loads from 100% to 25%, for motors about 10 kW,
      - from 0% to 1.6%, respectively, at loads from 100% to 25%, for motors below 1 kW.
2. On the basis of the investigation presented in this paper, it is confirmed that there are significant possibilities for energy savings by means of voltage magnitude setting, within values U_n±5%, in networks with induction motors which are light loaded (<70%).
3. Setting voltage within band 0.9 U_n - 0.95 U_n is not recommended, even if it leads to reduced power losses and reactive loads, because starting and maximal torque are decreased and it can also cause motor operation instability.

B) The most important conclusions regarding motor operation with the unbalanced voltage

4. It is explained that the rotor inverse resistance and rotor inverse reactance are higher for √2 times compared to the rotor resistance and rotor reactance in short-circuit mode, since current frequency of the negative sequence in the rotor winding is twice as high ($f_{r,NS} \approx 2f_1 = 2f_{r,SC}$), i.e. they are higher by 1.41 times than the corresponding values given in the literature.
5. Voltage unbalance causes increase of the motor heating, occurrence of inverse torques and a small increase in motor slip. Thus, for the voltage asymmetry of 2%, 3%, 4% and 5%, this causes an increase in power losses of 5.5%, 12%, 22% and 34% of motor nominal losses. Corresponding values of derating factors are 0.97, 0.94, 0.88 and 0.81, respectively, as noted in NEMA standards, so the acceptable voltage asymmetry is 2%.
6. Based on the actual calculation and analysis, it was found that the effects of an unbalance on power loss are smaller for motors of nominal power ≤ 10 kW. Thus, acceptable voltage asymmetry for these motors could be 3%.
7. Generally, motor operation is not allowed when voltage asymmetry is greater than 5%, because, in some cases, current and losses in one phase could be increased for 38% and 90%, respectively.

C) The most important conclusions regarding motor operation with the non-sinusoidal voltage

8. The given results show that, at the maximum permitted content of harmonics in supply voltage, $U_{h,i}$ = 5%, i = 1 to 37), the percentage of harmonic losses is relatively small:
   - about 2.109% $P_{Cu,N}$, for motors of lower power (< 5 kW), and
   - about 4.256%$P_{Cu,N}$, for motors of greater power (> 100 kW).
9. When the induction motors are supplied by rectangular shaped voltage with high levels of harmonic voltages, an increase in harmonic losses in stator and rotor windings are:
   - around 30-35% $P_{Cu,N}$, for high power motors (> 100 kW), and
   - around 15-20% $P_{Cu,N}$, for lower power motors (3–10 kW).

The increase in harmonic power losses for lower power motors is much higher than it was noted in the literature (5-10%) because it is (wrongly) believed that the resistance of the rotor does not change for higher harmonic frequencies.

Appendix

For deriving equations for electromagnetic torque and power, the equivalent Γ-circuit, shown in Fig. 13, is used (Kostic, 2010):

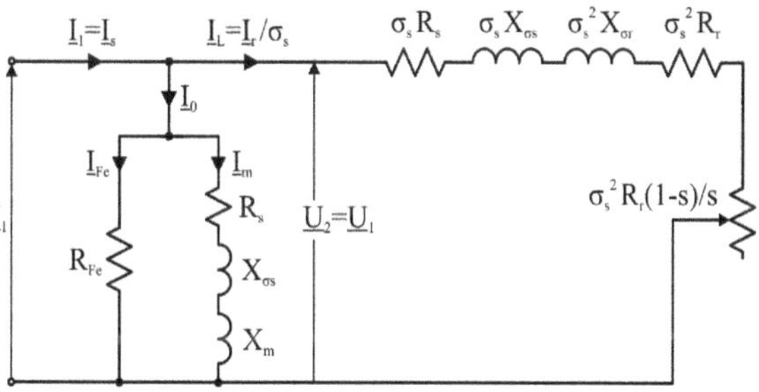

Figure 13. Equivalent Γ-circuit of induction machine

Equation (9) is completely derived in this Appendix.

1. Electromagnetic power ($P_{em,N}$) **at rated load**, i.e. at slip $s{=}s_N$, can be expressed as following:

$$P_{em,N} = T_{em,N} \cdot \Omega_1 = \frac{I_L^2 \cdot \sigma_s{}^2 R_r}{s_N} = \frac{U_1^2 \cdot \sigma_s{}^2 R_r / s_N}{(\sigma_s R_s + \sigma_s{}^2 R_r)^2 + (\sigma_s X_s + \sigma_s{}^2 R_r / s_N)^2} \tag{51}$$

For motors with power within the range of 1÷200kW, values for s_N are 0.05÷0.01, respectively, and therefore: $\sigma_s^2 R_r/s_N = (20\div100)\cdot\sigma_s R_s$ and $\sigma_s X_s + \sigma_s^2 X_r) \approx 0.20\cdot\sigma_s^2 R_r/s_m$.

$$P_{em,N} = T_{em,N} \cdot \Omega_1 = \frac{I_L^2 \cdot \sigma_s{}^2 R_r}{s_N} \approx \frac{U_1^2 \cdot \sigma_s{}^2 R_r / s_N}{(1.15 \div 1.05)^2 (\sigma_s{}^2 R_r / s_N)^2} = \frac{U_1^2}{(1.15 \div 1.05)(\sigma_s^2 R_r / s_N)} \tag{52}$$

2. **Regime with maximum input power,** i.e. at $s=s_{Pm}$, accrues when resistance $(\sigma_s X_{\sigma s}+\sigma_s^2 X_{\sigma r})$ and reactance in load branch $(\sigma_s R_s+\sigma_s^2 R_r/s_m)$ are equal, i.e., and when the load branch impedance is $Z_{2,m}=\sqrt{2}(\sigma_s X_{\sigma s}+\sigma_s^2 X_{\sigma r})$. Corresponding electromagnetic power $(P_{em,Pm})$ on the resistance $\sigma_s^2 R_r/s_m$ is:

$$P_{em,Pm} = T_{em,Pm}\Omega_1 = I_L^2 \cdot \sigma_s{}^2 \frac{R_r}{s_{Pm}} = \frac{U_1^2 \cdot \sigma_s{}^2 R_r / s_{Pm}}{2(\sigma_\sigma X_{\sigma s} + \sigma_\sigma{}^2 X_{\sigma r})^2} \tag{53}$$

Since for motors with power within the range of 1÷200kW, values for corresponding slip are s_{Pm} = 0.25÷0.05, respectively, the skin effect in the bars of the squirrel-cage is minor (the depth of penetration $\delta_r(s_m f_1) \geq H_b$-the bar (conductor rotor) height), so it is $\sigma_s^2 R_r/s_m$ = (5÷20) $\sigma_s R_s$. Consequently, it is:

$$\sigma_s^2 R_r s_{Pm} = (0.8 \div 0.95) \cdot (\sigma_s R_s + \sigma_s^2 R_r / s_{Pm}) = (0.8 \div 0.95) \cdot (\sigma_s X_{\sigma s} + \sigma_s^2 X_{\sigma r}) \tag{54}$$

and the electromagnetic power $(P_{em,m})$, in the regime with maximum input power, is:

$$P_{em,Pm} = T_{em,Pm} \cdot \Omega_1 = \frac{I_L^2 \cdot \sigma_s{}^2 R_r}{s_{Pm}} \approx \frac{U_1^2 \cdot (0.8 \div 0.95)(\sigma_s X_{\sigma s} + \sigma_s{}^2 X_{\sigma r})}{2 \cdot (\sigma_s X_{\sigma s} + \sigma_s{}^2 X_{\varsigma r})^2} = \frac{U_1^2 \cdot (0.8 \div 0.95)}{2 \cdot (\sigma_s X_{\sigma s} + \sigma_s^2 X_{\sigma r})} \tag{55}$$

3. If $\sigma_s^2 R_r/s_N$ is expressed from (A-4), and $(\sigma_s X_{\sigma s}+\sigma_s^2 X_{\sigma r})$ is expressed from (A-5), then it is:

$$\sigma_s{}^2 R_s + \sigma_s{}^2 R_r / s_N = \frac{U_1^2}{T_{em,N} \cdot \Omega_1 (1.15 \div 1.05)} \tag{56}$$

$$\sigma_s X_{\sigma s} + \sigma_s{}^2 X \sigma_r = \frac{U_1^2 \cdot (0.8 \div 0.95)}{2T_{em,Pn} \cdot \Omega_1} \tag{57}$$

On the base of (A-6) and (A-7), it is obtained:

$$\frac{\sigma_s X_{\sigma s} + \sigma_s{}^2 X_{\sigma r}}{\sigma_s{}^2 R_s + \sigma_s{}^2 R_r / s_N} = \frac{T_{em,N}}{2T_{em,Pm}} \cdot (0.8 \div 0.95) \cdot (1.15 \div 1.05) \tag{58}$$

Reactive power in the load branch of Γ-circuit, under rated condition, $Q_{2N} \approx Q_{LN}$ (Q_{LN} – load component of reactive power), can be expressed in terms of the electromagnetic power, $P_{em,N}$

$$Q_{2N} = P_{em,N} \cdot \frac{\sigma_s X_{\sigma s} + \sigma_s{}^2 X \sigma_r}{\sigma_s{}^2 R_s + \sigma_s{}^2 R_r / s_N} \approx Q_{LN} \tag{59}$$

Since the relation between the electromagnetic power $(P_{em,N})$ and the rating power (P_N) is:

$$P_{em,N} = P_N \cdot \frac{\sigma_s^2 R_s + \sigma_s^2 R_r / s_N}{(\sigma_s^2 R_r / s_N)} \cdot \frac{1}{1 - s_N} \tag{60}$$

then, based on equations (A-7), (A-9) and (A-10), it follows:

$$Q_{LN} = P_N \cdot \frac{T_{em,N}}{2T_{em,Pm}} \cdot \frac{(0.08 \div 0.95) \cdot (1.15 \div 1.05)}{0.95 \div 0.99} = \frac{T_{em,N}}{2T_{em,Pm}} \cdot (0.98 \div 1.01) \approx 0.5 P_N / (T_m / T_N) \quad (61)$$

$$Q_{LN} \approx P_N \cdot \frac{T_{em,N}}{2T_{em,Pm}} = 0.5 \cdot \frac{T_{em,N}}{T_{em,Pm}} \quad (62)$$

Since the maximum torque ($T_m \approx T_{em,m}$), which is catalogue data, is greater up to 2% from mentioned torque ($T_{em,Pm}$) in the regime with maximum input power, i.e. $T_{em,Pm} \leq 1.02\, T_m$, it might be concluded that the equation (9) sufficiently accurate for calculating the rating component of reactive power in load branch, $Q_{LN}=0.5 \cdot P_N/(T_m/T_N)$.

Author details

Miloje Kostic
Electrical Engineering Institute "Nikola Tesla", Belgrade University, Belgrade, Serbia

6. References

Aníbal, T. de Almeida Fernando J. T. E. Ferreira; João Fong & Paula Fonseca (December 2007). *EUP Lot 11 Motors, Final Report*, ISR-University of Coimbra, Lot 11-8-280408.

Boldea, I.S. & Nasar, A. (2002). *The Induction Machine Handbook*, 2002 by CRC Press LLC.

Bonnett, A.H. (2000). An overview of how AC induction motors' performance has been affected by the October 24, 1997 Implementation of the Energy Policy Act of 1992, *IEEE Transaction on Industry Applications*, Vol.36, No1, 2000, pp. 242-256.

Fei, R.; Fuch, E.F.& Huang, H. (December 1989). Comparison of two optimization techniques as applied to three-phase induction motor design, *IEEE Transactions energy Conversion*, Vol.4, pp. 651-660, December 1989.

Fink, D.G. (1983). *Standard Handbook for Electrical Engineers* (1983), 11th Edition McGraw-Hill Book Company, 1983, New York, pp. 2462, ISBN 0-07-020974-X

Hamer, P. S.; Love, D. M. & Wallace, S. E. (1997). Energy Efficient Induction Motors Performance Characteristic and Life Cycle Cost Comparison for Centrifugal Loads, *IEE Trans. Ind. Applications*, No. 5, 1997, pp. 1312-1320.

IEC 60034-30 (2008). *Efficiency classes of single speed three-phase cage induction motors*.

IEC 60034-31 (2010). *Guide for selection and application of energy-efficient motors including variable-speed applications*.

Ivanov-Smolensky, A. (1982). *Electrical Machines*, Vol. 2, Mir Publishers, 1982, pp. 464.

Kostic, M (2010). *Energy Efficiency Improvement of Motors in Drives*, Electrical Engineering Institute Nicola Tesla, Belgrade, 2010, pp.325 (in Serbian), ISBN 978-86-83349-11-1.

Kostic, M. & Kostic, B. (2011). Motor Voltage High Harmonics Influence to the Efficient Energy Usage, Invited paper for *15th WSEAS International Conference on Systems, Proc. pp.* 276-281, Corfu Island, Greece, July 2011.

Kostic, M. & Nikolic, A. (August 2010). Negative Consequence of Motor Voltage Asymmetry and Its Influence on the Inefficient Energy Usage, *Wseas Transaction On Circuits And Systems*, Issue 8, Volume 9, August 2010, pp. 547-556.

Kostic, M. (1998). Reduction of loads and electric energy consumption by setting voltage magnitude, *Elektroprivreda Magazine, No. 3, 1998*, pp 65-78 (in Serbian).

Kostic, M. (2001). Evaluation methods for load and efficiency of induction motor in the exploitation, *11th International Symposium Ee 2001*, Novi Sad, Serbia, pp.332-336.

Kostic, M. (2010). Equivalent circuit parameters of the squirrel-cage induction motors in short circuit regime, *Tehnika, separate Elektrotehnika* 5/2010), pp. 7E-13E (in Serbian).

Kostic, M.; Stanisavljevic, I.; Ivanovic, M.; Jankovic, R.; Mihajlovic, Lj. & Vasic, P. (2006). The Reduction of Own Electric Energy Consumption of Thermal Power Plants, *Symposium Power Plants 2006*, Vrnjacka Banja, Serbia, 2006, No paper 59.

Kravčik. A.E. (1982). *Induction Machines Handbook* (Moscow, 1982), p. 504, (in Russian).

Linders, J.R. (July/August 1972). Effects of Power Supply Variations on AC Motor Characteristics, *IEEE Transaction on Ind. Applic.* Vol. IA-8", No 4, 1972, pp. 383-400.

Radin, I.; Bruskin & Zorohovič, A.E. (1989). *Electrical machines: Induction machines* (Moscow, 1989), p. 328 (in Russian).

Vukic, DJ. (1985). Time Harmonics Influence on Operating of Induction Motors, *Tehnika*, separate Elektrotehnika 12/1985), pp.11E-13E (in Serbian).

Chapter 6

Optimization of Induction Motors Using Design of Experiments and Particle Swarm Optimization

Houssem Rafik El-Hana Bouchekara, Mohammed Simsim and Makbul Anwari

Additional information is available at the end of the chapter

1. Introduction

The level of prosperity of a community is related to its ability to produce goods and services. But producing goods and services is strongly related to the use of energy in an intelligent way. Energy can be exploited in several forms such as thermal, mechanical and electrical (Boldea & Nasar, 2002). Electrical energy, measured in kWh, represents more than 30% of all used energy and it is on the rise (Boldea & Nasar, 2002). The larger part of electrical energy is converted into mechanical energy in electric motors. Among electric motors, the induction motor is without doubt the most frequently used electrical motor and is a great energy consumer. About 70% of all industrial loads on a specific utility are represented by induction motors (Maljkovic, 2001). The vast majority of induction motor drives are used for heating, ventilation and air conditioning (Blanusa, 2010; Cunkas & Akkaya 2006).

The design of an induction motor aims to determine the induction motor geometry and all data required for manufacturing to satisfy a vector of performance variables together with a set of constraints (Boldea & Nasar, 2002). Because induction motors are now a well developed technology, there is a wealth of practical knowledge, validated in industry, on the relationship between their performance constraints and their physical aspects. Moreover, mathematical modeling of induction motors using circuit, field or hybrid models provides formulas of performance and constraint variables as functions of design variables (Boldea & Nasar, 2002).

The journey from given design variables to performance and constraints is called analysis, while the reverse path is called synthesis. Optimization design refers to ways of doing efficient synthesis by repeated analysis such that some single (or multiple) objective (performance) function is maximized and/or minimized while all constraints (or part of them) are fulfilled (Boldea & Nasar, 2002). The aim of this chapter is to present an optimal design method for induction motors using design of experiments (DOE) and particle swarm optimization (PSO) methods.

The outline of this paper is as follows. The current section is the introduction. Section 2 introduces and explains the DOE method. Section 3 gives an overview of the PSO method. In Section 4 the application of the DOE and PSO to optimize induction motors is explained and its results are also presented and discussed in detail. Finally, the conclusions are drawn in Section 5.

2. Design of Experiments (DOE)

With modern technological advances, the design and optimization of induction motors or any other electromechanical devices are becoming exceedingly complicated. As the cost of experimentation rises rapidly it is becoming impossible for the analyst, who is already constrained by resources and time, to investigate the numerous factors that affect these complex processes using trial and error methods (ReliaSoft Corporation, 2008). Computer simulations can solve partially this issue. Rather than building actual prototypes engineers and analysts can build computer simulation prototypes. However, the process of building, verifying, and validating induction motor simulation model can be arduous, but once completed, it can be utilized to explore different aspects of the modeled machine. Moreover, many simulation practitioners could obtain more information from their analysis if they use statistical theories, especially with the use of DOE.

In this section the DOE method is explained in order to make its use in this chapter understandable. The aim here is not to explain the whole method in detail (with all the mathematical developments behind), but to present the basics to demonstrate its interesting capabilities.

2.1. Why DOE?

Compared to one-factor-at-a-time experiments, i.e. only one factor is changed at a time while all the other factors remain constant, the DOE technique is much more efficient and reliable. Though, the one-factor-at-a-time experiments are easy to understand, they do not tell how a factor affects a product or process in the presence of other factors (ReliaSoft Corporation, 2008). If the effect of a factor is altered, due to the presence of one or more other factors, we say that there is an interaction between these factors. Usually the interactions' effects are more influential than the effect of individual factors (ReliaSoft Corporation, 2008). This is because the actual environment of the product or process comprises the presence of many factors together instead of isolated occurrences of each factor at different times.

The DOE methodology ensures that all factors and their interactions are systematically investigated. Therefore, information obtained from a DOE analysis is much more reliable and comprehensive than results from the one-factor-at-a-time experiments that ignore interactions between factors and, therefore, may lead to wrong conclusions (ReliaSoft Corporation, 2008).

Let's assume, for instance, that we want to optimize an induction motor taking into account, for simplicity, only two factors: the length and the external radius. Hence, the length is the first factor and is denoted by x_1 while the external radius is the second factor and it is denoted by x_2. Each factor can take several values between two limits, i.e. $\{x_{1_{min}}, x_{1_{max}}\}$ and $\{x_{2_{min}}, x_{2_{max}}\}$. We desire to study the influence of each of these factors on the system response or output for example the torque called Y. The classical or traditional approach consists of studying the two factors x_1 and x_2, separately. First we put x_2 at the average level $x_{2_{average}}$ and study the response of the system when x_1 varies between $x_{1_{min}}$ and $x_{1_{max}}$ using for example 4 steps (experiments) as shown in Fig. 1. Similarly, we repeat the same procedure to study the effect of x_2. Accordingly, the total number of tests is 8. However, we should ask a paramount question here, are these 8 experiments sufficient to have a good knowledge about the system? The simple and direct answer to this question is no. To get a better knowledge about the system, we have to mesh the validity domain of the two factors and test each node of this mesh as shown in Fig. 2. Thus, 16 experiments are needed for this investigation. In this example only two factors are taken into account. Therefore, if for example 7 factors are taken into account, the number of tests to be performed rises to $4^7 = 16384$ experiments, which is a highly time and cost consuming process.

Knowing that it is impossible to reduce the number of values for each factor to less than 2, the designer often reduces the number of factors, which leads to incertitude of results. To reduce both cost and time, the DOE is used to establish a design experiment with less number of tests. The DOE, for example, allows identifying the influence of 7 factors with 2 points per variable with only 8 or 12 tests rather than 128 tests used by the traditional method (Bouchekara, 2011; Uy & Telford, 2009).

Recently, the DOE technique has been adopted in the design and testing of various applications including automotive assembly (Altayib, 2011), computational intelligence (Garcia, 2010), bioassay robustness studies (Kutlea, 2010) and many others.

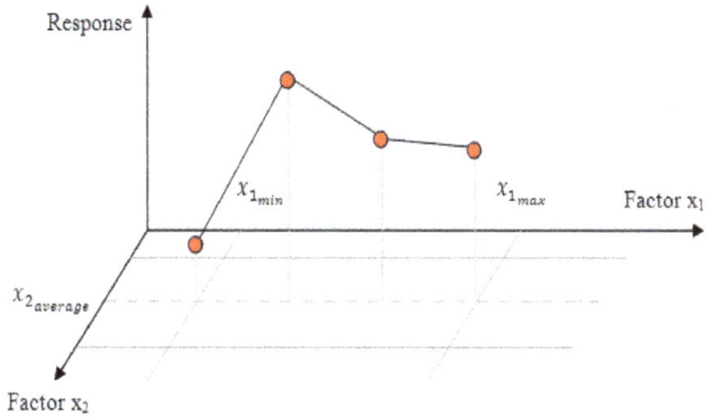

Figure 1. Traditional method of experiments.

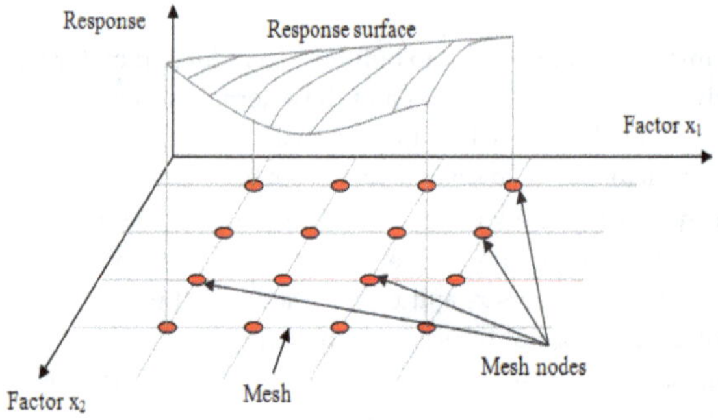

Figure 2. One experiment at each node of the mesh.

2.2. Methodology

The design and analysis of experiments revolves around the understanding of the effects of different variables on other variable(s). The dependent variable, in the context of DOE, is called the response, and the independent variables are called factors. Experiments are run at different values of the factors, called levels. Each run of an experiment involves a combination of levels of the investigated factors. The number of runs of an experiment is determined by the number of levels being investigated in the experiment (ReliaSoft Corporation, 2008).

For example, if an experiment involving two factors is to be performed, with the first factor having n_1 levels and the second having n_2 levels, then $n_1 \times n_2$ combinations can possibly be run, and the experiment is an $n_1 \times n_2$ factorial design. If all $n_1 \times n_2$ combinations are run, then the experiment is a full factorial. If only some of the $n_1 \times n_2$ combinations are run, then the experiment is a fractional factorial. Therefore, in full factorial experiments, all factors and their interactions are investigated, whereas in fractional factorial experiments, certain interactions are not considered.

2.3. Mathematical concept

Assume that y is the response of an experiment and $\{x_1, x_2, x_3, \dots, x_k\}$ are k factors acting on this experiment where each factor has two levels of variation x_{i-} and x_{i+}. The value of y, is approximated by an algebraic model given in the following equation:

$$y = a_0 + a_1 x_1 + a_2 x_2 + \cdots + a_k x_k + \cdots + a_1 x_1 x_2 + \cdots a_1 x_1 x_k + a_{1\dots k} x_{1\dots k} \tag{1}$$

where a_j are coefficients which represent the effect of factors and their interactions on the response of the experiment.

2.4. Full factorial design

As mentioned above, the study of full factorial design consists of exploring all possible combinations of the factors considered in the experiment (Kleijnen et al., 2005). Note that the design X^k means that this experiment concerns a system with k factors with x levels. Usually, two levels of the x's are used. The use of only two levels implies that the effects are monotonic on the response variable, but not necessarily linear (Uy & Telford, 2009). For each factor, the two levels are denoted using the "rating Yates" notation by -1 and +1 respectively to represent the low and the high levels of each factor. Hence, the number of experiments carried out by a full factorial design for k factors with 2 levels is $n = 2^k$. For example, Table 1 shows the design matrix of a full factorial design for 2 factors while, Fig. 3 shows the mesh of the experimental field where points correspond to nodes.

Run	Factor x_1	Factor x_2	Response Y
1	-1	-1	Y_1
2	-1	+1	Y_2
3	+1	-1	Y_3
4	+1	+1	Y_4

Table 1. Design Matrix for a full factorial design for 2 factors with 2 levels.

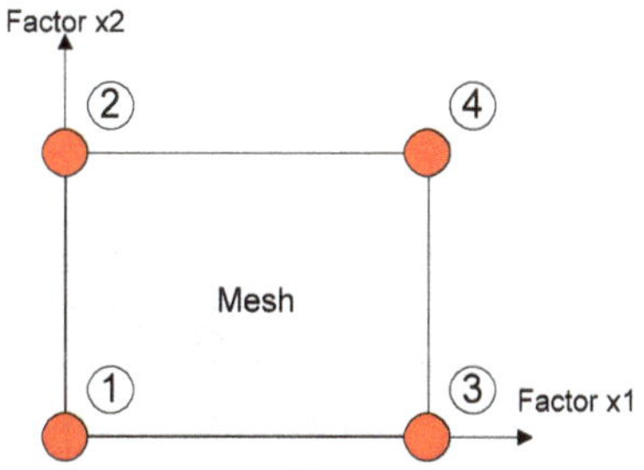

Figure 3. Strategy of experimentation; points corresponding to nodes in the mesh of the experimental field for a full factorial design for 2 factors with 2 levels.

2.5. Fractional factorial design

The advantage of full factorial designs, is their ability to estimate not only the main effects of factors, but also all their interactions, i.e. two by two, three by three, up to the interaction involving all k factors. However, when the number of factors increases, the use of such design leads to a prohibitive number of experiments. The question to be asked here is: is it necessary to perform all experiments of the full factorial design to estimate the system's response? In other words, is it necessary to conduct a test at each node of the mesh?

It is not necessary to identify the effect of all interactions because the interactions of order ≥ 2 (like $x_1x_2x_3$) are usually negligible. Therefore, certain runs specified by the full factorial design can be used instead of using all runs. To illustrate this phenomenon, an analogy can be made with a Taylor series approximation where the information given by each term decreases when its order increases. So, fractional factorial designs can be used to estimate factors effect and interactions that influence the experiments more with a reduced number of runs (Bouchekara, 2011). Taguchi Tables (Pillet, 1997), or Box generators (Demonsant, 1996), can be used to generate the fractional factorial design matrix of experiments.

To illustrate fractional factorial designs let's take an example. If $k = 3$, the design matrix of these three factors is given by Box generators in a way that the third factor is the product of the two other factors. The factor x_3 and interaction x_1x_2 are either confused or aliased, and there is a confusion of these aliases because only their sums are reachable (Pillet, 1997; Costa, 2001).

Table 2 shows a full factorial design for 3 factors with 2 levels. The number of runs is $2^3 = 8$. This number is reduced to 4 using a fractional factorial design as shown in Table 3 where the third factor is generated using Box generator for 3 factors given in Table 4. The comparison of the 2 designs is shown in Fig. 4.

Run	Factor x_1	Factor x_2	Factor x_3	Response Y
1	-1	-1	-1	Y_1
2	-1	-1	+1	Y_2
3	-1	+1	-1	Y_3
4	-1	+1	+1	Y_4
5	+1	-1	-1	Y_5
6	+1	-1	+1	Y_6
7	+1	+1	-1	Y_7
8	+1	+1	+1	Y_8

Table 2. Design Matrix for a full factorial design for 3 factors with 2 levels.

Run	Factor x_1	Factor x_2	Factor x_3	Response Y
1	-1	-1	+1	Y_1
2	-1	+1	-1	Y_2
3	+1	-1	-1	Y_3
4	+1	+1	+1	Y_4

Table 3. Design Matrix for a fractional factorial design for 3 factors with 2 levels.

Resolution	Design name	Number of Runs	Generators
3	2^{3-1}	4	$x_3 = x_1 \times x_2$

Table 4. G. Box generator of fractional factorial design for 3 factors.

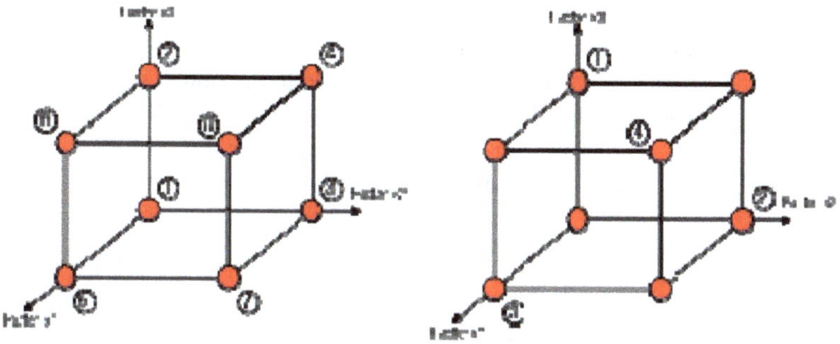

(a) (b)

Figure 4. Comparison between the design experimental field of full and fractional factorial designs with 3 factors. (a) full factorial design; (b) fractional factorial design

2.6. Estimation of model coefficients

The coefficient a_0 of (1) is estimated from the arithmetic average of all observed responses and it is given by:

$$a_0 = \bar{y} = \frac{1}{n}\sum_{i=1}^{n} y_i \tag{2}$$

where y_i is the response observed for the experiment i and n is the total number of experiments.

The effect of a factor x_j at the level x_{j+} can be calculated thus, the coefficient associated with this effect can be identified using the following equations:

$$a_j = e_{a_j} = y_{x_j}^{+} - a_0 \tag{3}$$

and

$$y_{x_j}^{+} = \frac{1}{n^{+}}\sum_{i=1}^{n} y_i^{+} \tag{4}$$

where $y_{x_j}^{+}$ is the response observed for experiment i when x_j is at level x_{j+}, n^{+} is the number of experiments when x_j is at level x_{j+} and e_{a_j} is the effect of coefficient a_j.

Once the method of how to calculate the coefficients of the model and how to identify the existing confusion between these factors has been presented, we can evaluate the contributions of contrasts (the sum of confusions) and therefore the most significant factors (affecting the response).

In (Demonsant, 1996) the identification of the significant factors has been proposed by evaluating the coefficients contribution (or contrasts, for fractional designs) on the model response from the normalization of their values compared to the sum of squared responses, such as given in the following equations:

$$C_{a_j} = \frac{SCE(a_j)}{SCE(y)} \, [\%] \tag{5}$$

with

$$SCE(y) = \sum_{i=1}^{n} (y_i - \bar{y})^2 \tag{6}$$

$$SCE(a_j) = \frac{n}{s} \sum_{j=1}^{s} \left(e_{a_j}\right)^2 \tag{7}$$

where s is the number of levels (equals to 2 in this case), e_{a_j} is the effect of coefficient a_j, and C_{aj} is the contribution of the contrast associated with the coefficient a_j.

According to (Demonsant, 1996):

- The contribution given by (5) is significant if it is higher than 5%.
- The interactions of order higher than two are negligible.
- If a contrast is negligible, all effects composing this contrast are negligible also.
- Two significant factors can generate a significant interaction. On the other side, two insignificant factors do not generate a significant interaction.

3. Particle Swarm Optimization

3.1. Introduction

PSO (Kennedy & Eberhart, 1995; Kennedy et al., 2001; Clerc, 2006) is an evolutionary algorithm for the solution of optimization problems. It belongs to the field of Swarm Intelligence and Collective Intelligence and is a sub-field of Computational Intelligence. PSO is related to other Swarm Intelligence algorithms such as Ant Colony Optimization and it is a baseline algorithm for many variations, too numerous to list (Brownlee, 2011). PSO was developed by James Kennedy and Russell Eberhart in 1995 (Kennedy & Eberhart, 1995).

PSO has similar techniques to traditional stochastic search algorithms, but the difference is that PSO is not totally stochastic. PSO can avoid trapping on suboptimal and provide a highly adaptive optimal method. Because of fast convergence, PSO has gradually been applied in identification of graphics, optimization of clustering, scheduling assignment, network optimization and multi-objective optimization. For an analysis of the publications on the applications of particle swarm optimization see (Poli, 2008).

3.2. Strategy

The goal of the algorithm is to have all the particles locate the optima in a multi-dimensional hyper-volume. This is achieved by assigning initially random positions to all particles in the space and small initial random velocities. The algorithm is executed like a simulation, advancing the position of each particle in turn based on its velocity, the best known global position in the problem space and the best position known to a particle. The objective function is sampled after each position update. Over time, through a combination of exploration and exploitation of known good positions in the search space, the particles cluster or converge together around an optimum, or several optima (Brownlee, 2011).

3.3. Procedure

The Particle Swarm Optimization algorithm is comprised of a collection of particles that move around the search space influenced by their own best past location and the best past location of the whole swarm or a close neighbor (Brownlee, 2011). In each iteration a particle's velocity is updated using:

$$v_i(t+1) = v_i(t) + \; c_1 \times rand() \times \left(p_i^{best} - p_i(t)\right) + c_2 \times rand() \times (\mathrm{p_{gbest}} - p_i(t)) \tag{8}$$

where $v_i(t+1)$ is the new velocity for the ith particle, c_1 and c_2 are the weighting coefficients for the personal best and global best positions respectively, $p_i(t)$ is the ith particle's position at time t, p_i^{best} is the ith particle's best known position, and $\mathrm{p_{gbest}}$ is the best position known to the swarm. The $rand()$ function generates a uniformly random variable $\in [0, 1]$.

Variants on this update equation consider best positions within a particles local neighborhood at time t. A particle's position is updated using:

$$p_i(t+1) = \; p_i(t) + v_i(t) \tag{9}$$

3.4. PSO algorithm

It is important to mention here that PSO has undergone many changes since its introduction in 1995. As researchers have learned about the technique, they have derived new versions, developed new applications, and published theoretical studies of the effects of the various parameters and aspects of the algorithm. (Poli, 2007) gives a snapshot of particle swarming from the authors' perspective, including variations in the algorithm, current and ongoing research, applications and open problems. Algorithm 1 provides a pseudocode listing of the Particle Swarm Optimization algorithm for minimizing a cost function used in this chapter.

Algorithm 1: Pseudocode for PSO (Brownlee, 2011).

```
Input: ProblemSize, Population_size
Output: P_g_best
Population ← ∅;
P_g_best ← ∅;
for     i = 1 to Population_size do
        P_velocity ← RandomVelocity();
        P_position ← RandomPosition(Population_size);
        P_cost ← Cost(P_position);
        P_p_best ← P_position ;
        if    P_cost ≤ P_g_best then
              P_g_best ← P_p_best;
        end
End
while   StopCondition() do
        foreach   P ∈ Population do
                  P_velocity ← UpdateVelocity(P_velocity, P_g_best , P_p_best);
                  P_position ← UpdatePosition(P_position, P_velocity);
                  P_cost ← Cost(P_position);
                  if    P_cost ≤ P_p_best then
                        P_p_best ← P_position ;
                        if    P_cost ≤ P_g_best then
                              P_g_best ← P_p_best ;
                        end
                  end
        end
end
return P_p_best;
```

According to (Brownlee, 2011):

- The number of particles should be low, around 20-40
- The speed a particle should be bounded.
- The learning factors (biases towards global and personal best positions) should be between 0 and 4, typically 2.
- A local bias (local neighborhood) factor can be introduced where neighbors are determined based on Euclidean distance between particle positions.
- Particles may leave the boundary of the problem space and may be penalized, be reflected back into the domain or biased to return back toward a position in the

problem domain. Alternatively, a wrapping strategy may be used at the edge of the domain creating a loop, torrid or related geometrical structures at the chosen dimensionality.

- An inertia coefficient can be introduced to limit the change in velocity.

4. Induction motor design: An optimization problem

Induction motors with power below 100 kW (Fig. 5) constitute a sizable portion of the global electric motor markets (Boldea & Nasar, 2002). The induction motor design optimization is a nature mixture of art and science. Detailed theory of design is not given in this chapter. Here we present what may constitute the main steps of the design methodology. For further information, see (Vogt, 1988; Boldea & Nasar, 2002; Murthy, 2008). The suitability of the DOE and the PSO techniques in induction motor design optimization will be demonstrated in this section.

Figure 5. Low power 3 phase induction motor with cage rotor (Boldea & Nasar, 2002).

4.1. The algorithm

The main steps in induction motor design optimization are shown in Fig. 6.

Step (1): Initialization

The design process may start with design specifications and assigned values of: rated power, nominal voltage, frequency, power factor, type (squirrel Cage or slip-ring), connection (star or delta), ventilation, ducts, iron factor, insulation, curves like B/H, losses, Carter coefficient, tables like specific magnetic loading, specific electric loading, density etc. Then, design constraints for flux densities, current densities are specified. After that, the computer program is formulated with imposing max & min limits for rotor peripheral speed, length/pole pitch, stator slot-pitch, number of rotor slots. Finally, suitable values for certain parameters are assumed and objective functions are defined.

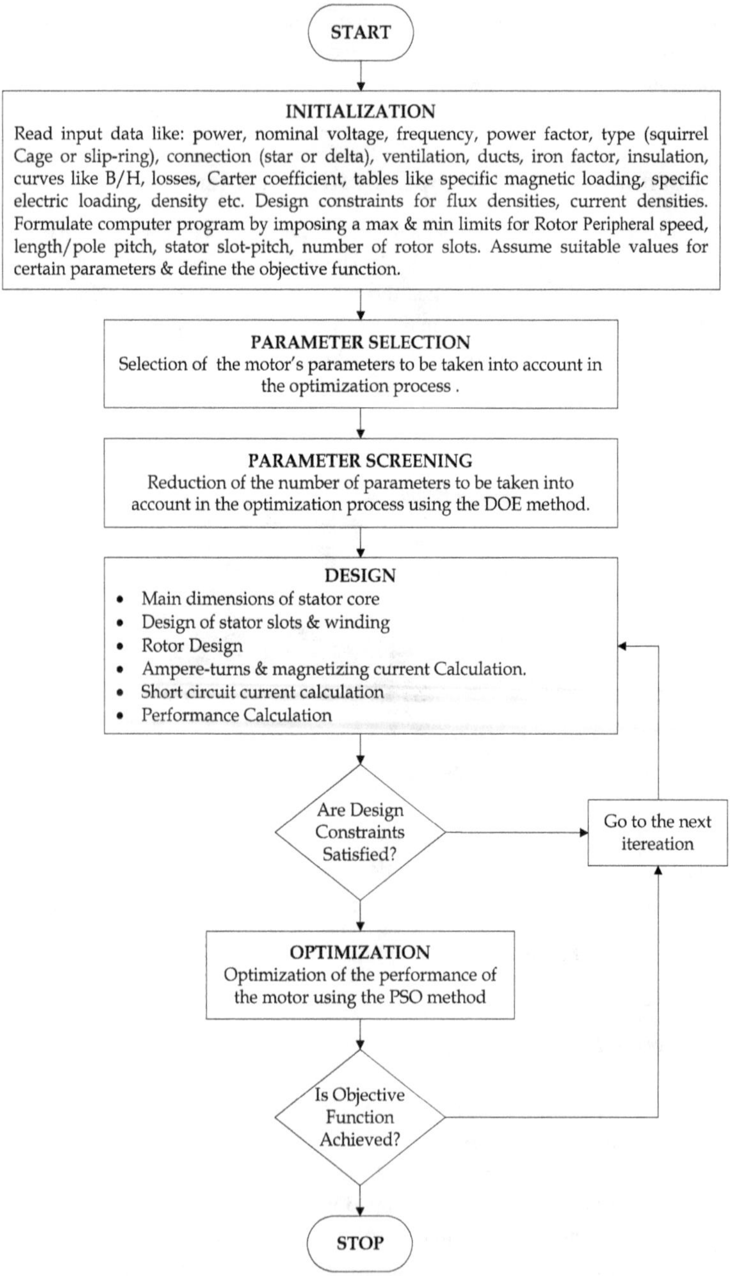
START
INITIALIZATION
Read input data like: power, nominal voltage, frequency, power factor, type (squirrel Cage or slip-ring), connection (star or delta), ventilation, ducts, iron factor, insulation, curves like B/H, losses, Carter coefficient, tables like specific magnetic loading, specific electric loading, density etc. Design constraints for flux densities, current densities. Formulate computer program by imposing a max & min limits for Rotor Peripheral speed, length/pole pitch, stator slot-pitch, number of rotor slots. Assume suitable values for certain parameters & define the objective function.
PARAMETER SELECTION
Selection of the motor's parameters to be taken into account in the optimization process .
PARAMETER SCREENING
Reduction of the number of parameters to be taken into account in the optimization process using the DOE method.
DESIGN
Main dimensions of stator core
Design of stator slots & winding
Rotor Design
Ampere-turns & magnetizing current Calculation.
Short circuit current calculation
Performance Calculation
Are Design Constraints Satisfied?
Go to the next itereation
OPTIMIZATION
Optimization of the performance of the motor using the PSO method
Is Objective Function Achieved?
STOP

Figure 6. Flowchart for computer-aided optimal design of 3-ph induction motor.

Step (2): Parameter selection

In this step the parameters to be taken into account in the optimization process are selected. The selection of parameters may be chosen by the designer or imposed by the user (for specific application for instance).

Step (3): Parameter screening

While there are potentially many parameters (factors) that affect the performance (objective functions) of the induction motor, some parameters are more important, *viz,* have a greater impact on the performance. The DOE provides a systematic & efficient plan of experimentation to compute the effect of factors on the performance of the motor, so that several factors can be studied simultaneously (Bouchekara, 2011). As said earlier, the DOE technique is an effective tool for maximizing the amount of information obtained from a study while minimizing the amount of data to be collected (Bouchekara, 2011). The DOE technique is used here to reduce the number of parameters (screening) to be taken into account in the optimization process. This goal is achieved by identifying the effect of each parameter on the objective function to be optimized. Only significant parameters (with contribution higher than 5%) are considered in the optimization step.

Step (4): Design

Total design is split into six parts in a proper sequence as shown in Fig. 6. The sequential steps for design of each part are briefly describes in the following sub sections. For more details see (Murthy, 2008).

Part I: Design of magnetic frame

In this part the output coefficient (C0) is calculated by:

$$\mathrm{C0} = 11 \times \mathrm{kW} \times \mathrm{Bav} \times \mathrm{q} \times \mathrm{EFF} \times \mathrm{pf} \times 10^{-3} \tag{10}$$

where: kW is the rating power, Bav is the specific magnetic loading, q is the specific electric loading, EFF is the efficiency and pf is the power factor.

Then the rotor volume that is (rotor diameter D)^2 × (rotor length L) is computed using the following formula:

$$\mathrm{D^2L} = \frac{\mathrm{kW}}{\mathrm{C0} \times \mathrm{ns}} \tag{11}$$

where: ns is the synchronous speed measured in rps.

Finally, the flux per pole ϕ is calculated by:

$$\phi = \frac{\tau_p \times \mathrm{L} \times \mathrm{Bav}}{10^6} \tag{12}$$

where: τ_p is the pole pitch and its is given by:

$$\tau_p = \frac{\pi \times D}{P} \tag{13}$$

Part II: Design of stator winding

The first step of this part consists of calculating the size of slots using the following equations:

$$\text{Slot Width(Ws)} = [\text{Zsw} \times (\text{Tstrip} + \text{insS}) + \text{insW}] \tag{14}$$

$$\text{Slot Height (Hs)} = [\text{Zsh} \times (\text{Hstrip} + \text{insS}) + \text{Hw} + \text{HL} + \text{insH}] \tag{15}$$

where: Zsw is the width-wise number of conductors, Tstrip is the assuming thickness of strip/conductor, insS is the strip insulation thickness, insW is the width-wise insulation, Zsh is the number of strips/conductors height-wise in a slot, Hstrip is the height of the strip, HL is height of lip, Hw is the height of wedge and insH is the height-wise insulation.

Then, the copper losses and the weight of copper are calculated by:

$$\text{Copper Losses (Pcus)} = 3 \times \text{Iph}^2 \times \text{Rph} \tag{16}$$

$$\text{Weight of Copper (Wcus)} = \text{Lmt} \times \text{Tph} \times 3 \times \text{As} \times 8.9 \times 10^{-3} \tag{17}$$

where: Iph is the current per phase, Rph is the resistance at 20°C, Lmt is the mean length of turn, Tph represents the turns per phase and As is the area of strip/conductor.

Finally, the iron losses are calculated by multiplying the coefficient deduced from the curve giving the losses in (W/kg) in function of the flux density in (T) by the core weight.

Part III: Design of Squirrel Cage Rotor

First, the air gap length is calculated by:

$$\text{Air} - \text{Gap Length (Lg)} = 0.2 + 2 \times \sqrt{\text{D} \times \text{L} \times 10^6} \tag{18}$$

Then, the rotor diameter is calculated using the following formula:

$$\text{Rotor Diameter (Dr)} = \text{D} - 2 \times \text{Lg} \tag{19}$$

Finally, the copper losses and the rotor weight are calculated using equations (20), (21) and (22).

$$\begin{aligned}&\text{Total Rotor Copper Loss (Pcur)}\\ &\qquad = \text{Copper Loss in the Bars} + \text{Copper Losses in the 2 End Rings}\end{aligned} \tag{20}$$

$$\text{Weight of Rotor Copper (Wcur)} = \text{Lb} \times \text{Sr} \times \text{Ab} \times 8.9 \times 10^{-6} \tag{21}$$

$$\text{Weight of Rotor End} - \text{Rings (Weue)} = \pi \times \text{Dme} \times 2 \times \text{Ae} \times 8.9 \times 10^{-6} \tag{22}$$

where: Lb is the length of bar, Sr is the number of Rotor Slots, Ab is the rotor bar area, Ae the area of cross sectional of end ring and Dme is mean diameter of end-ring.

Part IV: Total ampere turns and magnetizing current

First, the total ampere turns (ATT) for the motor are calculated using (23). Then, the magnetizing current (Im) is calculated using (24). Finally, the no load phase current (I0) and the no load power factor (pf0) are calculated using respectively (25) and (26).

$$\text{ATT} = \text{ATS} + \text{ATR} + \text{ATg} \tag{23}$$

$$\text{Im} = \frac{\text{P} \times \text{ATT}}{2 \times 1.17 \times \text{kW} \times \text{Tph}} \tag{24}$$

$$\text{I0} = \sqrt{\text{Iw}^2 + \text{Im}^2} \tag{25}$$

$$\text{pf0} = \frac{\text{Iw}}{\text{I0}} \tag{26}$$

where: ATS, ATR and ATg are the total ampere turns for the stator, the rotor and the air gap and Iw is the Wattful current.

Part V: Short-circuit current calculation

In this part the total reactance per phase, short-circuit current, and short-circuit power factor are calculated using the following formulas:

$$\text{Total Reactance/ph} = \text{Xs} + \text{X0} + \text{Xz} \tag{27}$$

$$\text{Short Circuit Current (Isc)} = \frac{\text{Vph}}{\text{Z}} \tag{28}$$

$$\text{Short Circuit pf} = \frac{\text{R}}{\text{Z}} \tag{29}$$

where: Xs is the slot reactance, X0 is the overhang reactance , Xz is the zig-zag reactance, R is the resistance and Z is the impedance.

Part VI: Performance calculation

In this last part of the design the performance of the induction motor are evaluated. The efficiency, the slip, the starting torque, the temperature rise and the total weight per kilo watt are calculated using the following formulas:

$$\text{Efficiency (EFF)} = \frac{\text{kW}}{\text{KW} + \text{Total Losses}} \tag{30}$$

$$\text{Slip at Full Load (SFL)} = \text{Total Rotor copper loss} \times \text{Rotor Input} \times 100 \tag{31}$$

$$\text{Starting Torque (Tst)} = \left(\frac{\text{Isc}}{\text{Ir}}\right)^2 \times \text{Slip at Full Load} \tag{32}$$

$$\text{Temperature Rise (Tr)} = 0.03 \times \frac{\text{Total Stator Losses}}{\text{Total Cooling Area}} \tag{33}$$

$$\text{kg/kW} = \frac{\text{Total Weight}}{\text{kW}} \tag{34}$$

where: Isc is the short circuit current and Ir is the equivalent rotor current.

At the end of step (4) an automatic check is performed. If the design constraints are satisfied we move to step (5) otherwise step (4) is restarted with new values of parameters.

Step (5): Optimization

In this step the motor's performances are checked and if found unsatisfactory, the process is restarted in step (4) with new values of parameters. The decision is made based on the PSO optimization method.

4.2. Design specifications

Design calculations are done for a given rating of an induction motor. Standard design specifications are:

- Rated power: P [kW] = 30.
- Line supply voltage: V [V] = 440.
- Supply frequency: f [Hz] = 50.
- Number of phases: 3.
- Phase connections: delta.
- Rotor type (squirrel cage or sling-ring): squirrel cage.
- Insulation class: F;
- Temperature rise: class B.
- Protection degree: IP55 – IC411.
- Environment conditions: standard (no derating).
- Configuration (vertical or horizontal shaft etc.): horizontal shaft.
- NEMA class: B.

4.3. Problem formulation

A very important problem in the induction motor design is to select the independent variables otherwise the problem would have been very much complicated using too many variables (Thanga, 2008). Therefore variables selection is important in the motor design optimization. A general nonlinear programming problem can be stated in mathematical terms as follows.

Find X = $(x_1, x_2 \ldots\ldots x_n)$ such that
$F_i(x)$ is a minimum or maximum
$g_i(x) \leq 0,\ i = 1, 2, \ldots m$

F_i is known as objective function which is to be minimized or maximized; g_i's are constants and x_i's are the variables. The following variables and constraints (Thanga, 2008) are considered to get optimal values of objective functions.

4.3.1. Variables

The variables considered are given in Table 5.

Name	Description	Minimum Value	Maximum Value	Type
P	Number of poles	4	6	Discrete
CDSW	Stator winding current density	3 [A/mm²]	5 [A/mm²]	Continuous
cdb	Current density in rotor bar	4 [A/mm²]	6 [A/mm²]	Continuous
Spp	Slots/pole/phase	3	4	Discrete
Tstrip	Stator conductor thickness	1 [mm]	2 [mm]	Continuous
Zsw	Number of conductors width-wise	1	2	Discrete

Table 5. Design optimization parameters with their domains.

4.3.2. Objective functions

Five different objective functions are considered while designing the machine using optimization algorithm. The objective functions are,

1. Maximization of efficiency; $F_1(x) = \max(EFF)$.
2. Minimization of kg/kW; $F_2(x) = \min(kg/kW)$.
3. Minimization of temperature rise in the stator; $F_3(x) = \min(Tr)$.
4. Minimization of I0/I ratio; $F_4(x) = \min(I0/I)$.
5. Maximization of starting torque; $F_5(x) = \max(Tst)$.

4.4. Fractional 2 levels factorial design

Here, the DOE is applied to analyze the objective functions. The proposed approach uses tools of the experimental design method: fractional designs, notably of Box generators to estimate the performance of the induction motor. The interest is to save calculation time and to find a near global optimum. The saving of time can be substantial because the number of simulations needed is significantly reduced.

Since six parameters define the shape of the motor, it is advisable to determine the effect of each parameter on the objective functions. Thus, it is very important to provide proper parameter ranges. The considered parameters are listed in Table 5. There are two types of parameters; continuous parameters and discrete parameters.

4.4.1. Results

Using two-level full factorial design needs 2^6=64 runs (simulations) to evaluate objective functions. However, using a 2^{6-2} fractional factorial design will significantly reduce the number of runs from 64 to 16. The 2^{6-2} design matrix and the simulation results obtained for this design are given in Table 6. This design has been generated using Box generators given in Table 7. The choice of a 2^{6-2} means that we have a 2 levels design with 6 factors where 2 of these factors are generated using the other 4 factors as shown in Table 7. Thus:

- The factor (5) will be generated using the product of factors (1), (2) & (3).
- The factor (6) will be generated using the product of factors (2), (3) & (4).

The contributions of obtained contrasts are given in Table 8. It shows in its first column contrasts and in the other columns their contribution or influences on objective functions. Keep in mind that a contribution is significant if it is higher than 5% and high order interactions (higher than 2) are considered negligible while only interactions of significant parameters are also significant.

N°	P	CDSW [A/mm²]	Cdb [A/mm²]	Spp	Tstrip [mm]	Zsw	EFF	Kg/kW	Tr [°]	I0/I	Tst [pu]
1	4	3	4	3	1	1	86.09	16.94	69.25	0.30	0.07
2	4	3	4	4	1	2	88.98	8.07	55.06	0.29	0.42
3	4	3	6	3	2	2	89.40	7.02	50.15	0.27	0.37
4	4	3	6	4	2	1	88.46	8.20	55.30	0.29	0.52
5	4	5	4	3	2	2	89.30	5.94	54.81	0.26	0.30
6	4	5	4	4	2	1	88.69	6.55	58.44	0.28	0.48
7	4	5	6	3	1	1	86.34	10.95	69.06	0.28	0.15
8	4	5	6	4	1	2	88.33	6.17	57.99	0.28	0.70
9	6	3	4	3	2	1	87.73	9.58	60.22	0.30	0.22
10	6	3	4	4	2	2	89.35	6.26	51.54	0.39	0.94
11	6	3	6	3	1	2	87.29	9.04	59.93	0.30	0.32
12	6	3	6	4	1	1	85.96	11.06	69.10	0.35	0.36
13	6	5	4	3	1	2	87.38	7.15	64.56	0.29	0.31
14	6	5	4	4	1	1	86.49	8.19	70.80	0.35	0.42
15	6	5	6	3	2	1	86.78	7.18	65.09	0.29	0.38
16	6	5	6	4	2	2	87.99	5.18	58.95	0.40	1.30

Table 6. Design matrix generated by the 2^{6-2} Box-Wilson fractional factorial design and the simulation results.

Resolution	Design name	Number of Runs	Generators
4	2^{6-2}	16	$x_5 = x_1 \times x_2 \times x_3$ $x_6 = x_2 \times x_3 \times x_4$

Table 7. Box generator of the fractional factorial design 2^{6-2}.

Contrasts	EFF	kg/kW	Tr	I0/I	Tst
P	13	2	9	45	7
CDSW	1	18	8	0	3
Cdb	3	1	0	0	5
Spp	4	10	2	25	41
Tstrip	34	24	36	0	14
Zsw	38	29	39	0	19
P × CDSW + Cdb × Tstrip	1	0	1	0	0
P × cdb + CDSW × Tstrip	2	2	1	0	0
P × Spp + Tstrip × Zsw	2	5	3	18	1
P × Tstrip + CDSW × Cdb + Spp × Zsw	1	3	0	4	5
P × Zsw + Spp × Tstrip	1	3	0	4	5
CDSW × Spp + Cdb × Zsw	0	1	0	0	1
Cdb × Spp + CDSW × Zsw	1	2	1	0	0

Table 8. Contrasts and contribution obtained.

The application of DOE identifies the effect of each parameter on each objective function. We can notice that for the efficiency Zsw, Tsrip, and P are the most significant factors with respectively 38% 34% and 13% of contribution on the objective function. Moreover, Fig. 7 gives more details. When P is low the efficiency is high and vice versa when P is high.

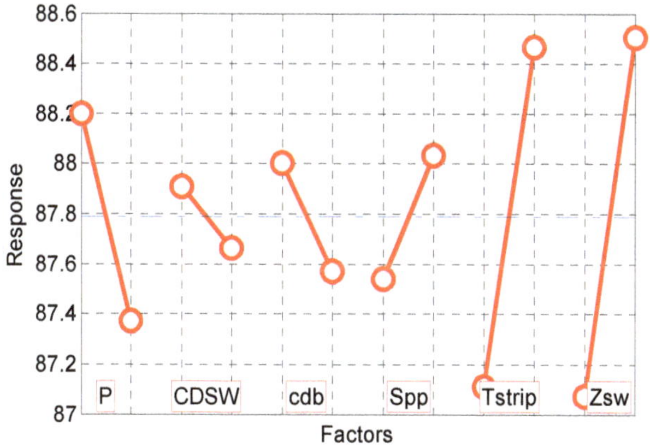

Figure 7. Plot of effects for the efficiency.

Contrariwise, when Tstrip and Zsw are low the efficiency is low, while it is high when Tstrip and Zsw are high.

For the objective function kg/kW the most important parameters are respectively Zsw (29%), Tstrip (24%), CDSW (18%) and Spp (10%). Fig. 8 shows that when each one of these parameters is low the kg/kW is high and inversely when they are high. Furthermore, for this objective function there is a significant interaction between some factors 'P × Spp + Tstrip × Zsw' (5%). Note that we have isolated all of the main effects from every 2-factors interaction. The two largest effects are Zsw and Tstrip, hence it seems reasonable to attribute this to the Tstrip × Zsw interaction.

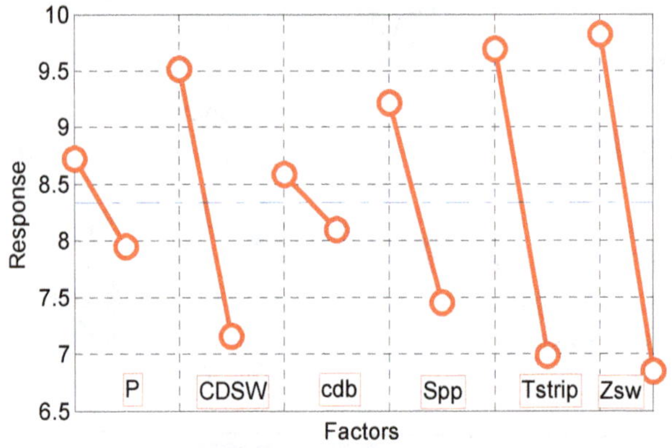

Figure 8. Plot of effects for kg/kW.

Concerning the temperature rise we can observe that, Zsw (39%), Tstrip (36%), P(9%) and CDSW (8%) are the most significant parameters. On the contrary, no significant interaction is discerned. Fig. 9 shows that the temperature rise is low when P and CDSW are low and it

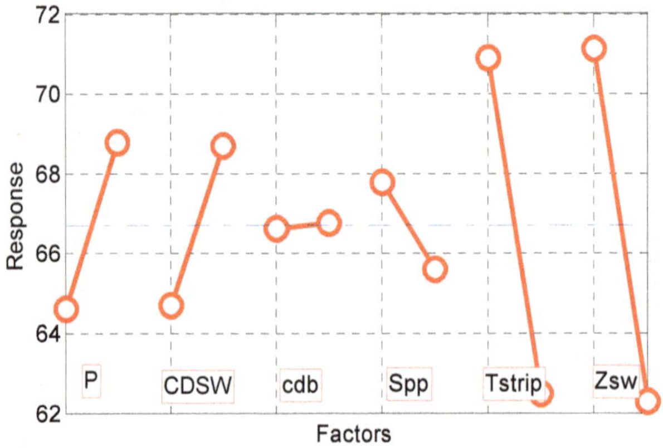

Figure 9. Plot of effects for temperature rise.

is high when they are high. Inversely, for Tstrip and Zsw the temperature rise is low when they are high.

For the objective function I0/I the significant parameters are P (45%) and Spp (25)%. Furthermore, there is a significant interaction between P and Spp included in the contrast 'P × Spp + Tstrip × Zsw'. Fig. 10 shows that I0/I is low when each parameter is low and vise versa.

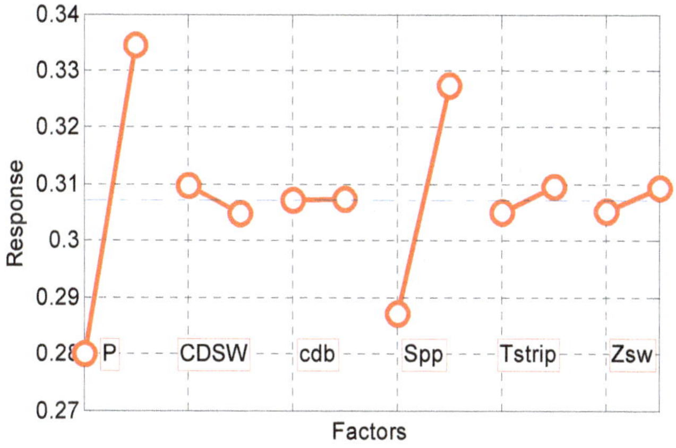

Figure 10. Plot of effects for I0/I.

Finally, for the starting torque the most significant parameters are given in this order: Spp (4%), Zsw (19%), Tstrip (14%), P (7%) and Cdb (5%). From Fig. 11 we can notice that when

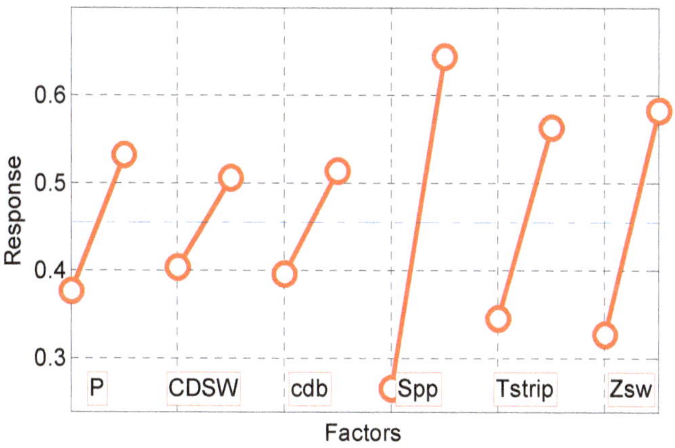

Figure 11. Plot of effects for starting torque.

each one of these parameters is low the starting torque is low. Likewise, when these parameters are high, the starting torque is high. Furthermore, for this objective function there is two significant interaction between some factors 'P × Tstrip + CDSW × Cdb + Spp × Zsw' (5%) and 'P × Zsw + Spp × Tstrip' (5%). Note that we have isolated all of the main effects from every 2-factor interaction. For the first contrast the two largest effects are Spp and Zsw. Thus, it seems reasonable to attribute this to the Spp × Zsw interaction. While, for the second contrast the two largest effects are Spp and Tstrip. Hence, it is appropriate to attribute this to the Spp × Tstrip interaction.

4.5. Optimization

Two optimization approaches can be achieved. The first one is to treat 1 of the 5 objective functions (defined in the Objective Function section) at a time. Thus, every time a single objective function is taken into account regardless of the 4 others. The second approach is to consider a multi objective function where the 5 objective functions are taken into account at the same time. The resulted complicated multiple-objective function can be converted into a simple and practical single-objective function scalarization. Among scalarization methods we can find the weighting method. In this method, the problem is posed as follows:

$$F_{\text{objective}} = \sum_{i=1}^{5} w_i f_i \tag{35}$$

where: $f_1 = \text{EFF}$, $f_2 = -\text{kg} / \text{kW}$, $f_3 = -\text{Tr}$, $f_4 = -\text{I0} / \text{I}$, $f_5 = \text{Tst}$ and w_i is a constant indicating the weight (and hence importance) assigned to f_i. By giving a relatively large value to w_i it is possible to favor f_i over other objective functions. Note that the condition $\sum_{i=1}^{k} w_i = 1$ can be posed in Eq.(35).

Nevertheless, since the 5 functions of the multi-objective function have different ranges, for instance f_1 varies from 85 to 91 and f_5 varies from 0.07 to 1.3. Thus, the values of these functions must be normalized between 0 and 1. The minimum of a given function is equal to 0 and the maximum is equal to 1. The normalization operation is given by:

$$\text{Normalized}_{\text{Value}} = \frac{\left(\text{Actual}_{\text{Value}} - \min(f_i)\right)}{\max(f_i) - \min(f_i)} \tag{36}$$

and (35) becomes:

$$F_{\text{objective}} = \sum_{i=1}^{5} w_i f_{i\,\text{Normalized}} \tag{37}$$

For this chapter we have chosen the first approach i.e. the single objective one.

The PSO algorithm is implemented to optimize the design of induction motor whose specifications are given above. The results of PSO algorithm for the optimized motor are given in the Table 9. The algorithm has returned an acceptable solution every time, which is indicated by a good value for objective with no constraint violations.

Parameters	EFF	Kg/KW	Tr	I0/I	Tst
P	4	4	4	4	6
CDSW	3	5	3	5	5
Cdb	4	6	4	6	6
Spp	4	4	4	3	4
Tstrip	2	2	2	2	2
Zsw	2	2	2	2	2
Existing Motor	89.7	5.36	53.8	0.27	0.5
Optimized Motor	90.1	5.15	48.5	0.26	1.3

Table 9. Optimum design results for efficiency maximization, minimization of kg/kW, minimization of temperature rise, minimization of the ratio I0/I and starting torque maximization.

According to the results presented in Table 9, when the efficiency of the motor is considered as the objective function, we can see that it increased from 89.7 to 90.1 compared to the existing motor. We can notice also that the when Kg/KW is minimized, it reduced from 5.36 to 5.15. Moreover, the optimization process allowed to the temperature rise to decrease form 53.8 to 48.5 which is a important reduction. Likewise, the I0/I is slightly reduced from 0.27 to 0.26 when it is the objective function. Finally, Table 9, shows that the starting torque is higher for the optimized motor (1.3) compared to the existing one (0.5).

According to these results, we can say that PSO is suitable for motor design and can reach successful designs with better performances than the existing motor while satisfying almost every constraint.

5. Conclusion

This chapter investigated the optimal design of induction motor using DOE and PSO techniques with five objective functions namely, maximization of efficiency, minimization of kg/kW, minimization of temperature rise in the stator, minimization of I0/I ratio, maximization of starting torque. It has been shown that DOE and PSO based algorithms constitute a viable and powerful tool for the optimal design of induction motor. The main objective of the DEO here is to identify the effect of each parameter on the objective functions. This is of a paramount importance mainly because of two reasons. The first one and also the obvious one is the reduction of the number of parameters to be taken into consideration in the optimization stage called screening. This can be achieved by neglecting the parameters with less effect. This will reduce the computing time burden and simplify the analysis of the designed motor. The second reason is that among the influent parameters themselves we can classify the parameters in function of their calculated effect. This will help the designer to have a clear picture of the importance of each parameter. For instance, if two parameters having respectively 45% and 5% of influence on a given objective function are compared; it is obvious that even if both parameters have an effect on the given objective function, the first one is greatly more important than the second one.

The approach developed here is universal and, although demonstrated here for induction motor design optimization, it may be applied to the design optimization of other types of electromagnetic device. It can be used also to investigate new types of motors or more generally electromagnetic devices. MATLAB code was used for implementing the entire algorithm. Thus, another valuable feature is that the developed approach is implementable on a desktop computer.

Author details

Houssem Rafik El-Hana Bouchekara
Department of Electrical Engineering, College of Engineering and Islamic Architecture, Umm Al-Qura University, Makkah, Saudi Arabia
Electrical Laboratory of Constantine "LEC", Department of Electrical Engineering, Mentouri University – Constantine, Constantine, Algeria

Mohammed Simsim and Makbul Anwari
Department of Electrical Engineering, College of Engineering and Islamic Architecture, Umm Al-Qura University, Makkah, Saudi Arabia

6. References

Altayib K., Ali, A. (2011). Improvement for alignment process of automotive assembly plant using simulation and design of experiments. International Journal of Experimental Design and Process Optimization, vol.2, no.2, pp.145-160.

Blanusa, B. (2010). New Trends in Efficiency Optimization of Induction Motor Drives, New Trends in Technologies: Devices, Computer, Communication and Industrial Systems, Meng Joo Er (Ed.), ISBN: 978-953-307-212-8, InTech.

Boldea, I. Nasar, S.A. (2002). The Induction Machine Handbook. CRC Press LLC, ISBN 0-8493-0004-5.

Bouchekara, H., Dahman, G., Nahas, M. (2011). Smart Electromagnetic Simulations: Guide Lines for Design of Experiments Technique. Progress in Electromagnetics Research B, Vol. 31, 357-379.

Brownlee, J. (2011). Clever Algorithms: Nature-Inspired Programming Recipes, lulu.com; 1ST edition, ISBN-10: 1446785068.

Clerc, M. (2006). Particle Swarm Optimization, Hermes Science Publishing Ltd., ISBN 1905209045, London.

Costa, M. C. (2001). Optimisation De Dispositifs Electromagnétiques Dans Un Contexte D'analyse Par La Méthode Des Eléments Finis. PhD thesis, National polytechnic institute of Grenoble.

Cunkas, M. Akkaya, R. (2006). Design Optimization of Induction Motor by Genetic Algorithm and Comparison with Existing Motor. Mathematical and Computational Applications, Vol. 11, No. 3., pp. 193-203.

Demonsant, J. (1996). Comprendre et Mener des Plans d'Expériences. Afnor, ISBN 2-124-75032-1.

Kennedy, J. Eberhart, R. C. (1995). Particle swarm optimization. Proceedings IEEE international conference on neural networks, Vol. IV, pp. 1942– 1948.

Kennedy, J., Eberhart, R. C., Shi, Y. (2001). Swarm Intelligence, San Francisco: Morgan Kaufmann Publishers.

Kleijnen, J. P. C., Sanchez S. M, T.W. Lucas, Cioppa, T. M. (2005). State-of-the-Art Review: A User's Guide to the Brave New World of Designing Simulation Experiments. Journal on Computing 17(3): 263–289.

Kutlea, L., Pavlovića, N., Dorotića, M., Zadroa, I., Kapustića, M., Halassy, B. (2010). Robustness testing of live attenuated rubella vaccine potency assay using fractional factorial design of experiments. Vaccine, vol.28, no.33, 2010, pp.5497-5502.

Maljkovic, Z., Cettolo, M. (2001). The Impact of the Induction Motor on Short-Circuit Current", IEEE Ind. Application Magazine, pp. 11-17.

Murthy, K.M.V. (2008). Computer-Aided Design of Electrical Machines', ISBN: 978-81-7800-146-3, Bs Publications/bsp Books.

Pillet, M. (1997). Les Plans d'Expériences par la Méthode TAGUCHI. Les Editions d'Organisation, ISBN 2-70-812031-X.

Poli, R. (2008). Analysis of the publications on the applications of particle swarm optimisation. Journal of Artificial Evolution and Applications, 1:1–10.

Poli, R., Kennedy, J., Blackwell, T. (2007). Particle swarm optimization an overview. Swarm Intelligence, 1:33–57.

ReliaSoft Corporation, 2008, Chapter 2: Overview. Available from: http://www.weibull.com/DOEWeb/experiment_design_and_analysis_reference.htm#introduction.htm>.

Thanga Raj, C., Srivastava, S. P., Agarwal, P. (2008). Realization on Particle Swarm Optimized Induction Motor Via SPEED/PC-IMD", IAENG Int. J. of Computer Science, Vol. 16, No. 4, pp. 486-492.

Uy, M., Telford, J.K. (2009). Optimization by Design of Experiment techniques. Aerospace conference, IEEE, pp.1-10, 7-14.

Vogt. V. (1988). Electrical Machines. Design of Rotary Electric Machines, Fourth edition (in German), VEB Verlag Technik Berlin.

Chapter 7

Role of Induction Motors in Voltage Instability and Coordinated Reactive Power Planning

Venkat Krishnan and James D. McCalley

Additional information is available at the end of the chapter

1. Introduction

As the power system is being operated in an economic and environment friendly fashion, there is more emphasis on effective resource utilization to supply the ever increasing demand. Consequently system experiences heavy power transaction, and one of the very important stability phenomena, namely voltage stability, is capturing the attention of many power system engineers, operators, researchers, and planners. Concerns for voltage instability and collapse are prompting utilities to better understand the phenomenon so as to devise effective, efficient and economic solutions to the problem.

Past studies have investigated the intricate relationship that exist between insufficient reactive power support and unreliable system operation including voltage collapses [1, 2, 3, 4], as was observed in 2003 blackout of USA [5]. It is not just the amount of reactive support, but also the quality and placement of reactive support that matters. For instance, it is found that due to the presence of electric loads that are predominantly induction motors the voltage recovery of the system following a severe disturbance is delayed due to lack of fast responding reactive support, thereby threatening to have secondary effects such as undesirable operation of protective relays, electric load disruption, and motor stalling [6, 7]. While a number of techniques have been developed in the past to address the problems of voltage instability [8], there has been little work towards a long term reactive power (VAR) planning (RPP) tool that addresses both steady state as well as dynamic stability issues.

The available reactive power devices can be classified into static and dynamic devices [9]. The static devices include mechanically switched shunt capacitors (MSCs) and series capacitors that exert discrete open-loop control action and require more time delay for correct operation. The dynamic devices are more expensive power electronics based fast-acting devices such as static VAR compensators (SVC), Static Synchronous Compensator (STATCOM), Unified Power Flow Controller (UPFCs) that exert continuous feedback control action and have better controllability and repeatability of operation. While MSCs are able to strengthen a power system against long term voltage instability issues [1], transient voltage dip and slow voltage recovery issues (influenced by electric load dynamics) is most effectively addressed by fast responding dynamic VAR sources [10, 11, 12]. While, past methods allocate static and dynamic VAR sources sequentially for one contingency at a time, the key to solving transmission system problems in the most cost-effective way will be to coordinate reactive power requirements simultaneously under many contingencies for both static and dynamic problems. The RPP should therefore identify the right mix of VAR devices, good locations and appropriate capacities for their installation.

In this chapter, we present a long term VAR planning algorithm, which coordinates between network investments that most effectively address steady-state voltage instability and those which most effectively address transient voltage recovery problems under severe contingencies. The study takes into account the induction motor dynamic characteristics, which influence the transient voltage recovery phenomenon. The algorithm is applied on a portion of a large scale system consisting of 16173 buses representing the US eastern interconnection. The planning method is a mixed integer programming (MIP) based optimization algorithm that uses sensitivity information of performance measures with respect to reactive devices to plan for multiple contingencies simultaneously.

The remaining parts of this chapter are organized as follows. In section 2, we discuss the very important role played by induction motor loads in the voltage stability phenomena. Section 3 sheds focus on the models used to build a base case for voltage stability assessment, and on appropriate performance criteria and solution strategies used in this chapter to devise the proposed coordinated planning. A summary of proposed RPP algorithm that considers both static as well as dynamic reactive resources in a coordinated planning framework, together with the various stages of planning are presented in Section 4. Section 5 illustrates the influence of induction motors on voltage stability phenomenon under severe contingencies, and demonstrates application of the coordinated VAR planning method on a large scale system to effectively avoid induction motor trips. Section 6 presents the conclusions.

2. Voltage instability phenomena

Several theories have been proposed to understand the mechanism of voltage instability. Voltage instability leading to collapse is system instability in that it involves many power system components and their variables at once. There are several system changes that can contribute to voltage collapse [4] such as increase in loading, SVC reaching reactive power limits, action of tap changing transformers (LTCs), load recovery dynamics, line tripping and generator outages. Most of the above mentioned system changes have a large effect on reactive power production or transmission. To discuss voltage collapse some notion of time scales is needed that accounts for fast acting variables in time scales of the order of seconds such as induction motors, SVCs to slow acting variables having long term dynamics in hours such as LTCs, load evolution etc.

2.1. Role of induction motors

A major factor contributing to voltage instability is the voltage drop that occurs when active and reactive power flow through inductive reactance of the transmission network; which limits the capability of the transmission network for power transfer and voltage support [13]. The power transfer and voltage support are further limited when some of the generators hit their field or armature current time-overload capability limits. Under such stressed conditions, the driving force for voltage instability is usually the inductive loads that try to recuperate after a disturbance. For instance, in response to a disturbance, power consumed by the induction motor loads tends to be restored by the action of motor slip adjustment, distribution voltage regulators, tap-changing transformers, and thermostats [6]. Restored loads increase the stress on the high voltage network by increasing the reactive power consumption and causing further voltage reduction. A run-down situation causing voltage instability occurs when load dynamics attempt to restore power consumption beyond the capability of the transmission network and the connected generation to provide the required reactive support.

The publication [14] corroborates this voltage instability phenomenon by means of a power-voltage (PV) curve, as shown in Figure 1. For a particular system and loads considered, the normal system can be stable with both resistive and motor loads at points where load curves and system curves intersect. However, when the system becomes stressed, with increased system reactance, it can only have a stable operating point with a resistive load. Due to lack of reactive power support that limits the transfer capability or loadability of the system, there is no intersection of system and load curves for the induction motor load since there is no stable operating point.

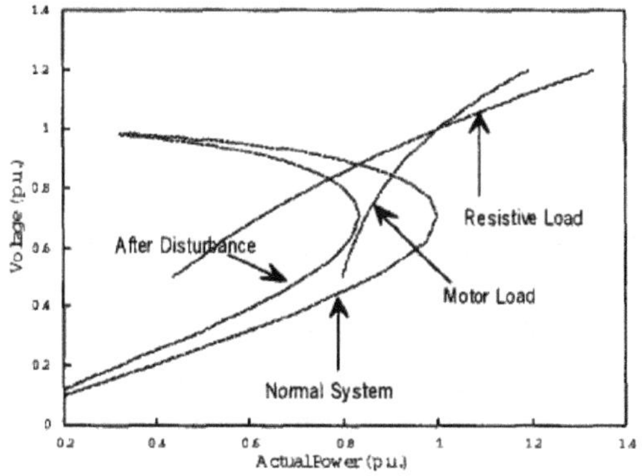

Figure 1. Stability and Load Characteristics

2.2. A typical scenario of slow voltage recovery leading to collapse

Heavily loaded transmission lines during low voltage conditions can result in operation of protective relays causing some transmission lines to trip in a cascading mode. A common scenario is a large disturbance such as a multi-phase fault near a load center that decelerates induction motor loads. Following fault clearing with transmission outages, motors draw very high current while simultaneously attempting to reaccelerate, as discussed in previous section, thereby making slowing down the voltage recovery process. A typical slow voltage recovery phenomenon following a disturbance is indicated in Figure 2 [15]. While trying to recover, if the voltage drops to a very low point for a sustained duration due to system's inability to provide reactive support, some motors may stall. Such drastic stalling of motor further exacerbates the conditions by increasing the reactive power requirements quickly, and the rate of voltage decline can accelerate catastrophically [6, 7, 16]. Massive loss of load and possibly area instability and voltage collapse may follow.

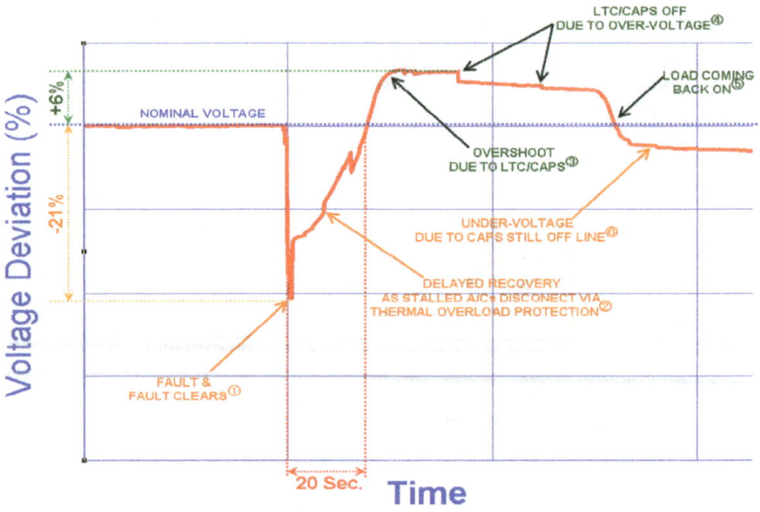

Figure 2. Post-Fault Transient Voltage Characteristics

There are several works [10] that have documented many short-term (few seconds) voltage collapse incidents with loss of load. In all cases, adequate dynamic reactive power support was not available which resulted in a large loss of load.

All the above discussions give a physical sense of how the problem of voltage instability occurs, and shed importance on the requirement of study techniques and good models, especially of induction motor dynamics as they are particularly hazardous from the viewpoint of voltage stability.

3. Voltage stability assessment and criteria

Traditionally, voltage stability investigations have been based on steady-state analyses, which involve solving conventional or modified power flow equations [17]. In such studies system P-V curve and the sensitivity information derived from the power flow jacobian are used to assess and plan for voltage instability. But the realization that voltage stability is a dynamic phenomenon has led to dynamic formulations of the problem and application of dynamic analysis tools [18] that uses time-domain simulations to solve nonlinear system differential algebraic equations.

A pre-contingency steady state base case is required for the voltage stability study to be performed, which is usually generated from real-time sequence control (State estimator solution), or via an already recorded power flow solution. In the case of a base case for dynamics study, as discussed in previous section, the dynamic phenomenon of voltage stability is largely determined by load characteristics and the available means of voltage control. The response speeds of these loads may be comparable to the speed of response of the dynamic voltage control equipment. So in such studies involving dynamic phenomena, using static models will give forth dubious results. So it is very important to properly model dynamic behavior of such large, small and trip induction motor loads; along with that of relevant voltage controls. Investigating the post-fault dynamic system response and effective planning to prevent a voltage collapse depends on inclusion of relevant system component models.

3.1. System component models

This section presents the various components and their relevant models that are required to build a suitable base case for comprehensive voltage stability assessment and planning.

3.1.1. Static device models

1. **Transmission lines** are represented as pi-sections, possibly with unsymmetrical line charging; accompanying data include line pi-section impedances/admittances data; line thermal limit both normal and emergency.
2. **Transformers** represented as pi-sections whereby the various impedance/admittance components may be explicit functions of tap settings; three winding transformers must be properly modeled. The data also include transformer limits under normal/emergency cases. **Phase-shifting transformers** are represented by complex tap ratios, allowing both shift in angle and change in voltage magnitude;
3. **Generators** as real-power source together with a reactive power capability curve as a function of terminal voltage; The required generator static data include minimum and maximum ratings, nominal terminal voltage and reactive power capability curve as a function of terminal voltage
4. **Shunt elements** by their impedance/admittance and Static Var compensators by static gain and maximum/minimum limits
5. **Loads** by ZIP model, i.e., as a combination of constant impedance (Z), constant current (I), and constant real/reactive injection (P) components; The data necessary are default ZIP load partition ratios at nominal voltage, load limits and default power factors

3.1.2. Dynamic device models

1. **Machine** mechanical dynamic equation (swing with damping) and machine electrical dynamic equations; machine mechanical parameters such as inertia constant and damping co-efficient and machine electrical parameters such as transient/sub-transient reactances and time constants etc are required. Saturation model data is also very vital.
2. **Excitation systems** of various types; the data for each model available in standard power system stability analysis programs such as EPRI's ETMSP, PTI's PSS/E etc are used in most cases.
3. **Governor systems** of various types; Again the necessary data for each model are usually available in standard power system stability analysis programs.
4. **Load modeling** is very vital for performing a voltage stability study.
5. Models for selected prime mover, power system stabilizers, and control devices such as SVC etc. are also required.

Induction motor and SVC modeling will be discussed further in this section. In addition to all the system/device data, other system data include convergence parameters such as threshold and maximum iteration counts for static power flow studies, and also various other solution parameters used for the dynamic time domain simulation.

3.1.2.1. Induction motor modeling

As mentioned earlier large induction motor loads generally affect the voltage recovery process after voltage sag has been incepted due to system faults, and in many occasions due to extended voltage sag secondary effects such as stalling or tripping of sensitive motors might happen leading to massive load disruption. So, it is very vital to represent large, small and trip induction motor loads in various combinations in the system, so that we capture the stalling phenomenon of induction motor load, the real and reactive power requirements in the stalled state, and the tripping caused by thermal protection.

The induction motor load must be modeled such that it is sensitive to dynamic variations in voltage and frequency, and emulates the typical characteristic of consuming more power at increased speeds. Equation (1) shows the modeling of mechanical torque (T_{load}) as a function of speed deviation from nominal and motor load torque at synchronous speed (T_{nom}).

$$T_{load} = T_{nom}(1+\Delta\omega)^{D} \tag{1}$$

Some of the primary model parameters include stator and rotor resistances and inductances, mutual inductance, saturation components, MVA base, intertia (H), per unit voltage level below which the relay to trip the motor will begin timing (V_T), time in cycles for which the voltage must remain below the threshold for the relay to trip (C_T), breaker delay time in cycles, nominal torque (T_{nom}), load damping factor (D) etc.

There are new composite load models developed by WECC LMTF [19] that improves the representation of induction motor load dynamics, and thereby more closely captures the critical role played by such loads in delayed voltage recovery events. These composite loads are represented by CMLD models in PSS/E and CMPLDW models in PSLF [20].

3.1.2.2. Static VAR compensator modeling

In the study done in this chapter we employ SVC as an effective means to mitigate transient voltage dip problem by providing fast responding dynamic reactive power support. The SVC is modeled as shown in the figure 3, with a non-windup limit B_{svc} (in MVAR) on the SVC output, which constrains the SVC output B. At its limit, SVC output is non-controllable and functions as a shunt capacitor. Hence, the ability of an SVC to provide dynamic support for mitigating the transient voltage dip problems depends on the SVC's capacitive limit (size) B_{SVC}, which also increases the SVC cost. The RPP finds the optimum rating of SVC that is economical and enhances system reliability.

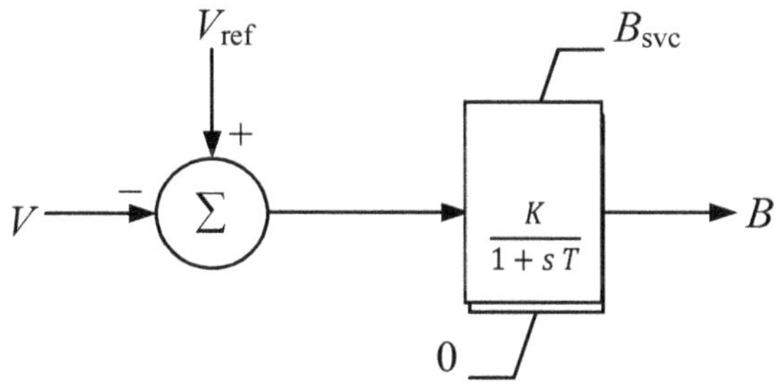

Figure 3. Static VAR compensator model

Some of the main parameters include voltage limits V_{MAX} and V_{MIN} to specify the active range of the voltage control, a time constant (of about 0.05 sec. or less) to model the delay in reactor's response, a steady-state voltage control gain K of about 100, and the time constant T of about 0.01 to provide transient gain reduction in the control loop.

Once the voltage stability base case is ready, next system analysis has to be performed to check the severity of the contingencies that need planning. So the next vital step in the planning procedure is voltage stability assessment of contingencies.

3.2. Post-contingency performance criteria for voltage stability assessment

In order to effectively plan against steady state and dynamic voltage stability problems under a certain set of contingencies, we need to identify proper performance criteria. Voltage stability of the power system should be assessed based on voltage security criteria of interest to, and accepted by, the utility.

As far as the steady state performance criteria are concerned, there are many criteria such as reactive reserve in different parts of the system, post-contingency voltage, Eigenvalues, etc. that enable to quantify the severity of a contingency with respect to voltage stability. In this work we utilize the most basic and widely accepted criteria, namely, post-contingency voltage stability margin [21, 22] for steady state performance assessments. Voltage stability margin, a steady state performance criterion, is defined as the amount of additional load in a specific pattern of load increase that would cause voltage instability. Contingencies such as unexpected component (generator, transformer, transmission line) outages often reduce the voltage stability margin, and control actions increase it [23, 24], as shown in figure 4 [25].

The disturbance performance table within the NERC (North American Electric Reliability Corporation)/WECC (Western Electricity Coordinating Council) planning standards [26] provides the minimum acceptable performance specifications for post-contingency voltage stability margin under credible events, that it should be atleast,

- greater than 5% for N-1 contingencies,
- greater than 2.5% for N-2 contingencies, and
- greater than 0% for N-3 contingencies.

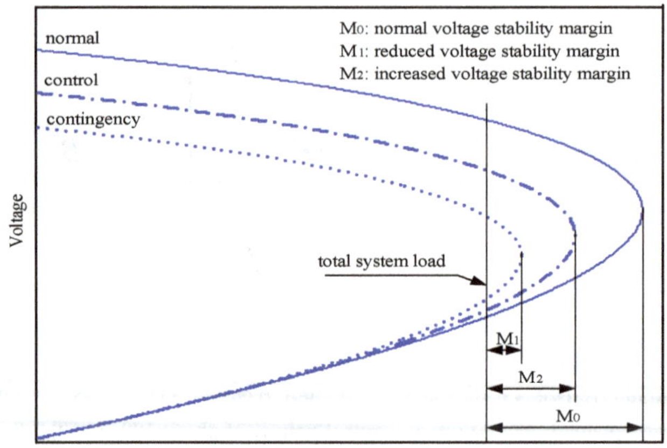

Figure 4. Voltage stability margin

As far as the performance measure for dynamic stability phenomena is concerned, in [11] it is stated that the needs of the industry related to voltage dips/sags for power system stability fall under two main scenarios. One is the traditional transient angle stability where voltage "swing" during electromechanical oscillations is the concern. The other is "short-term" voltage stability generally involving voltage recovery following fault clearing where there is no significant oscillations, for which much greater load modeling detail is required (specifically induction motor loads) with the fault applied in the load area rather than near generation. In [25], it is stated that many planning and operating engineers are insufficiently aware of potential short-term voltage instability, or are unsure on how to analyze the phenomena. In this work we focus on planning for the transient voltage recovery after a fault is cleared. For the transient voltage analysis, the minimum planning criteria is,

- **Slow voltage recovery**: The induction motor trip relay timer is actuated when the bus voltage dips below 0.7p.u and trips if voltage doesn't recover to 0.7p.u within next 20 cycles.

3.3. Sensitivities of post-contingency performance criteria

The objective of the reactive power planning problem is to satisfy these minimum performance criteria mentioned in the previous section under various credible contingencies. This is greatly dependent upon the amount, location and type of reactive power sources available in the system, which are all decision variables in the proposed coordinated RPP, as will be dealt in the next section of this chapter. If the reactive power support is far away, or insufficient in size, or too dependent on shunt capacitors (slow acting static device), a relatively normal contingency (such as a line outage or a sudden increase in load) can trigger a large system voltage drop. Hence, in order to properly allocate the reactive power support in terms of optimal type, location and amount, in this work we employ linear sensitivities of performance measure with respect to various control actions. These sensitivities enable obtaining the necessary information, i.e., the sensitivities of performance measure for the type of reactive support, the location and amount of reactive support, which are very useful to perform the proposed coordinated RPP. This section sheds light on these linear sensitivity measures.

3.3.1. Voltage stability margin sensitivity

The sensitivity of voltage stability margin refers to how much the margin changes for a small change in system parameters such as real power and reactive power bus injections, regulated bus voltages, bus shunt capacitance, line series capacitance etc [2, 27]. References [28, 29] first derived these margin sensitivities for different changing parameters.

Let the steady state of the power system satisfying a set of equations in the vector form be,

$$F(x, p, \lambda) = 0 \tag{2}$$

where x is the vector of state variables, p is any parameter in the power system steady state equations such as demand and base generation or the susceptance of shunt capacitors or the reactance of series capacitors, and λ denotes the system load/generation level called the scalar bifurcation parameter. The system reaches a state of voltage collapse, when λ hits its maximum value (the nose point of the system PV curve), and the value of the bifurcation parameter is equal to λ^*. For this reason, the system equation at equilibrium state is parameterized by this bifurcation parameter λ as shown below.

$$P_{li} = (1 + K_{lpi}\lambda)P_{li0} \tag{3}$$

$$Q_{li} = (1 + K_{lqi}\lambda)Q_{li0} \tag{4}$$

$$P_{gj} = (1 + K_{gj}\lambda)P_{gj0} \tag{5}$$

where P_{li0} and Q_{li0} are the initial loading conditions at the base case corresponding to λ=0. K_{lpi} and K_{lqi} are factors characterizing the load increase pattern (stress direction). P_{gj0} is the real power generation at bus j at the base case. K_{gj} represents the generator load pick-up factor. For a power system model using ordinary algebraic equations, the bifurcation point sensitivity with respect to the control variable p_i evaluated at the saddle-node bifurcation point is [27]

$$\frac{\partial \lambda^*}{\partial p_i} = -\frac{w^* F_{p_i}^*}{w^* F_\lambda^*} \tag{6}$$

where w is the left eigenvector corresponding to the zero eigenvalue of the system Jacobian F_x, F_λ is the derivative of F with respect to the bifurcation parameter λ and F_{Pi} is the derivative of F with respect to the control variable parameter p_i.

The sensitivity of the voltage stability margin with respect to the control variable at location i, S_i, is

$$S_i = \frac{\partial M}{\partial p_i} = \frac{\partial \lambda^*}{\partial p_i} \sum_{i=1}^{n} K_{lpi} P_{li0} \tag{7}$$

where M is the voltage stability margin given by

$$M = \sum_{i=1}^{n} P_{li}^* - \sum_{i=1}^{n} P_{li0} = \lambda^* \sum_{i=1}^{n} K_{lpi} P_{li0} \tag{8}$$

3.3.2. *Post-fault transient voltage recovery sensitivity*

The sensitivity of the voltage dip time duration to the SVC capacitive limit (B_{svc}) can be defined as the change of the voltage dip time duration (voltage recovery time) for a given change in the SVC capacitive limit. Let $\tau^{(1)}$ be the time at which the transient voltage dip begins after a fault is cleared, and $\tau^{(2)}$ be the time at which the transient voltage dip ends, as shown in figure 5 [25]. Then the time duration of the transient voltage dip τ_{dip} is given by

$$\tau_{\text{dip}} = \tau^{(2)} - \tau^{(1)} \tag{9}$$

Thus, the sensitivity of the voltage dip time duration to the capacitive limit of an SVC, S_τ, is

$$S_\tau \equiv \frac{\partial \tau_{dip}}{\partial B_{\text{svc}}} = \frac{\partial(\tau^{(2)} - \tau^{(1)})}{\partial B_{\text{svc}}} = \frac{\partial \tau^{(2)}}{\partial B_{\text{svc}}} - \frac{\partial \tau^{(1)}}{\partial B_{\text{svc}}} = \tau^{(2)}_{B_{\text{svc}}} - \tau^{(1)}_{B_{\text{svc}}} \tag{10}$$

where $\tau^{(1)}_{B_{\text{svc}}}$ and $\tau^{(2)}_{B_{\text{svc}}}$ are calculated based on trajectory sensitivity computations as derived in [25]. For a large power system, this method requires computing integrals of a set of high dimension differential algebraic equations. An alternative to calculate the sensitivities is using numerical approximation [25], as shown by equation (11). This procedure of sensitivity calculation by numerical approximation requires repeated simulations of the system model for the SVC capacitive limits B_{svc} and $B_{svc}+\Delta B_{svc}$. The sensitivities are then given by the change of the voltage recovery time divided by the SVC capacitive limit change

ΔB_{svc}. This procedure is easy to implement for a large power system using available simulation tools.

$$S_{\tau} = \frac{\partial \tau_{\text{dip}}}{\partial B_{\text{svc}}} \approx \frac{\Delta \tau_{\text{dip}}}{\Delta B_{\text{svc}}} = \frac{\tau_{\text{dip}}(B_{\text{svc}} + \Delta B_{\text{svc}}) - \tau_{\text{dip}}(B_{\text{svc}})}{\Delta B_{\text{svc}}} \tag{11}$$

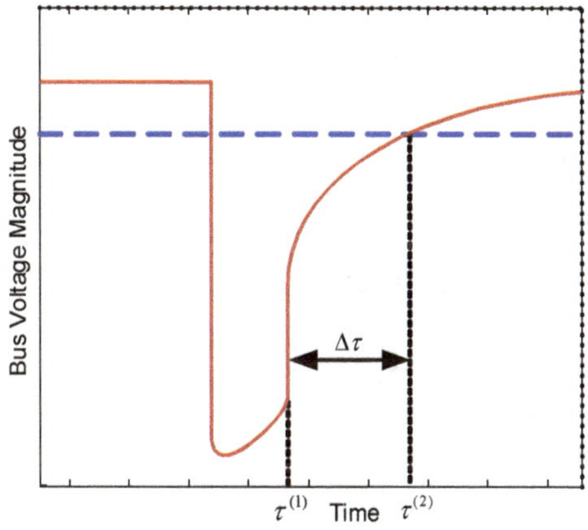

Figure 5. Post-fault clearance slow voltage recovery

4. Coordinated reactive power planning

The planning algorithm addresses the problems discussed in Table 1.

Problem	Planning Objective
A (Steady State)	To restore post-disturbance equilibrium and increase post contingency voltage stability margin beyond the minimum criteria after a severe contingency that can cause voltage instability
B (Dynamic)	To improve the characteristics of post-fault transient voltage recovery phenomenon satisfying the required minimum criteria, and prevent induction motor tripping

Table 1. Coordinated Reactive Power Planning Objectives

Figure 6 shows the general flow of the RPP procedure developed.

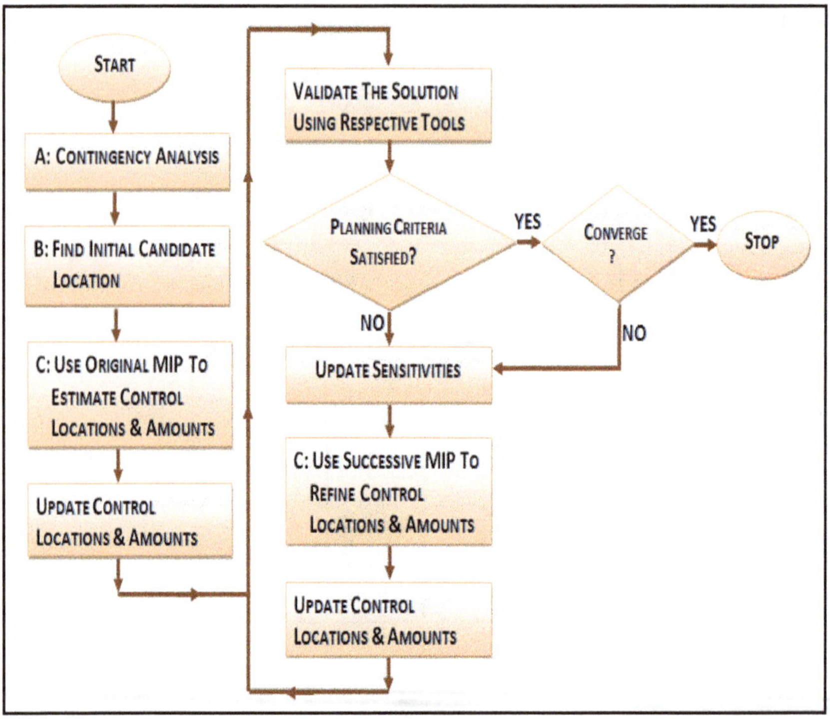

Figure 6. Flowchart of RPP procedure with successive MIP

4.1. Contingency analysis

Continuation power flow (CPF) [17] based tools and time domain simulation are used to perform the contingency analysis. The post-contingency state of the system is checked for performance criteria violations, and a list of critical contingencies is formed. In the case of steady state analysis the process of contingency screening using margin sensitivity information [29] is used to reduce the computational burden [30].

4.2. Formulation of coordinated control planning algorithm

This section presents the formulation of optimization developed to address the coordinated VAR planning problem for all the contingencies that have either one or both the planning problems shown in Table 1. The information required for the optimization algorithm are reactive device cost and maximum capacity limit at various voltage levels, performance measures and their sensitivities with respect to MSCs and SVCs under each critical contingencies, and an initial set of candidate locations for MSCs and SVCs. While many methodologies in the past determine static and dynamic VAR support sequentially, this method simultaneously determines the optimal allocation of static and dynamic VAR [31].

4.2.1. Original mixed integer programming

The objective of mixed integer program (MIP) is to minimize the total installation cost of MSCs and SVCs while satisfying the requirements of long-term voltage stability margin and short-term post-fault transient voltage characteristics.

Minimize

$$\sum_{i\in\Omega}[C_{vi\_MSC}B_{i\_MSC} + C_{fi\_MSC}q_{i\_MSC} + C_{vi\_svc}B_{i\_svc} + C_{fi\_svc}q_{i\_svc}] \quad (12)$$

Subject to

$$\sum_{i\in\Omega} S_{M,i}^{(k)}[B_{i\_MSC}^{(k)} + B_{i\_svc}^{(k)}] + M^{(k)} \geq M_{\mathrm{r}}, \forall k \quad (13)$$

$$\sum_{i\in\Omega_{svc}} S_{\tau,n,i}^{(k)}B_{i\_svc}^{(k)} + \tau_{\mathrm{dip},n}^{(k)} \leq \tau_{\mathrm{dip},n,\mathrm{r}}, \forall n,k \quad (14)$$

$$0 \leq B_{i\_MSC}^{(k)} \leq B_{i\_MSC}, \forall k \quad (15)$$

$$0 \leq B_{i\_svc}^{(k)} \leq B_{i\_svc}, \forall k \quad (16)$$

$$B_{i\min\_MSC}q_{i\_MSC} \leq B_{i\_MSC} \leq B_{i\max\_MSC}q_{i\_MSC} \quad (17)$$

$$B_{i\min\_svc}q_{i\_svc} \leq B_{i\_svc} \leq B_{i\max\_svc}q_{i\_svc} \quad (18)$$

$$q_{i\_MSC}, q_{i\_svc} = 0,1 \quad (19)$$

The decision variables are $B_i^{(k)}{}_{\_MSC}$, B_{i_MSC}, q_{i_MSC}, $B_i^{(k)}{}_{\_svc}$, B_{i_svc}, and q_{i_svc}.

C_{f_MSC} is fixed installation cost and C_{v_MSC} is variable cost of MSCs,

C_{f_svc} is fixed installation cost and C_{v_svc} is variable cost of SVCs,

B_{i_MSC}: size of the MSC at location i,

B_{i_svc}: size of the SVC at location i,

q_{i_MSC}=1 if the location i is selected for installing MSCs, otherwise, q_{i_MSC}=0,

q_{i_svc}=1 if the location i is selected for installing SVCs, otherwise, q_{i_svc}=0,

the superscript k represents the contingency causing insufficient voltage stability margin and/or slow voltage recovery problems,

Ω_{MSC}: set of candidate locations to install MSCs,

Ω_{svc}: set of candidate locations to install SVCs,

Ω: union of Ω_{MSC} and Ω_{svc},

$B_i^{(k)}{}_{\_MSC}$: size of the MSC to be switched at location i under contingency k,

$B_{i\_svc}^{(k)}$: size of the SVC at location i under contingency k,

$S_{M,i}^{(k)}$: sensitivity of the voltage stability margin with respect to the shunt susceptance of MSC at location i under contingency k,

$S_{\tau,n,i}^{(k)}$: sensitivity of the voltage recovery time duration at bus n with respect to the size of the SVC at location i under contingency k,

$M^{(k)}$: voltage stability margin under contingency k and without controls,

M_r: required voltage stability margin,

$\tau_{dip,n}^{(k)}$: time duration of voltage recovery at bus n under contingency k and without controls,

$\tau_{dip,n,r}$: maximum allowable time duration of voltage recovery at bus n,

B_{imin_MSC}: minimum size of the MSC at location i,

B_{imax_MSC}: maximum size of the MSC at location i,

B_{imin_svc}: minimum size of the SVC at location i, and

B_{imax_svc}: maximum size of the SVC at location i.

Note from (13) that SVCs can also be used to increase the voltage stability margin.

4.2.2. *Updated successive mixed integer programming*

The output of the mixed integer-programming problem in section 4.2.1 is the combined reactive compensation locations and amounts for all concerned contingencies. Now the network configuration is updated by including the identified reactive power support under each contingency. After that, the voltage stability margin is recalculated using CPF to check if sufficient margin is achieved for each concerned contingency. Also, time domain simulations are carried out to check whether the requirement of the transient voltage recovery performance is met. This step is necessary because the power system model is inherently nonlinear, and the mixed integer programming algorithm uses linear sensitivities to estimate the effect of variations of reactive support levels on the voltage stability margin and post-fault voltage recovery. So if need be, the reactive compensation locations and/or amounts can be further refined by re-computing sensitivities (with updated network configuration) under each concerned contingency, and solving a second-stage mixed integer programming problem.

This successive MIP problem based on updated sensitivity and system performance information is again formulated to minimize the total installation cost of MSCs and SVCs, subject to the constraints of the requirements of voltage stability margin and voltage recovery. It will terminate once the post-contingency performance criteria are satisfied for all concerned contingencies, and there is no significant change in decision variables from the previous MIP solution.

5. Numerical illustration

The control planning method described in this paper was applied to a particular portion of US Eastern Interconnection system consisting of about 16173 buses. This subsystem belonging to a particular utility's control area, henceforth will be referred to as the "study area." The study area within this large system consisted of 2069 buses with 30065.2 MW of loading and 239 generators producing 37946.7 MW.

The contingencies considered for the study are the more probable ones, i.e., N-1 and N-G-T. For N-1 and N-G-T contingencies, according to the WECC/NERC performance table, minimum steady state performance criteria is to have a post-contingency voltage stability margin of at least 5% of the sub-system's base load. For the slow voltage recovery problem, the minimum performance should be such that it avoids the induction motor tripping. The trip relay timer of the trip induction motor is actuated when the bus voltage dips below 0.7p.u and trips if voltage doesn't recover to 0.7p.u within next 20 cycles. Therefore, the objective of the coordinated RPP is to identify a minimum cost mix of static and dynamic Var resources that results in satisfactory voltage stability and transient voltage recovery performance for all considered contingencies.

The study area was grouped into 6 Market Zones (MZ), representing the 6 different load increase (stress) directions required to perform CPF analysis. For a particular stress direction, the sink is characterized by the set of loads inside a MZ, and the source is characterized by generators outside of that MZ, but within the study area.

For the transient study, dynamic models for generators, exciter, governor systems, and appropriate load and SVC models are used. Loads in the focus area were partitioned as 50% induction motor load (dynamic) and 50% ZIP load (static). Induction motor was modeled using CIM5 model in PSS/E, which is sensitive to changing voltage and frequency. SVCs were modeled using CSVGN1 and CSVGN3 family of SVC models. The PSS/E manual presents the block diagram, parameters and detailed description for each of these models. The dynamic models of induction motor loads were further split equally into three different kinds, i.e., large, small, and trip motors. Table 2 shows some of the important parameters of each of these motor loads as defined in section 3.1.2.1. The ZIP load is modeled as 50% constant impedance and 50% constant current for real power load and 100% constant impedance for reactive power load.

The steady state contingency analysis using CPF is performed in Matlab, while the dynamic voltage stability analysis using time domain simulation is performed in PTI PSS/E [32]. As part of this study, which required Matlab using the input files in PSS/E's "raw" data format, a data conversion module was built that converted the system raw data to a format that was understandable by Matlab. The conversion module includes tasks such as careful modeling of 3-winding transformer data, zero-impedance lines, switched shunt data, checking system topology, and checking for any islanding, so that the utility's base case is transported without any errors into Matlab.

	H (p.u. motor base)	V_T (p.u)	C_T (cycles)	*D*	T_{nom} (p.u)
Large Motor	1.5	0	20	2	1
Small Motor	0.5	0	20	2	1
Trip Motor	0.5	0.7	20	2	1

Table 2. Induction Motor Dynamic Model Parameters

Based on the pre-processing performed using these computational tools, respective sensitivity information of steady state and dynamic performance measures with respect to the reactive control device are computed, as explained in section 3.3. All the necessary input including candidate locations and control device cost information are fed as input to the coordinated planning algorithm, which is coded in Matlab and executed using CPLEX. Since candidate control locations and linear sensitivity information are used, the MIP optimization is faster even for a very large scale system. PSS/E's ability to store the results in *.csv format is utilized in post-processing the simulation results using MS Excel.

The cost is modeled as two components [24]: fixed cost and variable cost. For SVCs the fixed and variable costs were taken to be 1.5 M$ and 5 M$/100Mvar respectively. The maximum capacity limit for SVCs was fixed at 300 MVar. Table 3 shows the MSC cost information and the default maximum capacity constraints at every feasible location under different voltage levels. The maximum MSC limit at various voltage levels ensures avoiding over-deployment of MSCs at those voltage levels, which could degrade the voltage magnitude performance. At every solution validation step, if any bus has post-contingency voltage exceeding 1.06 p.u, a new maximum MSC (MVar) constraint is developed, and the optimization is re-run with the new constraint included.

Base KV	Fixed Cost (Million $)	Variable Cost (M $/100MVar)	Maximum MSC (MVar)
69	0.025	0.41	30
100	0.05	0.41	75
115	0.07	0.41	120
138	0.1	0.41	150
230	0.28	0.41	200
345	0.62	0.41	300
500	1.3	0.41	300

Table 3. MSC Cost Information and Maximum Compensation

5.1. Steady state reactive power planning and impact of induction motors

In this section we present only the steady state RPP, i.e., plan to satisfy post-contingency voltage stability margin only using MSCs, without including dynamic problem and SVCs in the overall planning methodology. This in effect means we consider only constraints (13), (15), and (17) in original MIP of section 4.2.1, and also in updated successive MIP of section 4.2.2. We also analyze the impact of induction motors on the transient voltage recovery performance, and demonstrate the ineffectiveness of steady state planning approach to counter the ill-dynamics caused by induction motors in voltage stability phenomena. This strongly motivates the need for considering the short-term voltage performance and fast-responding dynamic reactive resource within a coordinated planning approach.

Steady state voltage stability analysis using CPF-based contingency screening was performed for all credible N-1 (2100 branches (T) and 168 generators (G)) and N-G-T (all possible combinations of G and T) contingencies under 6 different stress directions. The results indicated that only the stress direction corresponding to the MZ1 region was found to have post contingency steady state voltage stability problems, as indicated by the 56 critical N-G-T contingencies (28 line outages under 2 N-G base case, where G = {14067, 14068}) in Table 4. The total base load in MZ1 zone (sink) is 2073 MW. The collapse point for basecase was at 2393 MW, which is equivalent to a stability margin of 15.44%. All the 56 contingencies in Table 4 have stability margin lesser than 5%, with the top 14 facing instability issues. Table 4 also shows the final optimal static VAR solution that just satisfies the minimum performance criteria under every critical contingency. The final optimal investments will be the maximum amount of MSCs required at each location. The total investment cost of MSCs to solve all the steady state voltage stability problems is 2.665 M$.

Cont. No	Transmission Line		Base KV	Stability Index (Margin in %)		Capacitor Allocation (p.u Mvar)			
	From	To		Gen#1–14067	Gen#2-14068	14073	14071	14080	14160
1	14079	14083	138	Unstable	Unstable	1.5	0	1.5	0.25
2	14094	14326	230	Unstable	Unstable	1.5	1.5	0.75	0
3	14175	14319	230	Unstable	Unstable	1.2	1.5	0	0.85
4	14296	14322	500	Unstable	Unstable	1	1.2	1.25	0
5	14319	14321	230	Unstable	Unstable	1.5	1.5	0.5	0
6	14319	14326	230	Unstable	Unstable	1.5	1.5	0.9	0
7	10983	14129	345	Unstable	Unstable	1.5	1.5	1.5	1
8	14077	14080	138	1.495417	1.784853	0.3	0	0	0
9	14130	14142	138	2.411963	2.556681	0.65	0	0	0
10	14129	14162	345	2.556681	2.701399	0.85	0	0	0
11	14294	14319	230	2.990835	3.135552	0.68	0	0	0
12	14295	14302	138	2.990835	3.135552	0.65	0	0	0
13	14071	14079	138	3.28027	3.424988	0.57	0	0	0
14	14126	14142	138	3.28027	3.424988	0.55	0	0	0
15	14109	14363	138	3.617945	3.617945	0.52	0	0	0
16	14148	14232	138	3.617945	3.762663	0.45	0	0	0

17	14302	14363	138	3.617945	3.907381	0.52	0	0	0
18	14124	14126	138	3.907381	3.907381	0.5	0	0	0
19	14148	14149	138	3.907381	3.907381	0.4	0	0	0
20	14232	14328	138	3.907381	4.052098	0.5	0	0	0
21	14237	14328	138	3.907381	4.052098	0.5	0	0	0
22	14322	14494	500	3.907381	4.052098	0.5	0	0	0
23	14106	14109	138	3.95562	4.052098	0.5	0	0	0
24	14297	14311	138	4.196816	4.341534	0.47	0	0	0
25	14134	14290	138	4.245055	4.486252	0.35	0	0	0
26	14103	14130	138	4.486252	4.63097	0.4	0	0	0
27	14106	14110	138	4.486252	4.63097	0.4	0	0	0
28	14238	14297	138	4.63097	4.823927	0.4	0	0	0

Table 4. Contingency List and Optimal Allocation of MSCs by Steady State Reactive Power Planning

Even though the system is planned against steady state voltage instability using MSCs, nevertheless the effect of induction motors in transient phenomena has not been investigated.

Figure 7 shows the voltage recovery phenomenon at a certain bus with and without dynamic modeling of induction motor. The figure shows the significance of modeling induction motor properly, which has important role in ascertaining post-fault-clearance short-term voltage stability of the system. Once this phenomenon is captured by appropriate modeling, we can counter it by the proposed coordinated RPP.

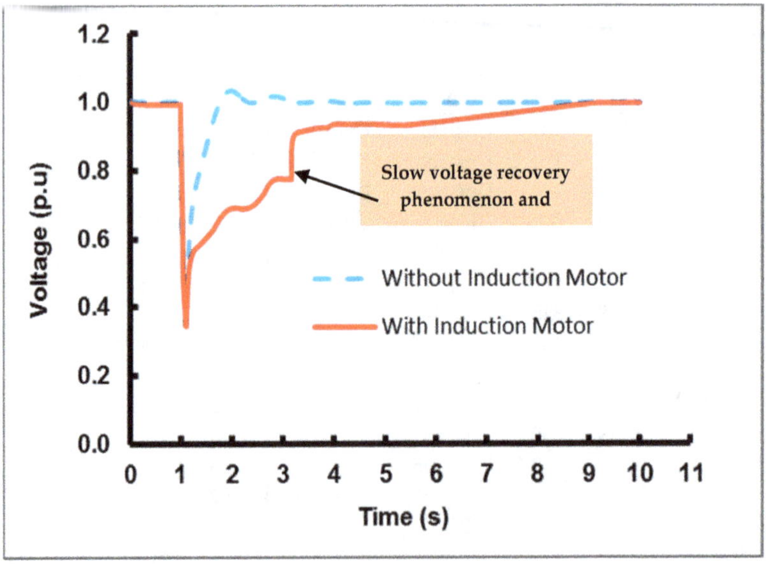

Figure 7. Voltage recovery phenomenon with and without Induction Motor modeling

With appropriate modeling of induction motor dynamics, time domain simulations were run for the top 7 contingencies by applying a 3-phase fault at t=0 at one end of the transmission circuit and then clearing the fault and the circuit at 6 cycles (t = 0.1s). All the contingencies lead to a slow voltage recovery due to the presence of induction motor loads that ultimately tripped. A list of all the buses having slow voltage recovery problem under each severe contingency was made. Table 5 shows the time domain simulation results under few of the contingencies of Table 4, wherein we notice the buses at which the minimum criteria for transient voltage recovery (more than 20 cycles below 0.7p.u) is violated. The induction motor loads connected to these buses also get tripped, in a similar manner as was shown in Figure 7. The transient voltage response under other contingencies was also very poor. For instance, under contingencies 2, 4, 5, and 6, about 76, 18, 72, 64 buses respectively violated the minimum post-fault voltage recovery criteria.

Bus Number	Contingency 1		Contingency 3	
	Recovery time	Cycles	Recovery time	Cycles
14079	0.841	50.46	0.344	20.64
14071	0.771	46.26	0.36	21.6
14084	0.694	41.64	0.36	21.6
14060	0.614	36.84	0.353	21.18

Table 5. Buses resulting in Slow Voltage Recovery and Induction Motor Tripping

Table 6 ranks the top 7 contingencies (under both gen. outages) shown in Table 4, based on their severity, which is quantified in terms of worst-case recovery times. It can be expected that the most severe contingencies will drive the amount of dynamic VARs needed.

Contingency No.	Bus Numbers		kV	Rank
	From	To		
1	14079	14083	138	2
2	14094	14326	230	4
3	14175	14319	230	6
4	14296	14322	500	7
5	14319	14321	230	3
6	14319	14326	230	1
7	10983	14129	345	5

Table 6. Contingency Ranking in terms of Worst-case Recovery Times

5.2. Coordinated reactive power planning

The proposed coordinated planning, as discussed in section 4.2, was performed to mitigate both transient as well as steady state problems under these contingencies. Sensitivities of post-contingency voltage stability margin and transient voltage recovery times are used as one of the inputs to find the optimal solution. The candidate locations were identified according to the following criteria:

1. Buses for which one or more contingencies result in:
   - the bus being among the top 5 worst voltage dips and
   - the bus has induction motor load that trips
2. Buses must have high steady state voltage stability margin sensitivity so that they can also efficiently address the static problems.

Table 7 shows the stage 1 MIP result, which selects two SVC locations to solve both static as well as dynamic problems. Investigation indicates the reason for this is that the transient voltage problems are so severe that the amount of SVC required to solve them is also sufficient to mitigate the steady state voltage stability problems.

No.	N-G-T Contingencies			SVC (p.u MVAR)		Steady State Stability Margin (%)
	Bus Number		KV	14071	14084	
	From	To				
1	14079	14083	138	3	0.85	9.64
2	14094	14326	230	3	0.8	9.16
3	14175	14319	230	3	0.7	9.4
4	14296	14322	500	1.7	0	8.68
5	14319	14321	230	3	1.5	9.4
6	14319	14326	230	3	1.53	8.92
7	10983	14129	345	3	0.95	5.1

Table 7. Stage 1 MIP Result of Coordinated RPP and Steady State Stability Validation

When stage I MIP solution was validated for steady state voltage stability problems for the top 7 contingencies (under both gen. outages) of Table 4, it was found that they not only attained equilibrium, but also had post contingency voltage stability margin more than the minimum requirement of 5% as shown in Table 7. The remaining contingencies in Table 4 were also validated with this stage I MIP solution and were found to have sufficient post contingency voltage stability margin.

When time domain simulations were done to validate stage 1 MIP result, it was found that contingencies 1, 2, 5, 6, and 7 still had buses that violated the minimum recovery time requirement, resulting in tripping of some induction motors. Figures 8 and 9 show the voltage profiles at the most severely-affected buses under contingency 2 before and after stage 1 MIP result implementation, respectively.

Figure 9 also shows that the SVC placed after stage 1 MIP at buses 14071 and 14084, which are just sufficient to provide voltage recover within 20 cycles at some buses. For the contingencies having slow voltage recovery problems at some buses even after stage 1 MIP solution, a successive MIP is performed with updated sensitivity information using stage 1 SVC solution.

Table 8 shows the final operational solution of the coordinated planning problem, which chooses only SVCs due to the nature of problems under contingencies.

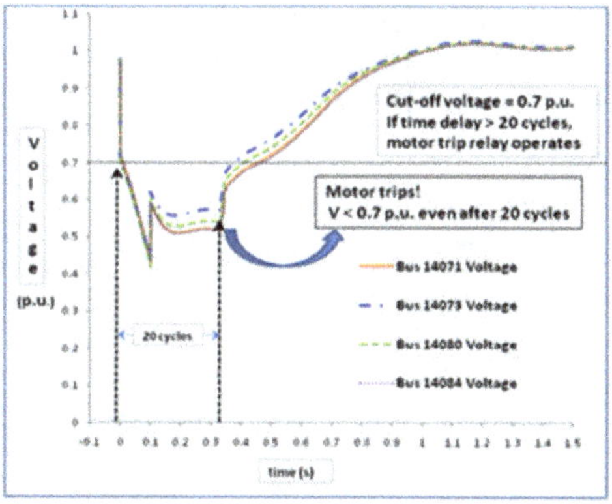

Figure 8. Voltage profiles of worst-hit buses under cont. 2 without SVC

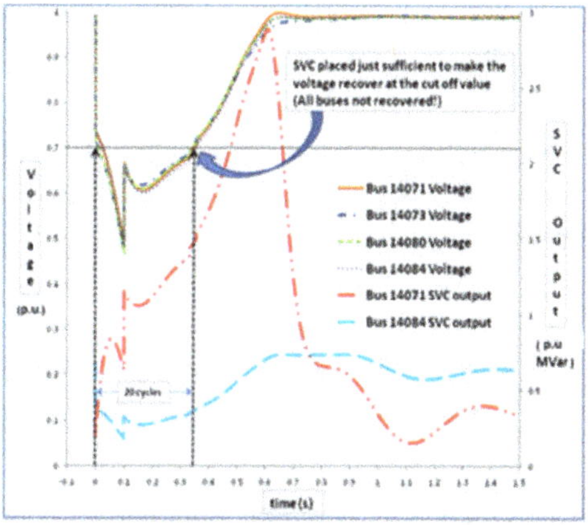

Figure 9. Voltage profiles of worst-hit buses under cont. 2 with SVC from stage 1 MIP result

N-G-T Contingencies under both the N-G				SVC (p.u MVAR)	
No.	Bus Number From	Bus Number To	KV	14071	14084
1	14079	14083	138	3	2.65
2	14094	14326	230	3	2.7
3	14175	14319	230	3	0.7
4	14296	14322	500	1.7	0
5	14319	14321	230	3	2.7
6	14319	14326	230	3	2.85
7	10983	14129	345	3	2.1

Table 8. Final Solution of Coordinated Optimal Planning

Figure 10 shows the improved voltage profile at Bus 14071 under contingency 1 after implementing the final SVC solution. Figure 10 also shows the output of the two SVCs. The SVC peak output at bus 14071 is 290 MVar, and that of bus 14084 is 270 MVar.

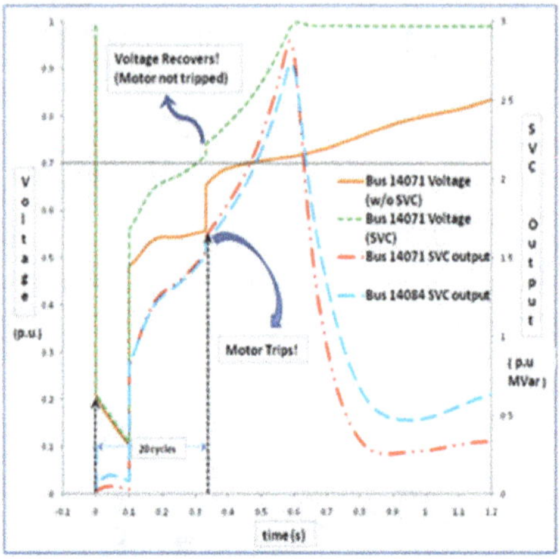

Figure 10. Bus 14071 voltage profile under cont. 1 with final allocation

The final investment solution for the coordinated planning problem includes SVC placements at buses 14071 and 14084 of maximum capacity 3.0 MVar and 2.85 MVar respectively[1]. The total cost is 32.25 M$ under the cost assumptions used for this study.

6. Conclusions

This chapter presented the critical role played by induction motors in short-term voltage stability phenomenon, and emphasizes the need for including proper dynamic models of motor loads in voltage stability related planning studies. The chapter also sheds light on the need of a coordinated framework for reactive power planning considering both static and dynamic reactive control devices, which mitigates both post-contingency steady state as well transient voltage recovery and motor stalling problems. The mixed integer optimization based VAR planning algorithm is formulated to satisfy the minimum post-contingency steady state voltage stability margin and post-fault transient voltage recovery performance criteria under many critical contingencies simultaneously. The developed method was illustrated in large-scale system with 16173 buses representing US Eastern Interconnection, wherein induction motor dynamic characteristics modeling and bus voltage magnitude limit enforcement were incorporated within the overall analysis and planning respectively. The results verify that the method works satisfactorily to plan an optimal mix of static and dynamic VAR devices under

[1] There may be cases, when the required SVC to mitigate slow voltage recovery problems might be less, and they may still not satisfy the post-contingency steady state voltage stability margin requirements. For such cases, the MIP may choose MSCs as an effective addition to the optimal solution.

many critical contingencies simultaneously and averts steady-state and dynamic voltage instability issues, including induction motor tripping.

Author details

Venkat Krishnan and James D. McCalley
Iowa State University, USA

7. References

[1] T. Van Cutsem, "Voltage instability: phenomena, countermeasures, and analysis methods," *Proc. IEEE*, vol. 88, pp. 208–227, Feb. 2000

[2] P. Kundur, *Power System Stability and Control*, EPRI Power System Engineering Series, New York: McGraw Hill, 1994

[3] NERC (North American Electric Reliability Corporation) Disturbance Analysis Working Group Database, "http://www.nerc.com/~dawg/database.html," 1984-2002

[4] Voltage Stability Assessment: Concepts, Practices and Tools, *Power systems stability subcommittee special publication*, sp101pss, IEEE-PES, Aug. 2002

[5] U.S.-Canada Power System Outage Task Force, (2004, Apr.), Final report on the August 14, 2003 blackout in the United States and Canada: Causes and recommendations. [Online]. Available: https://reports.energy.gov/BlackoutFinal-Web.pdf

[6] M. H. J. Bollen, *Understanding Power Quality Problems – Voltage Sags and Interruptions*, Wiley-IEEE Press September 24, 1999

[7] G.K. Stefopoulos and A.P. Meliopoulos, "Induction Motor Load Dynamics: Impact on Voltage Recovery Phenomena," *Proceedings of the 2005/2006 IEEE Transmission and Distribution Conference and Exposition*, pp.752 – 759, May 21-24, 2006

[8] "Criteria and countermeasures for voltage collapse," *CIGRE Publication*, CIGRE Task Force 38-02-12, 1994

[9] J.J. Paserba, "How FACTS controllers benefit AC transmission systems," *IEEE Power Engineering Society General Meeting, 2004.*, Vol.2, pp. 1257 – 1262, 6-10 June 2004

[10] J. A. Diaz de Leon II and C. W. Taylor, "Understanding and solving short-term voltage stability problems," *in Proc. 2002 IEEE/PES Summer Meeting*, Vol. 2, pp.745-752, July 2002

[11] D. J. Shoup, J. J. Paserba, and C. W. Taylor, "A survey of current practices for transient voltage dip/sag criteria related to power system stability," *in Proc. 2004 IEEE/PES Power Systems Conference and Exposition*, Vol. 2, pp. 1140-1147, Oct. 2004

[12] A. Hammad and M. El-Sadek, "Prevention of transient voltage instabilities due to induction motor loads by static VAr compensators," *IEEE Trans. Power Syst.*, Vol. 4, Issue 3, pp. 1182 – 1190, Aug. 1989

[13] IEEE/CIGRE joint task force on stability terms and definitions, "Definition and classification of power system stability," *IEEE Trans. Power Syst.*, vol. 19, pp. 1387-1401,2004

[14] Begovic. M et. al., "Summary of system protection and voltage stability", A summary of special publication prepared by Protection Aids to Voltage Stability Working Group of

the Substation Protection Subcommittee of the IEEE Power System Relaying Committee, *IEEE Trans. on Power Delivery*, Vol. 10, No. 2, April 1995

[15] NERC Transmission Issues Subcommittee, System Protection, and Control Subcommittee, "A technical reference paper fault-induced delayed voltage recovery," Technical report, NERC, June 2009

[16] P. Sakis Meliopoulos and George J. Cokkinides, "A Virtual Environment for Protective Relaying Evaluation and Testing", *IEEE Trans. of Power Syst.*, Vol. 19, pp. 104-111, 2004

[17] V. Ajjarapu and C. Christy, "The continuation power flow: A tool for steady state voltage stability analysis," *IEEE Trans. Power Syst.*, vol. 7, pp. 417-423, Feb. 1992

[18] B. Gao, G. K. Morison, and P. Kundur, "Toward the development of a systematic approach for voltage stability assessment of large-scale power systems," *IEEE Trans. Power Systems*, vol. 11, pp. 1314-1324, Aug. 1996

[19] D. Chassin et. al., "Load modeling transmission research," Technical report, CIEE, March 2010

[20] GE Positive Sequence Load Flow, User Manual Version 18.0, April 2011

[21] H. D. Chiang and H. Li, "CPFLOW for Power Tracer and Voltage Monitoring," *PSERC Publication 01-02*, May 2002

[22] C. A. Canizares and F.L. Alvarado, "Point of collapse and continuation methods for large ac/dc systems," *IEEE Trans. Power Syst.*, Vol. 8, No.1, pp. 1-8, Feb. 1993

[23] T. Van Cutsem and C. Vournas, Voltage Stability of Electric Power Systems, Boston: Kluwer Academic Publishers, 1998

[24] C. W. Taylor, *Power System Voltage Stability*, EPRI Power System Engineering Series, New York: McGraw Hill, 1994

[25] Haifeng Liu, "Planning reactive power control for transmission enhancement", PhD dissertation, Iowa State University, Ames, IA 2007

[26] Western Electricity Coordinating Council (2005, April). NERC/WECC planning standards. WECC, Salt Lake City, UT
Available: http://www.wecc.biz/documents/library/procedures/ CriteriaMaster.pdf

[27] V. Ajjarapu, *Computational Techniques for voltage stability assessment and control*, New York: Springer, 2006

[28] S. Greene, I. Dobson and F. Alvarado, "Sensitivity of the loading margin to voltage collapse with respect to arbitrary parameters," *IEEE Trans. Power Syst.*, vol. 12, 1997

[29] S. Greene, I. Dobson, and F. L. Alvarado, "Contingency ranking for voltage collapse via sensitivities from a single nose curve," *IEEE Trans. Power Syst.*, vol. 14, pp. 232-240, 1999

[30] V. Krishnan and J. D. McCalley, "Contingency assessment under uncertainty for voltage collapse and its application in risk based contingency ranking," *Int J Electr Power Energy Syst*, 43 (1), pp. 1025–1033, Dec 2012

[31] V. Krishnan, Coordinated Static and Dynamic Reactive Power Planning against Power System Voltage Stability Related Problems, M.S. Thesis, Iowa State University, Ames, 2007

[32] Siemens PTI Power Technologies Inc., PSS/E 33, Program Application Guide, Vol. II, May 2011

Chapter 8

Noise of Induction Machines

Marcel Janda, Ondrej Vitek and Vitezslav Hajek

Additional information is available at the end of the chapter

1. Introduction

Diagnostics of electric machines is very interesting and extensive. There are many methods used to detect properties of electrical machines. Between diagnostic methods include too the measurement and analyze of noise, which generates electrical machines.

Itself the noise of electric machines is byproduct of the machine operation. The generation of noise is involved in many physical principles.

Noise of electrical machinery is generated by the vibration of machine parts. Gradual spread of vibration from the engine to the surroundings causes pulses of air with certain frequencies. This creates a sound wave generator, which can be within a certain frequency range, audible to humans.

The main sources of noise in electrical machines are time change of the electromagnetic fields, noise of bearings and other mechanical sources. Finally, the unwanted noise is creating too due to coolant flow or parts that come into contact with coolant in electric machines. Level of noise sources in electrical machines depends on the structural arrangement and the accuracy of engine design.

A major problem in measurement noise is interference environment. For a perfect suppression ambient noise is necessary to have a specialized laboratory. It should be also measured machine have isolated from vibration, which it may be transferred from storage.

2. Basic concepts of noise

The sound wave is generated by vibrating objects and can be defined as mechanical interference with the finite speed of advancing through the media. These waves have small amplitude, adiabatic oscillation are characterized by a wave speed, wavelength, frequency and amplitude. Sound has the character of longitudinal sound waves in the direction of propagation in the environment. In other words, it is the movement of individual particles of the medium in a direction parallel to the transmission of energy. Sound waves spread in three-dimensional environment from the source. It is same in all directions, if is the environment homogeneous. Sound waves can be polarized, they cannot have orientation. Non-polarized waves can oscillate in any direction in the plane perpendicular to the direction of propagation.

Sound amplitude can be measured as sound pressure level (SPL), sound intensity (SIL), sound power level (SWL) and the intensity of the acoustic energy (SED). The human ear can perceive sound waves of sufficient intensity and frequencies are ranging from 20 to 20,000 Hz. The Minimum sound intensity is different for different frequency and it is called the threshold of audibility. Range of sound intensity, which can capture the human ear, is 10-12 to 1 W/m^2corresponding sound pressure of 20 MPa. Maximum sound level in which humans feel pain is called threshold of pain. Amplitude of sound about pressure 100 Pa is very loud.

2.1. Acoustic pressure

Concentration of particle of vibrating environment corresponds with increase or decrease pressure inside gasses and liquids. This means that the total pressure in the environment is changing and therefore fluctuates around the initial static value or barometric pressure. The acoustic pressure is then considered deviation of the total pressure from the static pressure. For acoustic pressure is valid relationship

$$p_c = p_b + p(t) \tag{1}$$

$$p(t) = p_0 . cos(\omega . t + \varphi) = p_0 . cos(2. \pi . f . t + \varphi) \tag{2}$$

$$p_c = p_0 + p_0 . cos(2. \pi . f . t + \varphi) \tag{3}$$

Where

- p_c... Acoustic Pressure [Pa]
- p_b... Barometric pressure [Pa]
- p_0...Amplitude of sound pressure [Pa]
- f... Frequency [Hz]
- t... Time [s]
- φ...phase shift

For effective sound pressure value is valid relationship

$$p_{ef} \frac{p_0}{\sqrt{2}} \tag{4}$$

Acoustic pressure is a variable and it describing the noise source quantitatively. The measured level depends on the observer's distance from the source and the quality of the transmission environment Acoustic pressure level gives us information on the total sound pressure across a entire audible band. For sound pressure level is valid relationship

$$L\_p = 20. log(p/p\_0) \tag{5}$$

Where

- p ... Static pressure [Pa]
- p_0 ...Minimum value of static pressure, which is able to capture the human ear [Pa]

2.2. Sound power level

Mechanical vibrations are transmitted in form of mechanical energy from the source through acoustic waves. Sound power level is called the energy that passes per unit time over surface. For sound output, we can write the relationship

$$P_{ac} = p_c . v . A \tag{6}$$

Where

P_{ac}... Sound power level [W]
p_c... Acoustic pressure [Pa]
v... Vibration velocity of particles[m/s]
A... Area [m^2]

The sound power level depend on the the environment parameters and distance from the measurement point. The sound power level can be expressed as

$$L_{Pac} = 20.\log\left(\frac{v}{10^{-9}}\right) + k \tag{7}$$

Where

k... constant

2.3. Acoustic intensity

Acoustic intensity is a vector quantity that describes the amount and direction of flow of acoustic energy in the environment. Vector of acoustic intensity is time-change of instantaneous sound pressure and it is corresponding instantaneous speed of vibrating particle environment in the same place

$$I = \overline{p(t).v(t)} \tag{8}$$

Where

I... Acoustic intensity$[W/m^2]$

3. Noise sources

From the physical point of view, mechanical sound is waves in a flexible environment. The Frequency range of sound audibly for human ear is from 20 Hz to 20 kHz. The sound spreads in all directions from resources by transmitting acoustic wave energy. Division by frequencies of sound waves:

- Infrasound - up to 20Hz
- Low frequency - 20Hz to 40Hz
- RF - 8kHz to 16kHz
- Ultrasound - 20kHz over

Dividing the sound by timing:

- Steady
- variable
- intermittent
- pulse

The interest noise frequency is over 1000 Hz for induction machines. Noise of Electrical Machines is characterized as a set of sounds that are caused by rapid changes in air pressure. These changes cause most commonly:

- Vibration of machine parts or the whole of its surface
- Aerodynamic phenomena that lead to pulsation of pressure near the machine

Basic sources of noise are induction motors (see diagram):

- Electromagnetic source
- The mechanical source
- The aerodynamic source

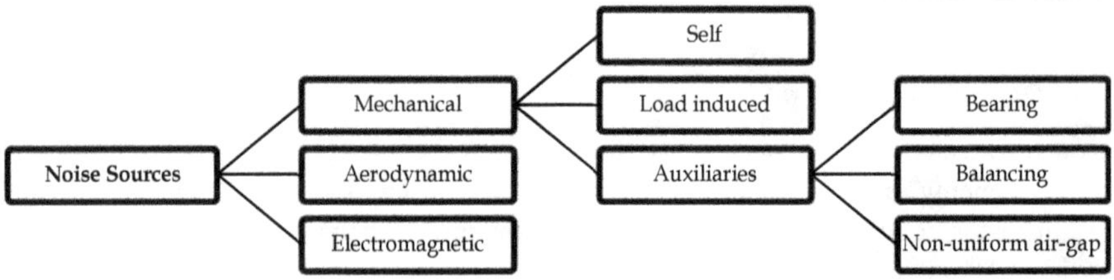

Figure 1. Division of noise sources in electrical machines

The noise from electromagnetic source is the most typical component noise of electrical machine. Its cause is the vibration of motor body, or other parts of the machine on which work the electromagnetic forces. Frequency Spectrum noise of the electromagnetic source has discrete character, while there is very distinct directional radiation characteristics of this component in many cases.

Determining the influence of this component on the overall noise of electric machine is often simply done so, that after switching off the machine from the network is observed decline in the acoustic signal in time. If is this decline immediately, then it is obviously a component of the noise of electromagnetic origin. Another method of investigation is the measurement of electromagnetic noise spectrum for different values of power - or even frequency.

Noise origin of ventilation is crucial observe especially in machines with high rotational speed. Detailed analysis of the fan noise shows that the main source in this case is very fan with its nearest surroundings. It is the decisive exceeds other sources of noise, which can be, for example rotor wings, radial or axial cooling channels in the machine, input and output caps and the like.

Frequency analysis of noise ventilation origin shows that the spectrum has a broadband character, either discrete or vice versa. In the first case, the aerodynamic noise is created from turbulent airflow near fan blade and near the entrance, but also the output edges of blades. These pulsations are uneven both in space and in time, so the frequency spectrum created of wind noise is broadband and contains all components of the audible band.

In contrast, discrete nature of the spectrum, sometimes the siren phenomenon can arise. This phenomenon arise if the fan or behind obstacles (such as a blade with these obstacles) is not profile of velocity uniform air flow around the wheel circumference, leading to periodic pulsation of pressure. Then the siren noise is produced naturally.

The noise of mechanical origin is primarily inflicted on roller bearings and unbalance of rotating machine parts. Rolling bearings can create multiple frequency components, which have their origin mainly in inequality as part of rolling paths of the bearing rings. In principle, the noise of mechanical origin has a mixed character.

3.1. Electromagnetic noise

The influence of magnetic induction in the air gap formed magnetic forces; these forces operate across various directions. They may also have various amplitude and frequency. Their work is split between the rotor and stator of electric machine. Their characteristics depend on the size and shape of the air gap and a number of other factors.

The construction of the rotor is the main radiator noise machine. If the frequency is close to the radial force or equal to one natural frequency of the stator system, resonance occurs which leads distorted stator system with vibration and noise. Magnetostriction noise electric machine can be neglected in most cases due low and high frequency 2f arrangement r = 2p of radial forces, where f is the fundamental frequency and p is the number of pole pairs. However, the radial forces due magnetostriction can reach up to 50% the radial forces produced in the air gap magnetic field.

Magnetic flux density wave

$$\text{Stator: } B_{m1}.cos(\omega_1.t + k.\alpha + \Phi_1) \tag{9}$$

$$\text{Rotor: } B_{m2}.cos(\omega_2.t + l.\alpha + \Phi_2) \tag{10}$$

Where

- B_{m1} … Amplitude of magnetic flux density in stator [T]
- B_{m2}… Amplitude of magnetic flux density in rotor [T]
- ω_{Φ_1} … Angular frequency of stator magnetic fields
- ω_{Φ_1} … Angular frequency of rotor magnetic fields
- k,l … Variable (values 1,2,3,4,….)

For total wave of magnetic flux density can be write relationship

$$P_{mr} = 0{,}5.B_{m1}.B_{m2}.cos[(\omega_1 + \omega_2).t + (k + l).\alpha + (\Phi_1 + \Phi_2)] +$$

$$+0{,}5.B_{m1}.B_{m2}.cos[(\omega_1 + \omega_2).t + (k - l).\alpha + (\Phi_1 - \Phi_2)] \quad (11)$$

The magnetic stress wave has worked in radial directions on the stator and on active surfaces of rotor. This causing the deformation and subsequently cause the vibration and noise.

The mixed product of stator and rotor winding space harmonic create forces at frequencies

$$f_r = f_1.\left[\frac{n.Z_r}{p}.(1 - s) + 2\right]$$

$$f_r = f_1.\left[\frac{n.Z_r}{p}.(1 - s)\right] \quad (12)$$

Where

- f_1… Supply frequency [Hz]
- n…. value n=0, ±1, ±2,… [-]
- p… number of pole pairs [-]
- N_{rs}…Number of rotor slots [-]
- s… slip

The mixed product of stator winding and rotor eccentricity space harmonics create forces with frequencies

$$f_r = f_1.\left[\frac{n.N_{rs}}{p}.(1 - s) + 2\right]$$

$$f_r = f_1.\left[\frac{n.N_{rs}}{p}.(1 - s)\right]$$

$$f_r = f_1.\left[\frac{n.N_{rs}}{p}.(1 - s) + \frac{1 - s}{p}\right]$$

$$f_r = f_1.\left[\frac{n.N_{rs}}{p}.(1 - s) + 2 + \frac{1 - s}{p}\right] \quad (13)$$

The mixed product of stator winding and rotor saturation harmonics create forces at frequencies

$$f_r = f_1.\left[\frac{n.N_{rs}}{p}.(1-s)+4\right] \tag{14}$$

$$f_r = f_1.\left[\frac{n.N_{rs}}{p}.(1-s)+2\right] \tag{15}$$

3.2. Rotor eccentricity

The air gap width depends only on position (no on time) in the static eccentricity. We conclude that the magnetic field in the air gap is rotating synchronous speed. That is given by the mains frequency and with the number of pole pair's induction machine. Modulation of magnetic field in one period is function, which is represented by a variable air gap, i.e. a function of its conductivity. Static eccentricity is defined as the rotor axis offset from the axis of the stator. The air gap has a variable character. There is stronger interaction of stator and rotor magnetic field at the point where the gap is smaller. Influence of the static eccentricity manifests as the emergence of side frequency bands, which are shifted from the mains frequency f_1 of the synchronous frequency f. For static eccentricity is the angular frequency $\Omega_\varepsilon = 0$.

Static eccentricity is straight-line. The frequency for static eccentricity is twice power frequency

$$f_{stat} = 2.f_1 \tag{16}$$

The relative eccentricity ε is defined as

$$\varepsilon = \frac{e}{g} = \frac{e}{R-r} \tag{17}$$

Where

- R... Inner stator core radius
- r... Outer rotor radius
- e... Rotor eccentricity
- g... Ideal uniform air-gap for e=0

Dynamic eccentricity occurs when the rotor failure or its affiliates. Ratios are complicated by the fact that the width of air gap is not just a function of position, but is also a function of time. The variable air gap is changing at the rotation of the rotor. There is emergence of side bands that appears in the frequency range of vibrations of electric machine.

Angular frequency for dynamic eccentricity

$$\Omega_\varepsilon = \Omega.(1-s) = \frac{\omega}{p}.(1-s) = 2.\pi.\frac{f}{p}.(1-s) \tag{18}$$

The frequency generated by the dynamic eccentricity

$$f_{DYN} = f_1 \pm (1-s).f_{SO} \tag{19}$$

For frequency generated by eccentricity is true also relationship

$$f_{exc}\left[(n_{rt}.R \pm n_d).\frac{1-s}{p}.n_{\varpi s}\right].f \quad (20)$$

Where

- R...Number of grooves engine
- s... Chute
- p... Number of pole pairs

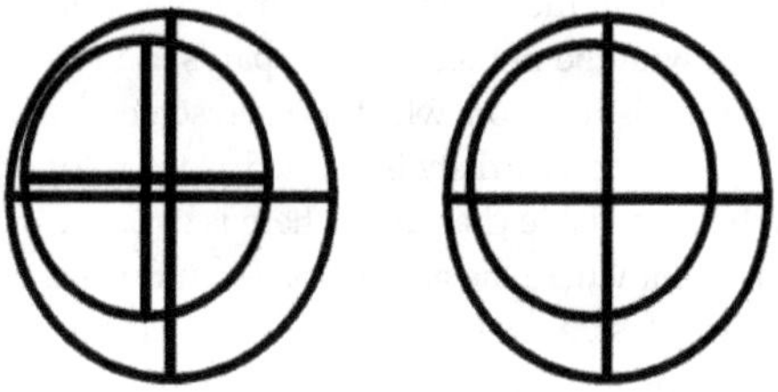

Figure 2. Rotor eccentricity

3.3. Aerodynamic noise

Aerodynamic noise arises most often around the fan, or in the vicinity of the machine that behaves like a fan. Noise can be created too on the necks stator slot windings or rotor. The aerodynamic noise sources can also include the noise produced by air flow inside and outside the design of electrical machines.

The main reason for the fan noise is formation of turbulent air flow around the blades. This noise is characterized by spectrum in a wide range, which has continuous character. Acoustic performance is increasing with the square of velocity. Siren noise can be eliminated by increasing the distance between the impeller and the stationary obstacle.

For the fan noise can write the relationship

$$L_A = 60.\,logU_2 + 10.\,logD_2.\,b_2 + \sum k_1 \quad (21)$$

Where

- U_2...Outer speed of fan on the circuit $[m.s^{-1}]$
- D_2...Outer diameter of the fan [m]
- B_2... Fan width [m]
- k_1... Constants for the correction

The vortex frequency is expressed by

$$f_v = 0{,}185.\frac{v}{D_2} \quad (22)$$

The frequency of the pure tone due to the fan blades is given by relationship

$$f_f = N_b.\frac{N}{60} \tag{23}$$

Where

- N...speed [rev/min.]
- N_b... Number of fan blades [-]

Sound power level of aerodynamic noise is

$$L_w = 67 + 10.log_{10}(P_{out}) + 10.log_{10}(p) \tag{24}$$

$$L_w = 40 + 10.log_{10}(Q) + 20.log_{10}(p) \tag{25}$$

$$L_w = 94 + 20.log_{10}(P_{out}) - 10.log_{10}(Q) \tag{26}$$

Where

- P_{out}... Motor rated power [kW]
- p... Fan static pressure [Pa]
- Q...Flow rate $[m^3.s^{-1}]$

Reducing aerodynamic noise in electrical machines can be use the following ways:

- Reducing the required amount of coolant used for ventilation of electrical machines
- Optimal design of fan. Especially the number and shape of the fan blades has an impact on the noise generated by the electric machine.
- To minimize the noise is needed to prevent vibration machine parts, which come into contact with a cooling medium.

3.4. Mechanical noise sources

Mechanical noise is mainly due with bearings, their defects, ovality, sliding contacts, bent shaft, rotor unbalance, shaft misalignment, couplings, U-joints, gears etc. In principle, the mechanical source of noise has a mixed character. The noise caused by unbalance of rotating parts and noise of bearing is spread after machine constructions very well. Dynamic balancing in production serves to reducing the noise of mechanical source. Especially for machines with high speed is necessary to perfect balance. Also, compliance with the manufacturing tolerances and technological processes, especially in the manufacture of small machines is the best solution to reduce the noise of mechanical source. Any change in noise from this source can mean failure of the mechanical parts inside the motor. For example, the bearings failure (damaged ball) is appear in the noise spectrum. There are specific frequencies by individual damage. The very faults of bearings and their effect on the noise spectrum of the electric machines are now well mapped.

Design of bearings can be either a sliding or rolling bearings. Rolling bearings can create multiple vibration frequencies, which have their origin mainly in the uneven parts or rolling themselves paths to the bearing rings. If bearing has mechanical damage, there is uneven movement of the whole system and thus increasing vibration and noise of the electric machine.

The main mechanical sources of the noise

- Alignment
- Inaccurate machining of parts
- Running speed
- Number of rolling elements carrying the load
- Mechanical resonance frequency of the outer ring
- Lubrication conditions
- Temperature

3.4.1. Rolling bearings

The noise of rolling bearings depends on the type of bearing and its construction and accuracy of bearing parts. The increase in vibration and noise level of bearings, when the rotational speed changes from n_1 ton_2 can be expressed as

$$\Delta L_v = 20.log\frac{n_2}{n_1} \quad (27)$$

Ball pass frequency – outer race

$$f_{or} = \frac{N_b}{2}.n_m.\left(1 - \frac{d_b}{D}.cos\alpha\right) \quad (28)$$

Where

- D…Pitch diameter [m]
- n_m… Rotation speed [rev/s]
- N_b…Number of balls [-]
- d_b…Diameter of balls]m]
- α…Contact angle of balls

Ball pass frequency – inner race

$$f_{ir} = \frac{N_b}{2}.n_m.\left(1 + \frac{d_b}{D}.cos\alpha\right) \quad (29)$$

3.4.2. Sleeve bearings

- Uneven journal

$$f_{ov} = k.n_m \quad (30)$$

k=1, 2, 3, …

- Axial groves

$$f_{gr} = N_g.n_m \quad (31)$$

N_g…number of groove

3.4.3. Load induced noise

In certain cases, the vibrations and thus noise transmitted from the load, which is connected to the induction motor. In most cases, this occurs with wrong balance or bad connects of couplings. Uneven distribution load acting on the motor shaft or inappropriate use of gears may also affect noise machine. The only possible protection against these effects is the perfect balance of the whole set and if possible an even distribution of forces acting on the connecting elements. Noise arises too due to coupling of the machine with a load, e.g., shaft misalignment. Next noise arises from belt transmission, from cogwheels and couplings. It may also arise to noise due to mounting the machine on foundation or other structure.

4. Noise measurement

For measurement noise of induction machines can be used several techniques. The basic method for the measurement noise is the sound meter. It is a device which measures sound pressure.

4.1. Measurement process

Measurement of noise can be divided into three main parts. The first part is data capture. For this purpose, the most commonly used microphones, or specialized equipment to measure noise (sound level meter). Their output is usually an analog signal, which must be further processed. When choosing of microphone is needed careful heed on certain parameters that can affect measurement accuracy. One of the most important parameter is the sensitivity of frequency. Worse microphones not recorded of the entire spectrum of the measured noise. Thanks to this complicates achieve it of accurate analysis results. Other parameters include the microphone sensitivity, which indicates the size of the output voltage (mV / Pa), depending on the pressure acting on the membrane. In addition, the structural dimensions of the measurement microphone and also the type of sound field that which is measured. Computers are most frequently use for Signal processing. For this reason it is necessary to convert from analog signal to digital form.

Large numbers of types A/D converters is on the market. Some are stand-alone converters; others are integrated to the specialized measurement cards. In both cases, the measurement depends on the three main parameters. The first is the measuring range of the converter. It gives the minimum, respectively maximum, measurable value. Because the signal is weak from a microphone, there should be used an amplifier for its amplification. Another parameter of the A/D converter is the bit depth conversion. This parameter defines the limitations of this device.

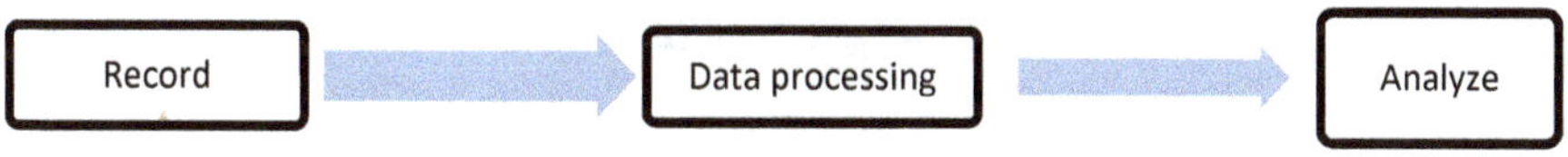

Figure 3. Block diagram of measurement process

Factors to selecting a suitable type of microphone are as follows

Characteristics of sound field	Required accuracy	Environmental conditions
Freely field for a closed chamber	Tolerance sensitivity	Noise level background
An important range of sound pressure levels	Frequency distortion tolerance	Humidity
An important frequency range	Phase distortion tolerance	Atmospheric pressure
	Tolerance of non-linear distortion	wind
	Own noise tolerance	Strong electromagnetic fields
		Mechanical shock

Table 1. Selecting factors of microphone

4.2. Sound level meter

Sound level meter is an essential instrument for measuring sound pressure levels. This device consists of the following components: Microphone, preamp, overload detector, central Unit, weighing Network, filters, amplifier, RMS detector, Output and Display.

One of the basic parameters of sound level meter is range of frequency. The sound intensity I has broad frequency range. The dispersion of the frequencies is from lower f_1 to higher f_2. The immediate value is indicated by I (f). For sound intensity is valid the relationship

$$I = \int_{f_1}^{f_2} I(f)df \tag{32}$$

Where $I(f) = \frac{\Delta I}{\Delta f}$ is intensity in the frequency interval f = 1 Hz.

Spectral intensity level (ISL) L_{Is} is defined

$$L_{Is} = 10log\left[\frac{I(f)}{I_{ref}}\right] \tag{33}$$

Where I_{ref} is the reference intensity levels (for air $\frac{10^{-12}W}{m^2}$).

$$L_I = L_{Is} + 10log(\Delta f) \tag{34}$$

Similarly, the sound pressure level L_p is related to the level of spectral noise L_{ps}as follows:

$$L_p = L_{ps} + 10log(\Delta f) \tag{35}$$

$$\Delta f = f_u - f_l \tag{36}$$

Where f_l and f_u are the lower and upper frequency to half power.

5. Fast Fourier Transformation (FFT)

Fast Fourier Transformation is one of the most common mathematical functions, which is used for noise analysis of electrical machines. The Fast Fourier Transformation is applied in an increasing scale in science, engineering, and technology. The use of complex exponentials has often been convenient rather than fundamental. Most signals and functions used in real applications are real rather than complex. In areas such as digital filtering, convolution, correlation, image processing, and partial differential equations, the actual signals or functions, are real, but they are considered to be the real part of a complex quantity in order to be able to use the complex formulation of Fourier series and transforms. The complex Fourier transform (CFT) of a signal$x(t) - \infty \leq t \leq -\infty$ with finite energy, is defined as

$$x_c(f) = \int_{-\infty}^{\infty} x(t).e^{-j2.\pi.f.t}dt \quad (37)$$

The inverse complex Fourier transform (ICFT) is given by

$$x(t) = \int_{-\infty}^{\infty} x_c(f).e^{j.2.\pi.f.t}dt \quad (38)$$

The real Fourier transform (RFT) of $x(t)$ can be defined as

$$x(f) = 2\int_{-\infty}^{\infty} x(t).cos(2.\pi.f.t + \Theta(f))dt \quad (39)$$

$$\text{Where: } \Theta(f) = \begin{cases} 0, & f \geq 0 \\ \frac{\pi}{2}, & f < 0 \end{cases}$$

The inverse real Fourier transform (IRFT) is given by

$$x(t) = \int_{-\infty}^{\infty} x(f).cos(2.\pi.f.t + \Theta(t))df \quad (40)$$

Equation (3) and (5) can be written for $f \geq 0$ as follows

$$x_1(f) = 2\int_{-\infty}^{\infty} x(t).cos(2.\pi.f.t)dt \quad (41)$$

$$x_0(f) = 2\int_{-\infty}^{\infty} x(t).sin(2.\pi.f.t)dt \quad (42)$$

And

$$x(t) = \int_{-\infty}^{\infty} [x_1(t).cos(2.\pi.f.t) + x_0(t).sin(2.\pi.f.t)]df \quad (43)$$

Thus $x(f)$ equals $x_1(f)$for$f \geq 0$, and $x_0(f)$ for$f < 0.x_1(f)$and$x_0(f)$ will be referred toas the cosine and the sine parts. The relationship between the CFT and the RFT can be expressed for $f \geq 0$ as$x_c(0) = x_1(0)$

$$\begin{bmatrix} x_c(f) \\ x_c(-f) \end{bmatrix} = \frac{1}{2}.\begin{bmatrix} 1 & -j \\ 1 & j \end{bmatrix}.\begin{bmatrix} x_1(f) \\ x_{10}(f) \end{bmatrix}, f \neq 0 \quad (44)$$

Equation (44) reflects the fact that $x_1(f)$ and $x_0(f)$are even and odd functions, respectively.

The inverse of (44) for $x_1(f)$ and $x_0(f)$ is

$$\begin{bmatrix} x_1(f) \\ x_0(-f) \end{bmatrix} = \begin{bmatrix} 1 & 1 \\ 1 & -j \end{bmatrix} \cdot \begin{bmatrix} x_c(f) \\ x_c(-f) \end{bmatrix} \tag{45}$$

Equations (44) and (45) are very useful to convert from one representation to the other. When $x(t)$is real, $x_1(f)$ and $x_0(f)$are also real. Then, (44) shows that $x_c(f)$ and $x_c(-f)$ are complex conjugates of each other. Equations (44) and (45) are also valid in the case of the discrete time Fourier transformation. In addition, they are valid for Fourier series and the discrete Fourier transforms with the replacement of fby the frequency index n. The RFT relations given by (43) can be proven by using (44), and writing (38) as

$$x(t) = \int_0^{\infty} \frac{1}{2} . [x_1(f) - jx_0(f)]. e^{j.2.\pi.f.t} df + \int_0^{-\infty} \frac{1}{2} . [x_1(-f) - jx_0(-f)]. e^{j.2.\pi.f.t} df \tag{46}$$

Then

$$x(t) = \int_0^{\infty} [x_1(f). cos(2.\pi.f.t) + x_0(f). sin(2.\pi.f.t)] df \tag{47}$$

6. Measurement noise of induction machines

6.1. Disturbed surroundings

Surrounding noise sources have an impact on the measurement of electrical machinery. It is not always possible to perform measurements in specialized laboratories, which are perfectly sound-insulated. To laboratory measurement can penetrate the noise from nearby sources (see Fig. 4), which is inaudible to the human ear. The interference from other sources can be created undesirable frequencies in the frequency band.

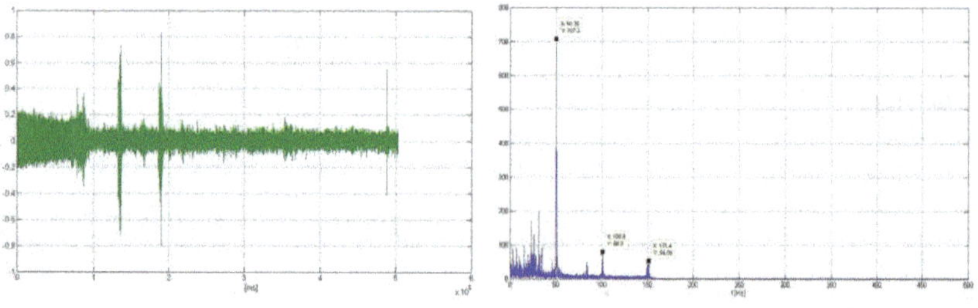

a) Noise detected in the laboratory b)Fast Fourier Transformation of laboratory noise

Figure 4. Noise measurement in the laboratory when the machine is switched off

Interference of other sources in the neighborhood of workplace cannot be directly prevented, but you can minimize their impact on analysis of the measured signal. Before the measurements it must be made measurement ambient noise before the main measurements. It is necessary to determine whether the background noise is random, or it is periodically repeated. In the case of random noise is preferable to wait to other time of measurement or it must count with errors in the measurement. In the event that can be measurement of noise repeated. Can be recorded the extent of spectral interference with which will be calculate when evaluating the measured results. From Spectral analyses of interference is possible to determine the proportion of individual harmonics. These harmonic then they can be the "subtracted" from the noise levels of electrical machines.

The next part of the measurement was performed on the induction motor which worked without a load. The electric motor was loosely placed on a foam board. This board was for suppression the transmission of vibrations from the surroundings. External vibrations are not desirable for accurate measurements.

Measurement noise of electric machine, that is run, is shown in Fig. 5. As seen from the measured values, that the noise level is constantly fluctuating.

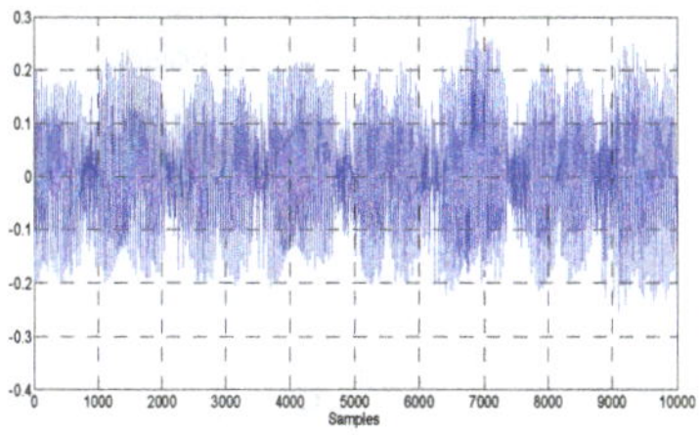

Figure 5. Noise of induction machines - no load

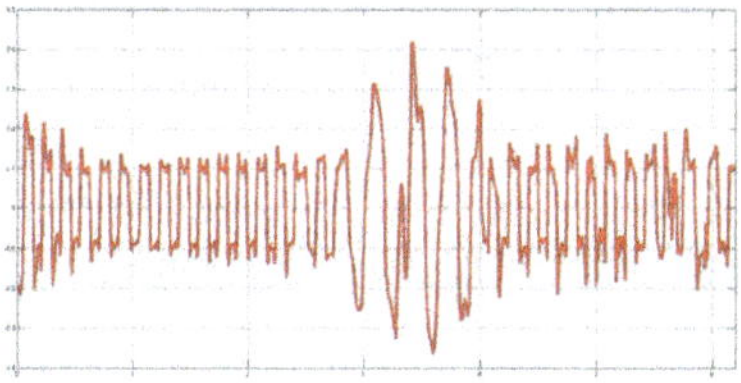

Figure 6. Noise of induction machine – 1 rotation

6.2. Noise of induction machine

On Fig. 7 is an analysis of the measured noise using MATLAB. Specifically, was carried Fast Fourier Transform (FFT). Dominant frequency is 600 Hz. This frequency is multiple of power supply frequency. It is a frequency of radial forces. In measurement signal can be involved many harmonics frequencies of radial forces. Than we can write equation

$$f_v = 6.k.f \tag{48}$$

Where

- f_v ... Frequency of radial force [Hz]
- f ... Power supply frequency [Hz]
- k ... Number (k=1, 2, 3, ...)

For $f = 50Hz$ are frequencies of radial forces $f_v = 300, 600, 900, ... Hz$.

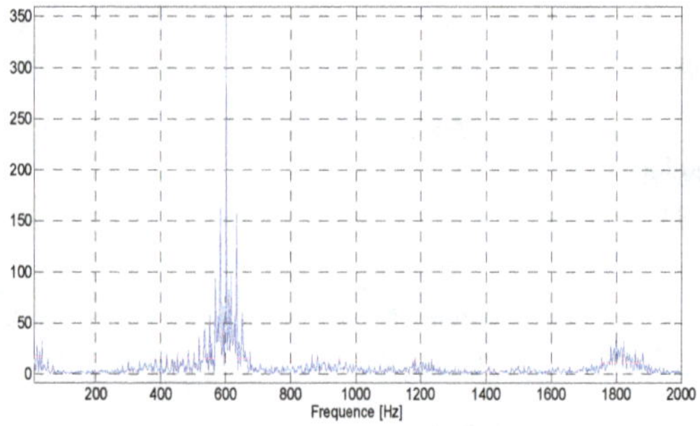

Figure 7. Fast Fourier Transformation of induction machine noise

It was done measurements eccentricity of rotor. Eccentricity of rotor is shown in Fig. 9. From the measured values it was found that the largest deviations occur in the range of approximately 120 degrees.

When comparing the noise of induction machines recorded on one rotation and values of rotor eccentricity can see a connection. In both cases (Fig. 8 and Fig. 9) appeared larger deflection in the range of 120 degrees. Extreme deviation is in a different quadrant in each graph. This is due to the different measurement principles. Noise measurements done digitally, while measuring the eccentricity was used mechanical method. It was therefore not possible to accurately determine the initial rotor position in both measurements.

it can be argued that the noise of induction machines is generated of the rotor who has eccentricity. Given that the, that machine is equipped with a ventilator, there are two sources of noise. The influence of the fan but will not cause displacement of only a specific part of one rotation.

Given that the measured induction motor was not equipped with cooling system (fan) can be assumed, that the vibration and thus the noise are produced only by electromagnetic source and mechanical source.

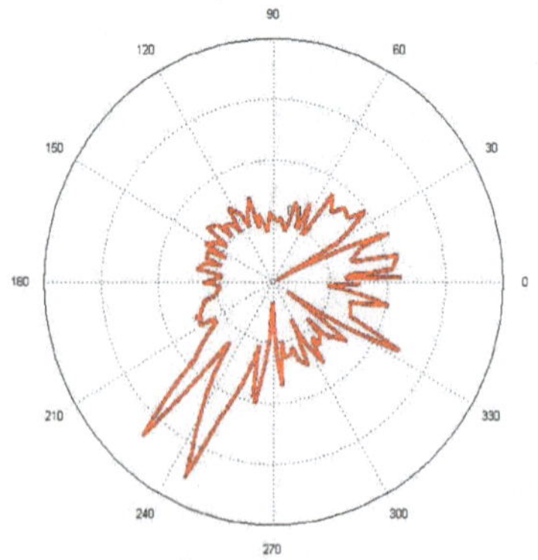

Figure 8. Noise envelope – 1 rotation

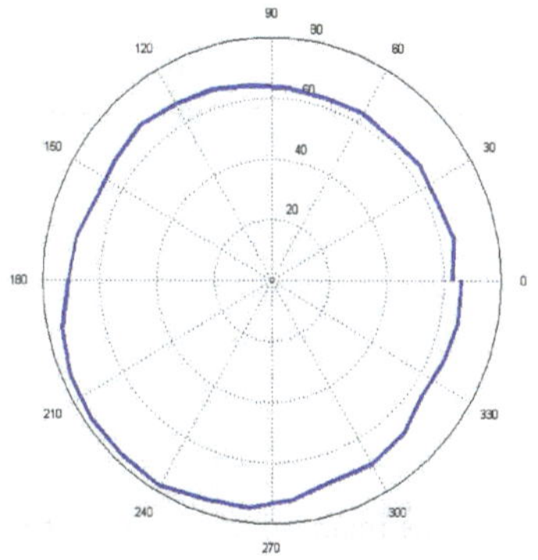

Figure 9. Rotor Eccentricity - Mechanical measurement

Analyze of noise was performed on the one rotation of rotor. The Fig 5 shows the noise levels depending on the position of the rotor. As the graph shows it is to generate greater levels of noise in the position of the rotor from 300 to 60 degrees (about 120 degrees).

7. Conclusion

Diagnosis of induction motors is a very complex issue that has many components. One of them is the analysis of motor noise. Noise measurement asynchronous machines are the commonly used diagnostic method. This method is relatively simple. You need to be near an electrical machine quality microphone and recording equipment. Analysis itself can be the performed on specialized software, either on the spot or later in laboratory.

Subsequent analysis of the signal can then indicate whether the machine operates as required, or whether there was damage to electrical equipment. Based on the fast Fourier analysis of noise can be determined which components of the signal are dominant. Based on knowledge of layout design of the engine is then possible to determine what is causing individual harmonics. According to the frequency it is possible to determine which there the main sources of noise are.

A major problem in measuring the noise may be interference from nearby sources. To avoid the external influence of external noise is possible only in specialized laboratories.

During measurements realized appeared possible link between noise and rotor eccentricity of electrical machinery. In the analysis of noise is dominant skew in the range of 120 ° in one rotation. In the same range (120 °) was measured the dominant deflection of rotor eccentricity this rotor machine. Given that the machine has not a cooling system, there is not source of aerodynamic noise; there are only two possible causes of this deviation. Source of electromagnetic noise would not cause deviation only at certain rotor position, but in the whole rotation. Displacement of noise in a certain position the rotor it cannot assign too resources source of mechanical noise. This group includes vibration bearings. During the measurement was verified that the bearings are not damaged. There are not larger deviations of movement in rotation of bearing.

As a source of noise is impact of rotor eccentricity on the running of the induction motor. Unfortunately, the verification of this theory would require accurate measurement with recording of the rotor position and size of air gap. This measurement is very difficult.

Author details

Marcel Janda, Ondrej Vitek and Vitezslav Hajek
Brno University of Technology, Czech Republic

Acknowledgement

Research described in this paper was financed by the Ministry of Education of the Czech Republic, under project FR-TI3/073 Research and development of small electric machines; and the project of the Grant agency CR No. 102/09/1875 - – Analysis and Modeling of Low Voltage Electric Machines Parameters. The work was supported by Centre for Research and utilization of renewable energy - CZ.1.05/2.1.00/01.0014

8. References

Vijayraghavan, P.; Krishnan, R.; "Noise in electric machines: a review," Industry Applications Conference, 1998. Thirty-Third IAS Annual Meeting. The 1998 IEEE ,vol.1, no., pp.251-258 vol.1, 12-15 Oct 1998 doi: 10.1109/IAS.1998.732298

Ellison, A.J.; Yang, S.J.; "Effects of rotor eccentricity on acoustic noise from induction machines," Electrical Engineers, Proceedings of the Institution of, vol.118, no.1, pp.174-184, January 1971doi: 10.1049/piee.1971.0028

Hamata, Václav. Hluk elektrických strojů.Praha : Academia, 1987. 176 s.

Mišun, Vojtěch. Vibrace a hluk. 2. Brno :CERM, s.r.o., 2005. 176 s. ISBN 80-216-3060-5.

Gieras, Jacek F.; Wang, Chong; Cho Lai, Joseph. Noise of Polyphase Electric Motors. [s.l.] :CRC Press, 2005. 392 s. ISBN 978-0824723811.

Srinivas, K.N.; Arumugam, R.; "Analysis and characterization of switched reluctance motors: Part II. Flow, thermal, and vibration analyses", IEEE Transactions on Magnetics, vol.41, no.4, pp. 1321- 1332, April 2005 doi: 10.1109/TMAG.2004.843349. ISSN: 0018-9464

Srinivas, K.N.; Arumugam, R.; "Static and dynamic vibration analyses ofs witched reluctance motors including bearings, housing, rotor dynamics, and applied loads", IEEE Transactions on Magnetics, vol.40, no.4, pp. 1911- 1919, July 2004 doi: 10.1109/TMAG.2004.828034. ISSN: 0018-9464.

Ersoy, O.K.; "A comparative eview of real and complex Fourier-related transforms" Proceedings of the IEEE , vol.82, no.3, pp.429-447, Mar 1994 doi: 10.1109/5.272147. ISSN: 0018-9219

Chapter 9

Induction Motors with Rotor Helical Motion

Ebrahim Amiri and Ernest Mendrela

Additional information is available at the end of the chapter

1. Introduction

A demand for sophisticated motion control is steadily increasing in several advanced application fields, such as robotics, tooling machines, pick-and-place systems, etc. These kinds of applications require implementation of at least two or more conventional motors/actuators, often operating with different type of mechanical gear. Electric motors/actuators that are able directly perform complex motion (with multiple degrees of mechanical freedom – multi-DoMF) may provide appreciable benefits in terms of performances, volume, weight and cost.

This chapter is organized as follows. Section 2 provides a brief overview of the main typologies of induction motors with two degrees of mechanical freedom (IM-2DoMF) structure. Section 3 introduces the mathematical model for helical-motion induction motors. Section 4 discusses the phenomenon known end effect caused by finite length of the armature and its negative influences on the motor performance. Section 5 presents a construction of a twin-armature rotary-linear induction motor with solid double layer rotor, its design data and the performance prediction of the motor. The results obtained from FEM modeling are then verified by the test carried out on experimental model of the motor what validates the theoretical modeling of the motor.

2. Topologies of induction motors with two degrees of mechanical freedom

Several topologies of electromagnetic motors featuring a multi-DoMF structure were investigated in the technical literatures (Mendrela et al., 2003, Krebs, et al., 2008). Considering the geometry, three classes of motors can be distinguished:

- X-Y motors – flat structure
- Rotary-linear motor – cylindrical geometry
- Spherical motors – spherical geometry

2.1. X-Y motors

X-Y motors, also called planar motors, are the machines which are able to translate on a plane, moving in the direction defined by two space co-ordinates. They may be usefully employed for precision positioning in various manufacturing systems such as drawing devices or drive at switch point of guided road/e.g. railway. The representative of X-Y motors is shown schematically in Fig. 1. Primary winding consist of two sets of three phase windings placed perpendicularly to one another. Therefore, magnetic traveling fields produced by each winding are moving perpendicularly to one another as well. Secondary part can be made of non-magnetic conducting sheet (aluminum, copper) backed by an iron plate. The motor with a rotor rectangular grid-cage winding is another version that can be considered.

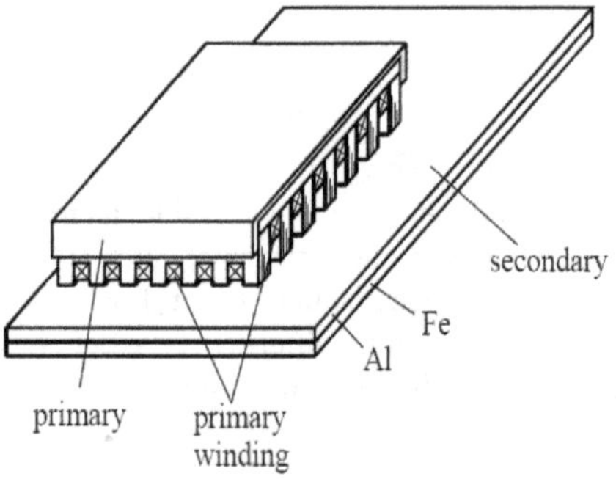

Figure 1. Construction scheme of X-Y induction motor (Mendrela et al., 2003).

The forces produced by each of traveling fields can be independently controlled contributing to the control of both magnitude and direction of the resultant force. This in turn controls the motion direction of the X-Y motor.

2.2. Rotary-linear motors

Mechanical devices with multiple degrees of freedom are widely utilized in industrial machinery such as boring machines, grinders, threading, screwing, mounting, etc. Among these machines those which evolve linear and rotary motion, independently or simultaneously, are of great interest. These motors, which are able effectively generate torque and axial force in a suitably controllable way, are capable of producing pure rotary motion, pure linear motion or helical motion and constitute one of the most interesting topologies of multi-degree-of freedom machines (Bolognesi et al., 2004). Some examples of such actuators have already been the subject of studies or patents (Mendrela et al., 2003, Giancarlo & Tellini, 2003, Anorad, 2001). A typical rotary-linear motor with twin-armature is shown in Fig. 2. A stator consists of two armatures; one generates a rotating magnetic field, another traveling magnetic field. A solid rotor, common for the two armatures is applied. The rotor consists of an iron cylinder covered with a thin copper layer. The rotor cage winding that looks like grid placed on cylindrical surface is another version that can be applied. The direction of the rotor motion depends on two forces: linear (axially oriented) and rotary, which are the products of two magnetic fields and currents induced in the rotor. By controlling the supply voltages of two armatures independently, the motor can either rotate or move axially or can perform a helical motion.

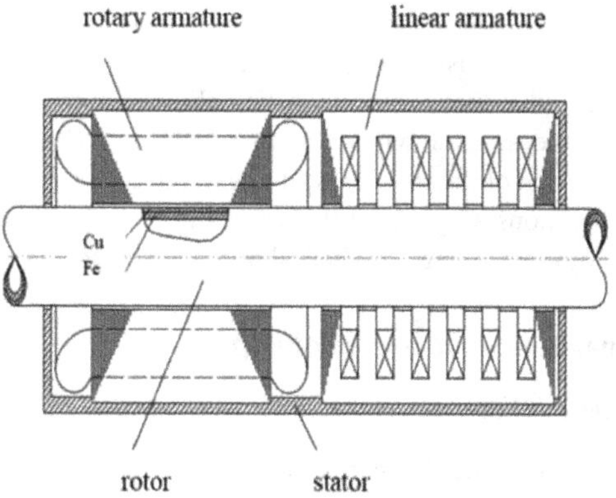

Figure 2. Scheme of twin armature rotary-linear induction motor (Mendrela et al., 2003).

2.3. Spherical motors

The last class of multi-DoMF motors has spherical structure. The rotor is able to turn around axis, which can change its position during the operation. Presently, such actuators are mainly proposed for pointing of micro-cameras and laser beams, in robotic, artificial vision, alignment and sensing applications (Bolognesi et al., 2004). In larger sizes, they may be also used as active wrist joints for robotic arms. Fig. 3 shows one of the designs in which the rotor driven by two magnetic fields generated by two armatures moving into two directions perpendicular to one another. This design is a counterpart of twin-armature rotary-linear motor.

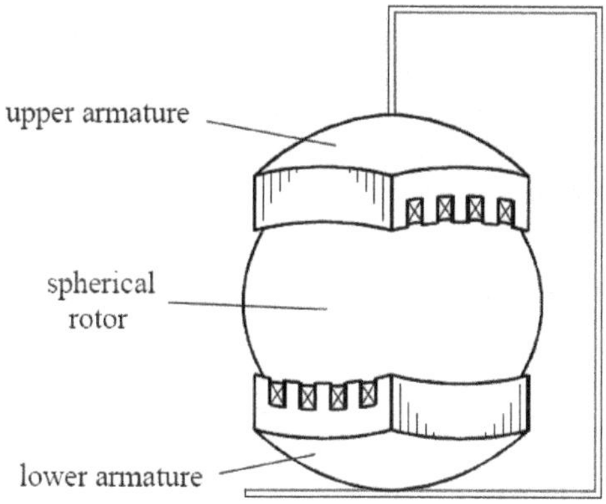

Figure 3. Construction scheme of twin-armature induction motor with spherical rotor (Mendrela et al., 2003).

3. Mathematical model of induction motor with magnetic field moving helically

The basic and most comprehensive research on IM-2DoMF is contained in the book (Mendrela et al., 2003). The analysis of these motors is based on theory of the induction motors whose magnetic field is moving in the direction determined by two space coordinates. According to this theory the magnetic field of any type of motor with 2DoMF can be represented by the sum of two or more rotating-traveling field what allows to consider the complex motion of the rotor as well as end effects caused by the finite length of stator. In the next subsections a sketch of theory of the motor with the rotating-traveling magnetic field whose rotor is moving with helical motion is presented.

3.1. Definition of magnetic field and rotor slip

3.1.1. Magnetic field description

The magnetic field moving helically in the air-gap is represented by the magnetic flux density *B* wave moving in the direction placed between two co-ordinates *z* and θ (Fig. 4).

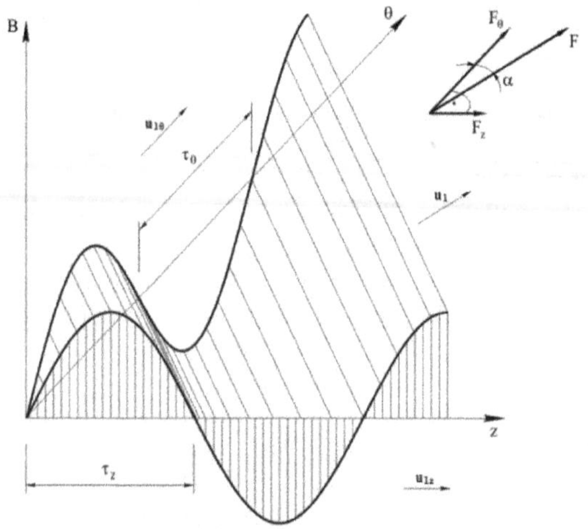

Figure 4. Magnetic field wave moving into direction between two spaces coordinates (Mendrela et al., 2003).

It can be expressed by the following formula (Mendrela et al., 2003):

$$B = B_m \exp\left[j\left(\omega t - \frac{\pi}{\tau_\theta}\theta - \frac{\pi}{\tau_z}z \right) \right] \tag{1}$$

where B_m - amplitude of travelling wave of magnetic flux density, ω - supply pulsation, τ_θ and τ_z - pole-pitch length along θ and z axes.

The physical model of the motor which could generate such a field is shown in Fig. 5.

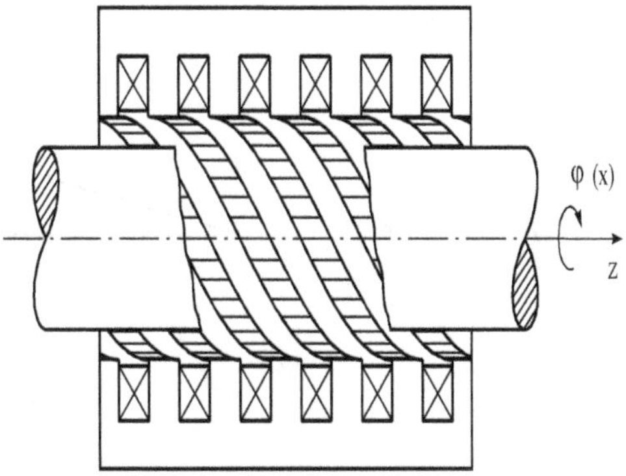

Figure 5. Rotary-linear induction motor with rotating-traveling magnetic field (Mendrela et al., 2003).

The electromagnetic force that exerts on the rotor is perpendicular to the wave front and can be divided into two components: F_z – linear force, F_θ - rotary component (Fig. 4). The relationships between force components are:

$$\frac{F_\theta}{F_z} = ctg\,\alpha\ , \qquad F_\theta = F\cos\alpha \tag{2}$$

where,

$$ctg\alpha = \frac{\tau_z}{\tau_\theta} \tag{3}$$

3.1.2. Rotor slip

To derive a formula for the rotor slip the motor is first considered to operate at asynchronous speed. Meaning, an observer standing on the rotor surface feels a time variant magnetic field. Therefore, magnetic field for a given point $P(\theta_1 z_1)$ (Fig. 6) on the rotor surface is varying in time and expressed by the following equation:

$$B(t,\theta_1,z_1) = B_m \exp\left[j\left(\omega t - \frac{\pi}{\tau_\theta}\theta_1 - \frac{\pi}{\tau_z}z_1\right)\right] = var \tag{4}$$

Eqn. (4) is true if:

$$\omega t - \frac{\pi}{\tau_\theta}\theta_1(t) - \frac{\pi}{\tau_z}z_1(t) = \psi(t) \tag{5}$$

where $\psi(t)$ is the angle between point P and the wave front of the magnetic field wave. Differentiating Eqn. (5) with respect to time t yields:

$$\omega - \frac{\pi}{\tau_\theta}\omega_\theta - \frac{\pi}{\tau_z}u_z = \omega_2 \tag{6}$$

where ω_2 (slip speed) is the angular speed of point P with respect to the stator field and ω_θ and u_z are the angular and linear speeds of the rotor, respectively.

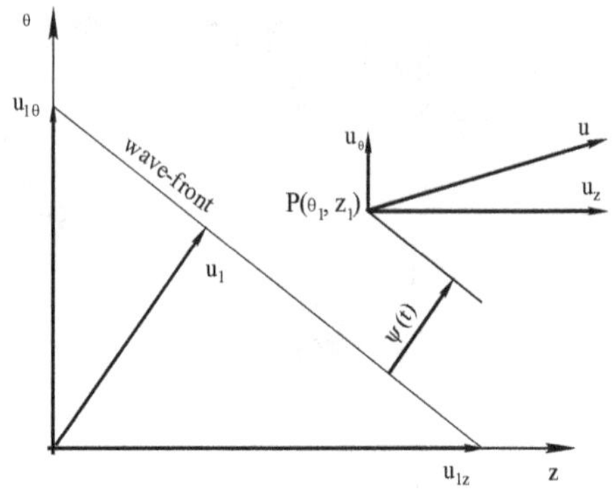

Figure 6. Rotary-linear slip derivation (Mendrela et al., 2003).

The, field velocities along θ and z axes are expressed by the equations:

$$\omega_{1\theta} = 2\tau_\theta f \ , \qquad u_{1z} = 2\tau_z f \tag{7}$$

Inserting (7) to (6), it takes the form:

$$\omega - \frac{\omega}{\omega_{1\theta}}\omega_\theta - \frac{\omega}{u_{1z}}u_z = \omega_2 \tag{8}$$

Similarly, as in the theory of conventional induction motors, it can be written:

$$\omega_2 = \omega s \tag{9}$$

From (8) and (9) the following equation for the rotor slip is finally derived:

$$s_{\theta z} = 1 - \frac{\omega_\theta}{\omega_{1\theta}} - \frac{u_z}{u_{1z}} \tag{10}$$

The two dimensional rotor slip obtained in Eqn (10) is a function of rotary and linear rotor speed components as well as the speeds of the magnetic field moving along two space coordinates. If the motion of rotor is blocked along one of the coordinates, this slip takes the form known for motors with one degree of freedom. For example: if the rotor is blocked in the axial direction, the u_z component drops to zero and the slip takes the form:

$$s_\theta = 1 - \frac{\omega_\theta}{\omega_{1\theta}} \tag{11}$$

which is the form of rotor slip in the theory of conventional induction motor.

3.2. Motor equivalent circuit

For the induction motor with rotating-travelling field the equivalent circuit is shown in Fig. 7, which corresponds to the well-known circuit of rotary induction motor. The only difference is in the rotor slip, which for rotary-linear motor depends on both rotary and linear rotor speeds.

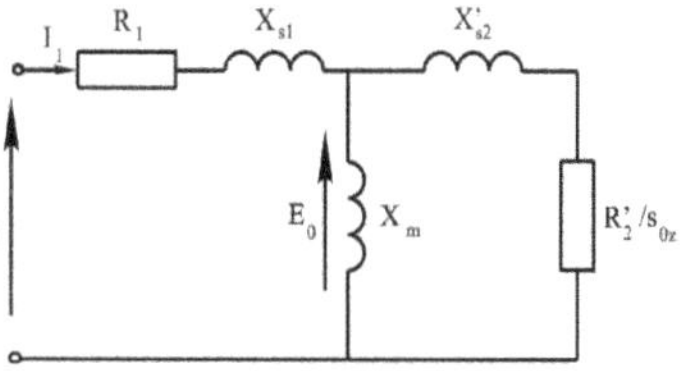

Figure 7. Equivalent circuit of induction motor with rotating-travelling magnetic field.

Similarly to the conventional rotary motors, the secondary resistance can be split into two resistances as shown in Fig. 8. R_2' represents the power loss in the rotor windings and the second part ($R_2' \frac{1-s_{\theta z}}{s_{\theta z}}$) contributes to mechanical power P_m.

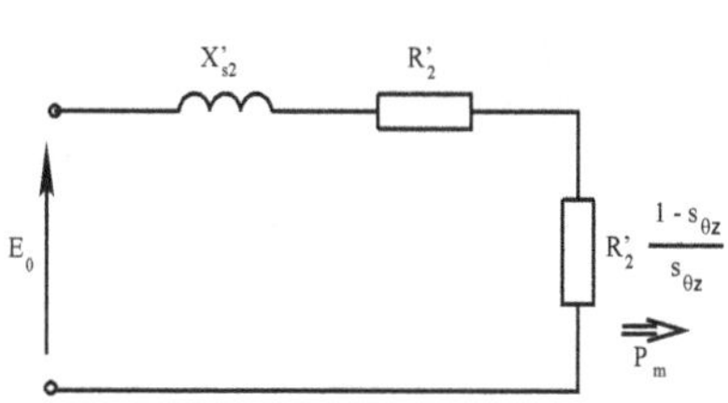

Figure 8. Equivalent circuit of rotor of rotary-linear induction motor.

The mechanical power of the resultant motion between θ and z axis is proportional to the resistance $R_2' \frac{1-s_{\theta z}}{s_{\theta z}}$ and is equal to:

$$P_m = mR_2' \frac{1-s_{\theta z}}{s_{\theta z}} I_2'^2 \tag{12}$$

where m is the number of phases.

Inserting Eqn (10) into Eqn (12), it takes the form:

$$P_m = mI_2'^2 \frac{R_2'}{s_{\theta z}} \left(\frac{\omega_\theta}{\omega_{1\theta}} + \frac{u_z}{u_{1z}} \right) \tag{13}$$

The resultant mechanical power P_m can be expressed in the form of its component in θ and z direction:

$$P_m = P_{m\theta} + P_{mz} \tag{14}$$

From (11), (13) and (14):

$$P_{m\theta} + P_{mz} = mR_2' \frac{1-s_\theta}{s_{\theta z}} I_2'^2 + mR_2' \frac{1-s_z}{s_{\theta z}} I_2'^2 \tag{15}$$

Therefore, Fig. 8 can be redrawn in terms of the resistance split into two components as shown in Fig. 9.

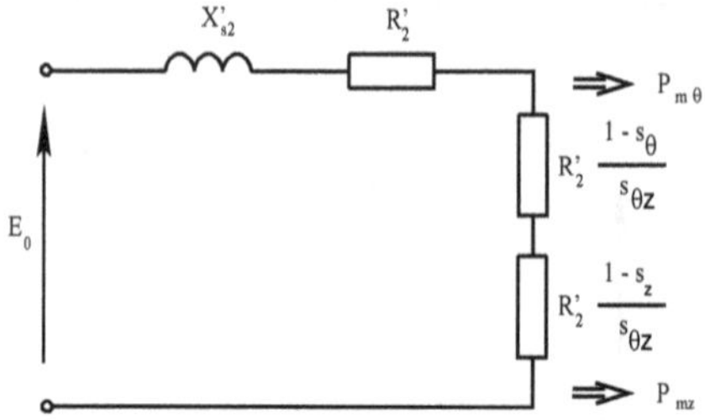

Figure 9. Equivalent circuit of rotor of rotary-linear induction motor with mechanical resistance split into two components.

3.3. Electromechanical characteristics

Unlike conventional rotary motors with the curvy characteristics of electromechanical quantities versus slip, electromechanical quantities in rotary-linear motor cannot be interpreted in one dimensional shape and should be plotted in a surface profile as a function of either slip $s_{\theta z}$ or speed components u_θ and u_z. The circumferential speed u_θ is expressed as follows:

$$u_\theta = R_r . \omega_\theta \tag{16}$$

where R_r is the rotor radius.

As an example, the force-slip characteristic of a typical rotary-linear motor is plotted in Fig. 10.

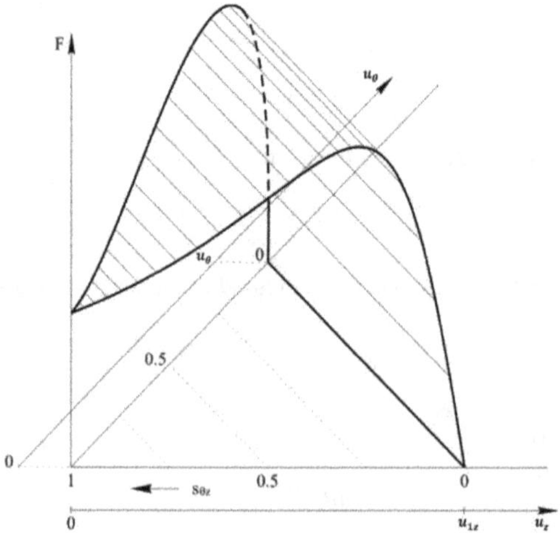

Figure 10. Fig. 10. Electromechanical characteristic of the induction motor with a rotating-travelling field (Mendrela et al., 2003).

In order to determine the operating point of the machine set, let the rotor be loaded by two machines acting independently on linear (axial) and rotational directions with the load force characteristics shown in Fig. 11.

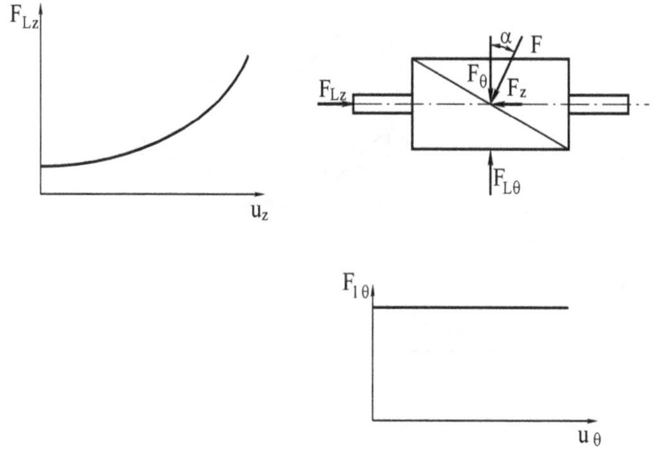

Figure 11. Load characteristics for IM-2DoMF, F_{lz}: load force in axial direction, $F_{l\theta}$: load force in rotary direction.

The equilibrium of the machine set takes place when the resultant load force is equal in its absolute value and opposite to the force developed by the motor. The direction of the electromagnetic force F of the motor is constant and does not depend on the load. Thus, at steady state operation both load forces $F_{l\theta}$ and F_{lz} acts against motor force components F_θ and F_z in the same direction if the following relation between them takes place:

$$\frac{F_\theta}{F_z} = \frac{F_{l\theta}}{F_{lz}} = ctg\alpha = \frac{\tau_z}{\tau_\theta} \tag{17}$$

To draw both load characteristics on a common graph, the real load forces $F_{l\theta}$ and F_{lz} acting separately on rotational and linear directions should be transformed into $F'_{l\theta}$ and F'_{lz}, forces acting on the direction of the motor force *F*. These equivalent forces are:

$$F'_{l\theta} = \frac{F_{l\theta}}{\cos\alpha} \quad , \quad F'_{lz} = \frac{F_{lz}}{\sin\alpha} \tag{18}$$

The transformed load characteristics drawn as a function of u_θ and u_z, as shown in Fig. 12, are the surfaces intersecting one another along the line segment $\overline{KA}$. This line segment that forms the $F_l(u)$ characteristic is a set of points where the following equation is fulfilled:

$$F'_{l\theta} = F'_{lz} \tag{19}$$

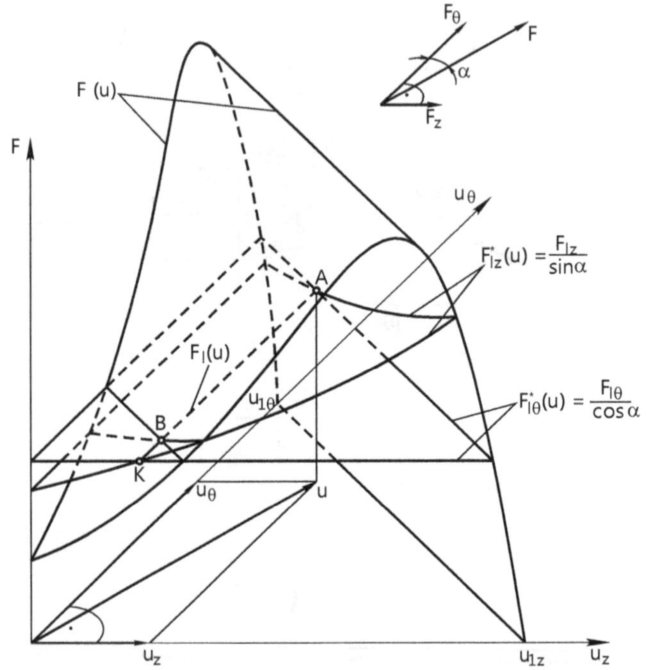

Figure 12. Determination of the operating point A of the machine set with IM-2DoMF (Mendrela et al., 2003).

The load characteristic $F_l(u)$ intersects with the motor characteristic at points A and B, where the equilibrium of the whole machine set takes place. To check if the two points are stable the steady state stability criterion can be used, which is applied to rotary motors in the following form:

$$\frac{dF}{du} < \frac{dF_l(u_\theta, u_z)}{du} \tag{20}$$

$$\frac{dF_{l\theta}}{du_\theta} > 0, \quad \frac{dF_{lz}}{du_z} > 0 \tag{21}$$

Applying this criterion, point A in Fig. 12 is stable and point B is unstable.

3.4. Conversion of mathematical model of IM-2DoMF into one of IM-1DoMF

The mathematical model of IM-2DoMF presented in previous subsections is more general than the one for linear or rotating machines. The rotating magnetic field wave of rotating machines and the traveling field of linear motors are in the mathematical description special cases of the rotating–traveling field. If the wave length remains steady, pole pitches along both axis (τ_θ and τ_z) will vary by changing the motion direction of field waves. For example: if the wave front (see Fig. 4) turns to θ axis, then $\tau_z = \infty$. This makes the formula (1) changes to:

$$B = B_m \exp\left[j\left(\omega t - \frac{\pi}{\tau_\theta}\theta \right) \right] \tag{22}$$

which is the flux density function for rotary motor. The $\alpha = 0$ and according to Eqn (2) the force $F = F_\theta$ what is the case for rotary motors. On the other hand, turning the wave completely toward z axis leads to infinity pole pitch value along θ axis ($\tau_\theta = \infty$). By inserting this into Eqn (1) and (10) the description of both field and slip expressed by two space coordinates turns to the description of such quantities in linear motors.

In other word, the mathematical model of the rotary-linear motor is a general form of conventional, one dimensional motors and can be reduced at any time to the model either of rotary or linear motors.

4. Edge effects in rotary-linear induction motors

The twin-armature rotary-linear induction motor, which is the object of this chapter consists of two armatures what makes this machine a combination of two motors: rotary and tubular linear, whose rotor are coupled together. This implies that the phenomena that take place in each set of one-degree of mechanical freedom motors also occur in the twin armature rotary-linear motor in perhaps more complex form due to the complex motion of the rotor. One of these phenomena is called end effects and occurs due to finite length of the stator at rotor axial motion. This phenomenon is not present in conventional rotating induction machines, but play significant role in linear motors.

These effects are the object of study of many papers (Yamamura, 1972, Greppe et all, 2008, Faiz & Jafari, 2000, Turowski, 1982, Gierczak & Mendrela, 1985, Mosebach et all, 1977, Poloujadoff et all, 1980). In the literature, end effects are taken into account in various ways. In the circuit theory a particular parameter can be separated from the rest of equivalent circuit elements, and it represents the only phenomena that are caused by finite length of primary part of linear motor. This approach has been done in (Pai et all, 1988, Gieras et all, 1987, Hirasa et all, 1980, Duncan & Eng, 1983, Mirsalim et all, 2002). Kwon et al, solved a linear motor (LIM) with the help of the FEM, and they suggested a thrust correction coefficient to model the end effects (Kwon et all., 1999). Fujii and Harada in (Fujii & Harada, 2000) modeled a rotating magnet at the entering end of the LIM as a compensator and reported that this reduced end effect and thrust was the same as a LIM having no end effects. They used FEM in their calculations. Another application of FEM in analysing LIMs is reported by (Kim & Kwon, 2006). A d-q axis equivalent model for dynamic simulation purposes is obtained by using nonlinear transient finite element analysis and dynamic end effects are obtained.

The end effect has been also included in the analysis of rotary-linear motors in the literature (Mendrela et al., 2003, Krebs et all, 2008, Amiri et all, 2011). This inclusion was done by applying Fourier's harmonic method when solving the Maxwell's equations that describe motor mathematically (Mendrela et al., 2003). This approach was also applied to study the linear motor end effects (Mosebach et all, 1977, Poloujadoff et all, 1980).

The edge effects phenomena caused by finite length of both armatures can be classified into two categories as follows:

- End effects: which occurs in the tubular part of the motor.
- Transverse edge effects: which exists in the rotary part.

4.1. End effects

One obvious difference between LIM and conventional rotary machines is the fact that in LIM the magnetic traveling field occurs at one end and disappears at another. This generates the phenomena called end effects. End effects can be categorized into two smaller groups called: static end effects and dynamic end effects.

4.1.1. Static end effects

This is the phenomenon which refers to the generation of alternating magnetic field in addition to the magnetic traveling field component. The process of generation of alternating magnetic field at different instances is shown in Fig. 13.

Instant $t_1 = 0$: The 3-phase currents are of the values shown by phasor diagram in Fig. 13.b and 13.c with maximum in phase A. The current distribution in the primary winding relevant to these values is shown in Fig. 13.a and its first space harmonic is represented by curve J in Fig. 13.b. The distribution of magneto-motive force F_m in the air-gap corresponding to this linear current density has the cosine form with the maximum value at both edges of the primary part shown in Fig. 13.b. This mmf generates the magnetic flux which consists of two components: alternating flux Φ_a shown in Fig. 13.a and traveling flux component. The distribution of these two components B_a and B_t as well as the resultant flux density ($B_a + B_t$) are shown in Fig. 13.c.

Instant $t_2 = ¼ T$ (where T is sine wave period): The 3-phase currents are of value shown by phasor diagram in Fig. 13.d with zero in phase A. These currents make the distribution of the first harmonic of mmf F_m as shown in Fig. 13.d. Since there is no mmf at primary edges, the alternating flux component Φ_a does not occur and the traveling flux B_t is the only available component.

Instant $t_3 = ½ T$: After a half period, the currents are of values shown by phasor diagram in Fig. 13.e with the maximum negative value in phase A. The relevant mmf distribution reveals its maximum negative value at primary edges which generates the magnetic flux Φ_a represented by its flux density B_a (see Fig. 13.e) which adds to the traveling flux component B_t.

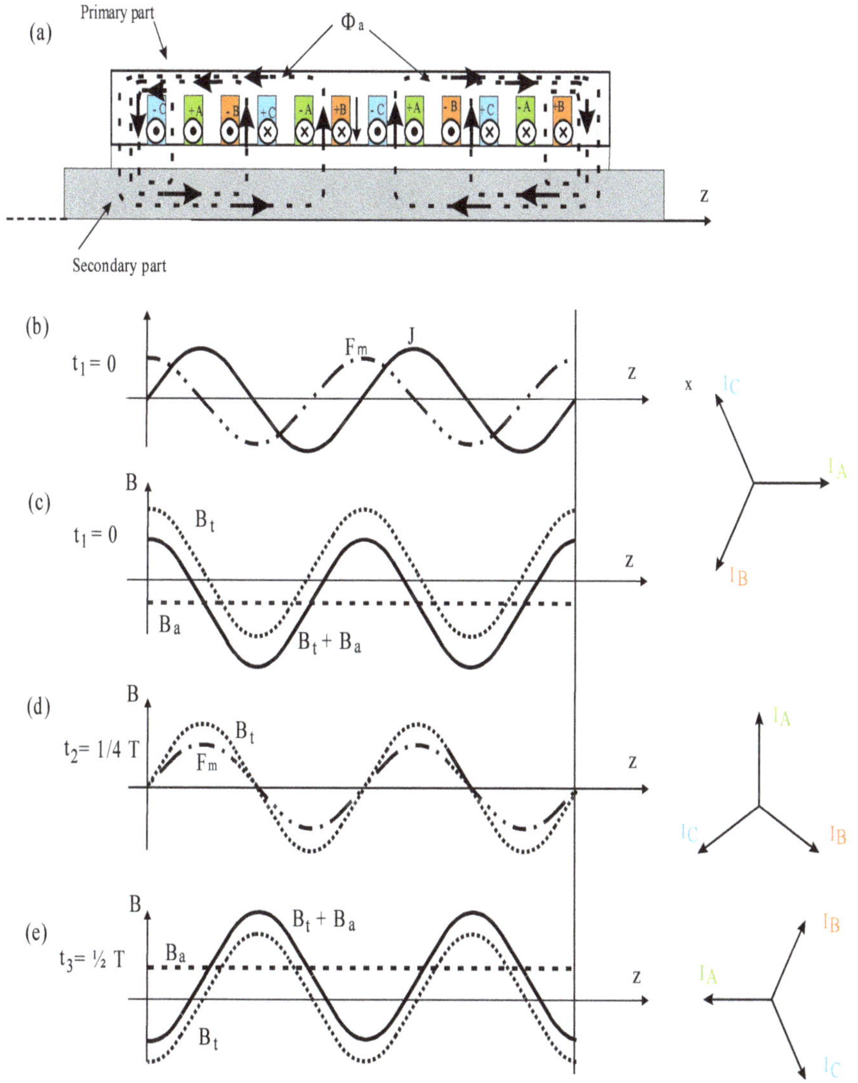

Figure 13. The process of generation of alternating magnetic field at different instances in 4-pole tubular motor: (Φ_a) – alternating component of magnetic flux, (B_a) – alternating component of magnetic flux density in the air-gap, (B_t) - travelling component of magnetic flux density in the air-gap, (J) – linear current density of the primary part, (F_m) – magneto-motive force of primary part in the air-gap.

Analysing the above phenomenon in time, one may find that magnetic flux density has two components: B_a which does not move in space but changes periodically in time called alternating component and B_t which changes in time and space is called traveling component. The first component B_a does not exist in motors with infinity long primary part, which is the case in conventional rotary machines.

Summarizing, the resultant magnetic flux density distribution is a combination of the traveling wave component $B_t(t)$ and alternating magnetic field $B_a(t)$ denoted by:

$$B(t,z) = B_t \cos\left(\omega t - \frac{\pi}{\tau_p} z + \delta_m\right) + B_a \sin\left(\omega t - \frac{\pi}{\tau_p} + \delta_a\right) \quad (23)$$

where τ_p is pole pitch and δ_m is phase angle.

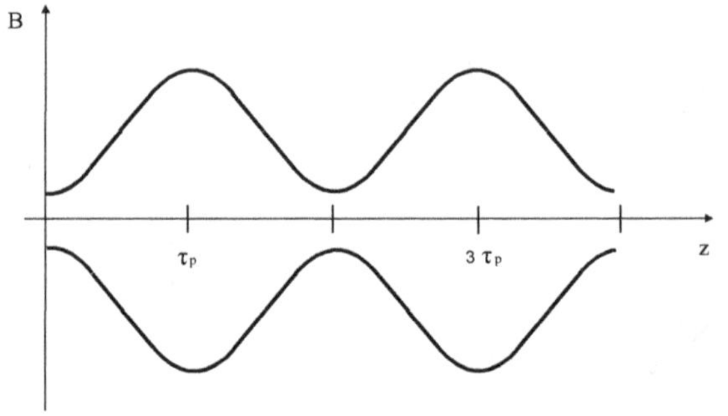

Figure 14. The envelope of the resultant magnetic flux density in the air-gap of four pole linear motor due to presence of the alternating magnetic field (τ_p - pole pitch).

When only the travelling wave exists, the envelop of flux density distribution in the air gap is uniform over the entire length of the primary core but the second term deforms the air gap field distribution to the shape shown in Fig. 14. The alternating flux contributes to the rising of additional power losses in the secondary and to producing of braking force when one part of LIM motor is moving with respect to the other one (Mosebach et all, 1977, Poloujadoff, et all, 1980, Amiri, et all, 2011). This component occurs no matter what is the value of the speed of the secondary part (Poloujadoff, 1980). The envelope of the resultant magnetic flux density for the four-pole motor is no longer uniform as shown in Fig. 14.

4.1.2. Dynamic end effects

The dynamic end effects are the entry and exit effects that occur when the secondary moves with respect to the primary part. This phenomenon will be explained in two stages:

Stage I: secondary part moves with synchronous speed

There are no currents induced in the rotor (within the primary part range) due to traveling magnetic field component (since the secondary moves synchronously with travelling field). However, the observer standing on the secondary (see Fig. 15) feels relatively high change of magnetic flux when enters the primary part region and when leaves this region at exit edge. This change contributes to rising of the eddy currents at both the entry and exit edges. These currents damp the magnetic field in the air-gap at entry in order to keep zero flux linkage for the secondary circuit. At the exit edge the secondary eddy currents tries to sustain the magnetic flux linkage outside the primary zone the same as it was before the exit. This leads to damping magnetic flux at the entry edge and to appearance of magnetic flux tail beyond the exit edge (Fig. 16.a). The distribution of the primary current J_1 is uniform over the entire region. The envelop of the eddy currents induced in the secondary J_2 shown in Fig. 16.a is relevant to the magnetic flux density distribution in the air-gap.

The eddy currents at the entry and exit edges attenuate due to the fact that the magnetic energy linked with these currents dissipate in the secondary resistance. Thus, the lower is the secondary resistance the more intensive is damping at entry and the longer is the tail beyond the exit.

Stage II: rotor (secondary part) moves with a speed less than the synchronous speed

The currents are induced in the secondary over the entire primary length due to slip of the secondary with respect to the travelling component of primary magnetic flux. These currents superimpose the currents that are due to the entry and exit edges. The resultant eddy current envelop is shown schematically in Fig. 16.b. The flux density distribution in the air-gap and current density in the primary windings are also shown in Fig. 16.b. As it is illustrated the primary current density is uniformly distributed along the primary length only if the coils of each phase are connected in series and the symmetry of 3-phase currents is not affected by the end effects. The magnetic flux density distribution has the same shape and changes in a same pattern in both stages, but due to the rotor current reaction, the second stage has a lower magnetic flux density (B). However, primary current density is higher at the second stage if the primary winding is supplied in these two cases with the same voltage.

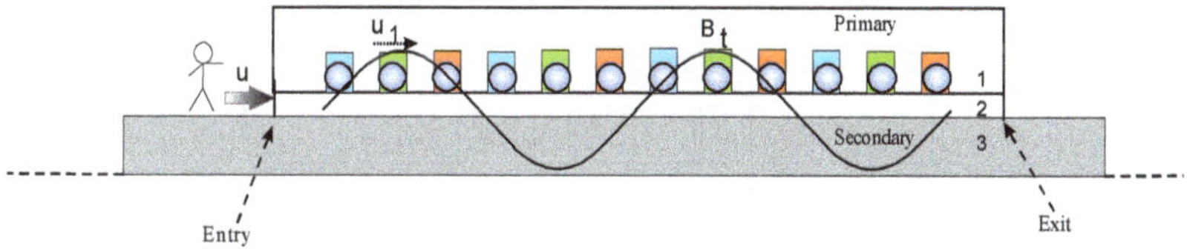

Figure 15. End effect explanation: (B_t) - travelling component of magnetic flux density in the air-gap (u_1) speed of traveling magnetic field (u) speed of the rotor.

In general, end effect phenomena leads to non-uniform distribution of:

- magnetic field in the air-gap,
- current in the secondary,
- driving force density,
- power loss density in the secondary.

Thus, this contributes to:

- lower driving force,
- higher power losses,

- lower motor efficiency,
- lower power factor.

Due to dynamic end effects, the resultant magnetic flux density in the air-gap can be expressed as a summation of three flux density components as follows (Greppe, et all, 2008):

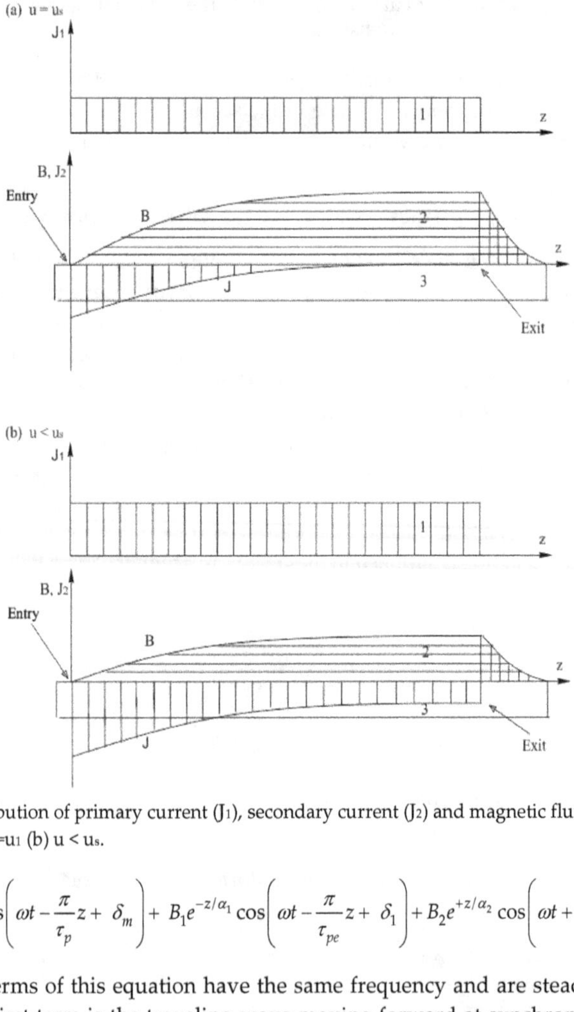

Figure 16. Distribution of primary current (J_1), secondary current (J_2) and magnetic flux density in the air-gap (B): (a) u =u_1 (b) u < u_s.

$$B(t,z) = B_t \cos\left(\omega t - \frac{\pi}{\tau_p} z + \delta_m\right) + B_1 e^{-z/\alpha_1} \cos\left(\omega t - \frac{\pi}{\tau_{pe}} z + \delta_1\right) + B_2 e^{+z/\alpha_2} \cos\left(\omega t + \frac{\pi}{\tau_{pe}} z + \delta_2\right) \quad (24)$$

All the three terms of this equation have the same frequency and are steady with respect to time t. The first term is the traveling wave moving forward at synchronous speed. The second term is an attenuating traveling wave generated at the entry end, which travels in the positive direction of z and whose attenuation constant is $1/\alpha_1$ and its half-wave length is τ_{pe}. The third term of Eqn (24) is an attenuating traveling wave generated at the exit end, which travels in the negative direction and whose attenuation constant is $1/\alpha_2$ and half-wave length is τ_{pe}. The B_1 wave is caused by the core discontinuity at the entry end and the B_2 wave is caused by the core discontinuity at the exit end, hence, both are called end effect waves. Both waves have an angular frequency ω, which is the same as that of power supply. They have the same half wave-length, which is different from half-wave length (equal to pole pitch) of the primary winding. The traveling speed of the end waves is given by $v_e = 2 f \tau_{pe}$ and is the same as the secondary speed if high speed motors is studied. However, in low speed motors, the speed of the end waves can be much higher than that of secondary (Yamamura, 1972). The length of penetration of entry end wave α_1 depends on motor parameters such as gap length and secondary surface resistivity. The impact of these parameters on α_1 are quiet different at high speed motors and low speed motors. As a result, α_1 is much longer at high speed motors with respect to low speed motors. Also, in the high-speed motors, half wave length τ_{pe} is almost linearly proportional to the speed of secondary and is independent from gap length and secondary surface resistivity while it is dependent to such parameters at low speed motors (Yamamura, 1972). Therefore, the speed of the end waves is equal to the secondary speed at high speed motors regardless the value of parameters such as supply frequency, gap length and surface resistivity, while in low speed motors, end wave's speed depends on such parameters and may reach to even higher than synchronous speed at low slip region. The super-synchronous speed of the end-effect wave at motor speed lower and close to synchronous speed occurs only in low speed motors (Yamamura, 1972).

The entry-end-effect wave decays relatively slower than the exit-end-effect wave and unlike exit-end-effect wave, is present along the entire longitudinal length of the air-gap and degrades the performance of the high speed motor. The exit-end-effect wave attenuates much faster due to the lack of primary core beyond the exit edge. Therefore, the influence of the exit field component B_2 on motor performance is less than that of the entry component B_1, and it may be disregarded for many applications Gieras et all, 1987, Hirasa et all, 1980, Greppe, et all, 2008).

For the motors with the number of magnetic pole pairs greater than 2 if the synchronous speed is below 10 m/s the end effects can be ignored. For the motors with higher synchronous speeds the influence of end effects can be seen even for the motors with higher number of pole-pairs (Mendrela, 2004).

4.2. Transverse edge effects

The transverse edge effect is generally described as the effect of finite width of the flat linear motor and is the result of x component of eddy current flowing in the solid plate secondary (Fig. 17.b). Since, there are no designated paths for the currents, as it is in cage rotors of rotary motors, the currents within the primary area are flowing in a circular mode (Fig. 17.b). These currents generate their own magnetic field, whose distribution is shown schematically in Fig. 17.a. This magnetic field shown schematically as B_r in Fig. 18 subtracts from the magnetic field B_s generated by the primary part winding. The resultant field has non-uniform distribution in transverse direction (x axis) (Fig. 19). This non-uniform distribution of the magnetic field and circular pattern of the secondary currents contribute to the increase of power losses, decrease of motor efficiency and reduction of maximum electromagnetic force (Boldea & Nasar, 2001).

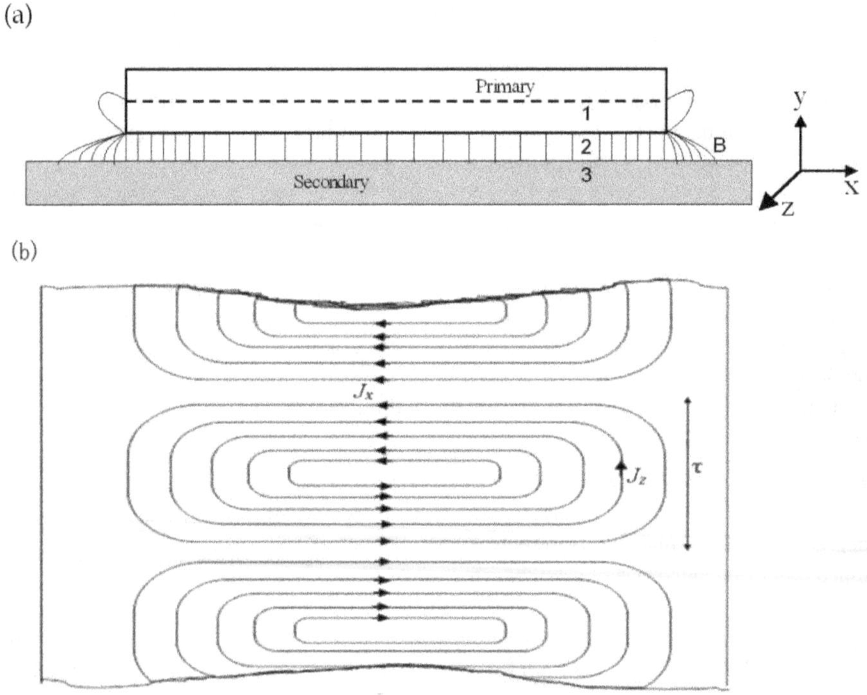

Figure 17. Transverse edge effect explanation: (a) The resultant magnetic flux distribution, (b) eddy current induced in the secondary.

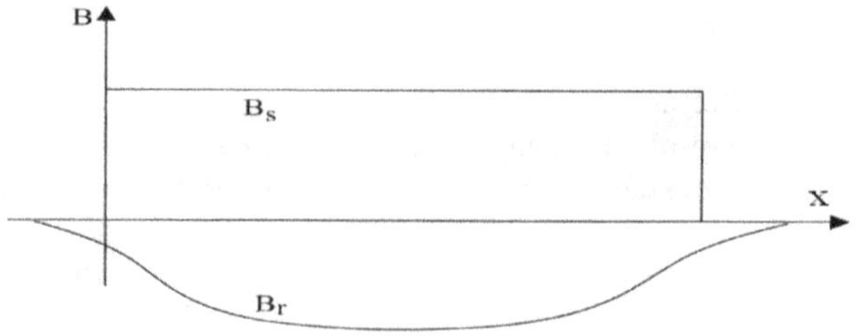

Figure 18. The distribution of magnetic flux density B_s produced by the primary current and B_r by the secondary currents.

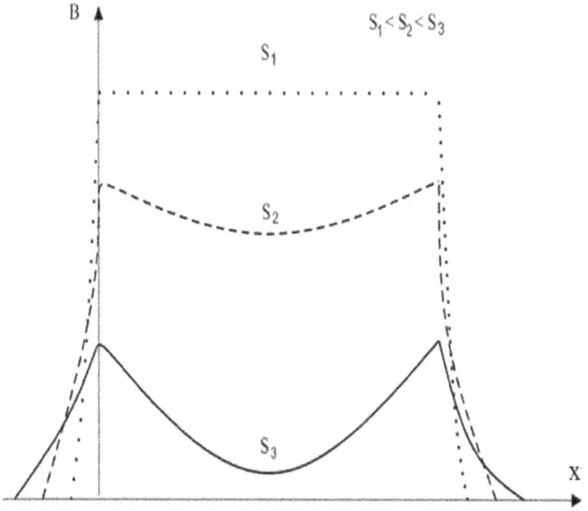

Figure 19. The resultant magnetic flux density distribution in the air-gap at different secondary slips.

As the rotary-linear motor is concerned, the transverse edge effect occurs for rotary armature. This effect has here more complex form due to the additional axial motion of the rotor. The above transverse edge effects superimpose on entry and exit effects whose nature is the same as discussed earlier for linear part of the rotary-linear motor. This motion makes the flux density distribution distorted as shown in Fig. 20. At the entry edge the flux density in the air-gap is damped, but at the exit edge it increases.

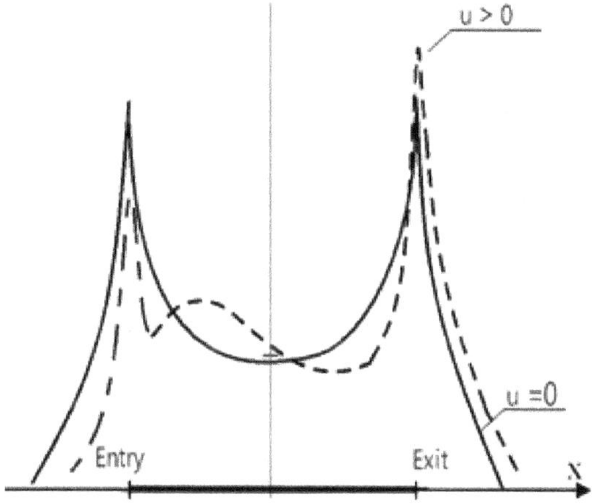

Figure 20. Resultant magnetic flux density in the air-gap of rotary part of the IM-2DoMF motor with linear speed greater than zero (u > 0).

5. Performance of twin-armature rotary-linear induction motor

5.1. Design parameter of the motor

One of a few design versions of rotary-linear motors is a Twin-Armature Rotary-Linear Induction Motor (TARLIM) shown schematically in Fig. 21. The stator consists of a rotary and linear armature placed aside one another. One generates a rotating magnetic field, another traveling magnetic field. A common rotor for these two armatures is applied. It consists of a solid iron cylinder covered with a thin copper layer. The direction of the rotor motion depends on two forces; linear and rotary, which are the products of two magnetic fields and currents induced in the rotor.

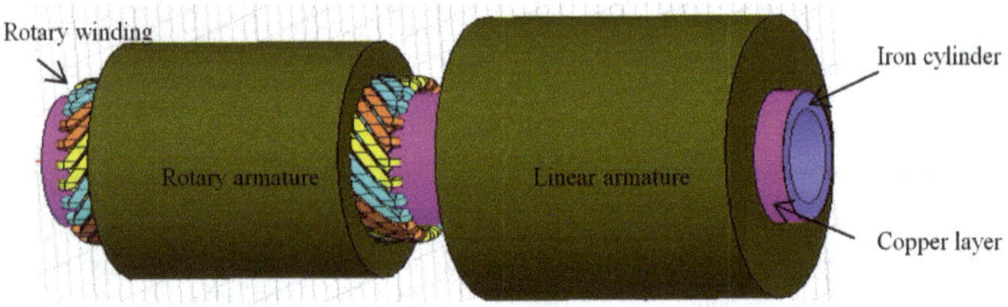

Figure 21. Schematic 3D-view of twin-armature rotary-linear induction motor.

The TARLIM in its operation can be regarded as a set two independent motors: a conventional rotary and tubular linear motor with the rotors joined stiffly. This approach can be applied only if there is no magnetic link between the two armatures, what practically is fulfilled due to the relatively long distance between the armatures and the low axial speed of the rotors. In case of the motor analysed here both conditions are satisfied and the analysis of each part of the TARLIM can be carried out separately as the analysis of IM 3-phase rotary and linear motors. The only influence of one motor on the other is during the linear motion of the rotor which will be considered at the end of this chapter.

To study the performance of TARLIM the exemplary motor has been chosen with the dimensions shown in Fig. 22. The dimensions of rotary armature are presented in details in Fig. 23.

The core of both armatures is made of laminated steel. The common rotor is made of solid steel cylinder covered by copper layer. Both armatures possess a 3-phase winding. The rotary and linear winding diagrams are shown in Figs. 24.a and 24.b, respectively. The winding parameters and the data of stator and rotor core material are enclosed in Table 1.

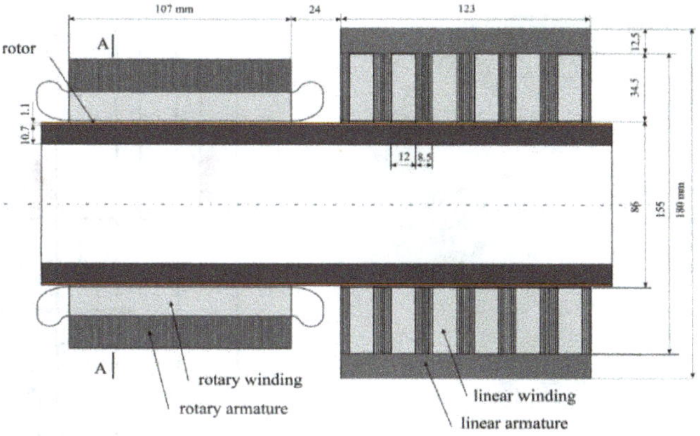

Figure 22. Dimensions of TARLIM chosen for analyse.

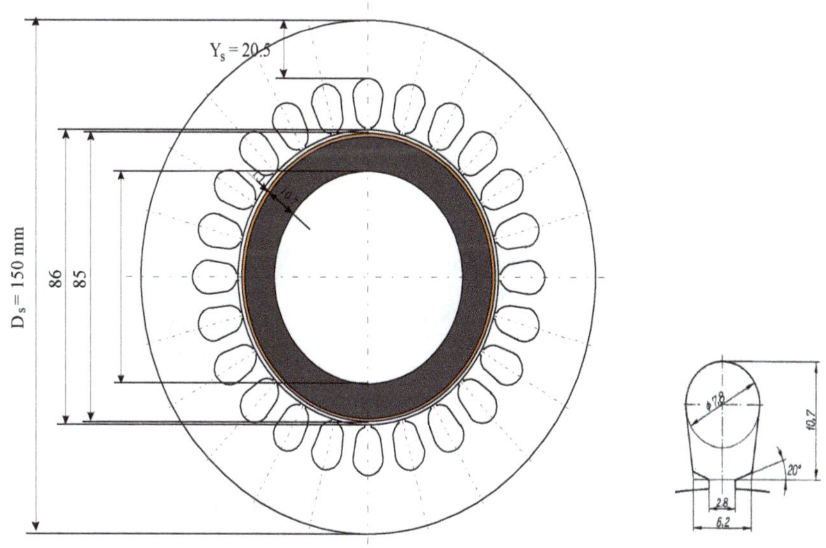

Figure 23. Rotary armature dimensions.

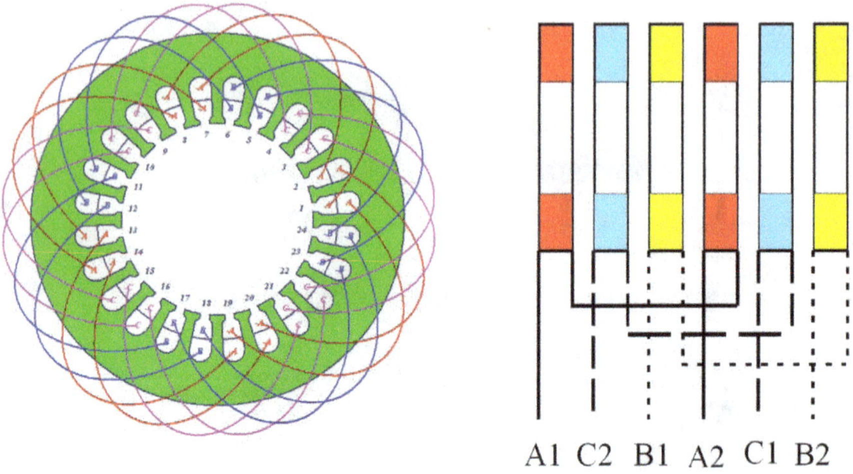

(a) (b)

Figure 24. Winding diagram of the TARLIM, (a) rotary winding, (b) linear winding.

Linear winding data:	
Number of phases	3
Number of poles	2
Number of slots per pole per phase	1
Number of wires per slot, N_w	215
Copper wire diameter	1.29 mm
Rotary winding data:	
Number of phases	3
Number of poles	4
Number of slots per pole per phase	2
Number of wires per slot, Nw	96
Copper wire diameter	0.7 mm
Armature Core	Laminated steel
Air gap length, mm	0.5
Rotor	
Copper layer	
Thickness mm	1.1 mm
Conductivity (γ_{Cu}, @ 20C)	57.00x106 S/m
Solid iron cylinder	
Thickness mm	10.7 mm
Conductivity (γ_{Fe} , @ 20C)	5.91x106 S/m

Table 1. Winding and materials data for TARLIM.

5.2. Experimental model

To verify the modeling results, a real prototype of the motor was built (see Figs. 25 and 26) and tested. The laboratory model of TARLIM has a relatively short secondary length. Therefore, measuring motor performances at linear speed greater than zero was practically difficult so the test was carried out only at zero linear speed. The TARLIM operates practically at low rotary slip and at linear slip close to one. Thus the dynamic end effects does not influence much the motor performance but the static end effect caused by finite length of each of the armatures has a large impact on the linear motor performance.

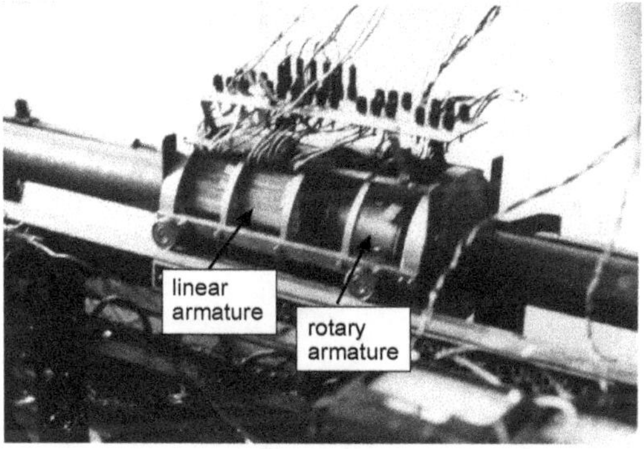

Figure 25. Laboratory model of twin-armature rotary-linear induction motor.

Figure 26. Measurement stand for testing of rotary-linear motors.

5.3. Motor performance

The analysis of each part of TARLIM performance is carried out separately as an independent tubular linear and rotary motor by 3-D FEM modelling.

The linear armature is being supplied from the constant voltage source of 86.6 V (rms), 50 Hz frequency. The results of simulation are shown in Figs. 27, 28 and 29 in form of the electromechanical force (F_{em}), mechanical power (P_m) and efficiency, respectively, versus linear slip of the rotor. These characteristics illustrate a significant impact of end effects on motor performance and are drawn as dashed line curves for the infinitely long motor when no end effects are considered, circles for the actual motor of finite armature length, when both static and dynamic end effects are taken into account and triangular sign for the experimental result.

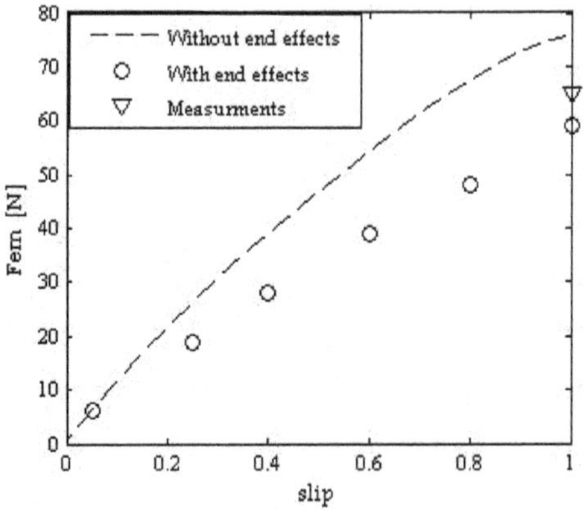

Figure 27. Characteristic of force vs linear slip.

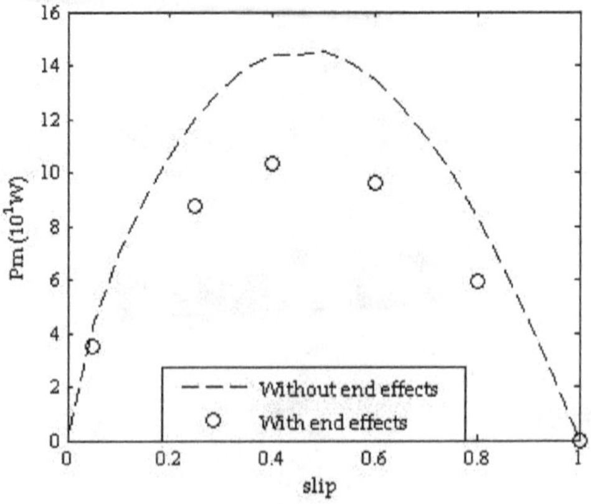

Figure 28. Characteristic of output power vs linear slip.

The motor under study has a linear synchronous speed equal to 6.15 m/s and is considered as low-speed motor. In low-speed motors, the speed of the end effect wave can be higher than the motor speed and even much higher than the synchronous speed, while in high-speed motors the speed of the end effect wave is about the same as the motor speed and cannot be higher the synchronous speed. In low-speed motors, the attenuation of the entry end-effect wave is quick, while in high-speed motors the attenuation is very slow and the entry-end-effect wave presents over the entire longitudinal length of the air-gap. As a consequence of the difference, the influence of the end-effect wave on motor performance is also quite different at high-speed motors and low-speed motors. In low-speed motors, the end effect wave may improve motor performance in low-slip region, the important motor-run region, increasing thrust, power factor and efficiency, and allowing net thrust to be generated even at synchronous and higher speeds. On the contrary, in high speed motors, thrust, power factor and efficiency are reduced to a large extent in the low-slip region, and it is not an over statement to say that high-speed applications of linear induction motors may not be feasible if the end effect is overlooked and is allowed to remain as an influence.

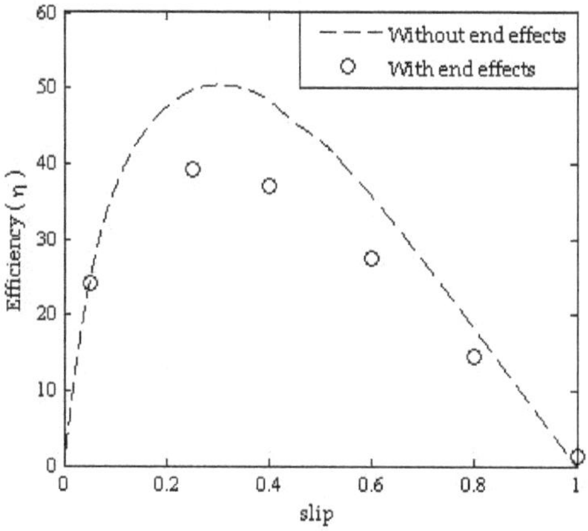

Figure 29. Characteristic of efficiency vs linear slip.

To study the performance of the motor at higher speeds, let us change the supply frequency to 3 times higher (synchronous speed, v_s= 3 * 6.15 m/s) and then recalculate the forces acted on rotor when all end effects are taken into account. Table 2 compares the output forces of the low-speed and high-speed motor at relatively low operational slip region.

Synchronous speed	Operational slip region	
	S=0.25	S=0.05
v_s= 6.15 m/s	19 N	6 N
v_s= 3*6.15 m/s	6 N	0 N

Table 2. Electromechanical force of low speed and high speed LIM.

The simulation of rotary part of the motor is done for the winding being supplied by the three-phase voltage of 150 V (rms), 50 Hz frequency.

Due to closed magnetic circuit in rotary armature, static end effect does not exist in rotational direction. However, the performance of rotary armature might be affected by dynamic end effects during rotor axial motion. This is the only influence of linear part of TARLIM on rotary motion and, as stated earlier, both armatures have no more influence on one another due to the relatively long distance and the lack of magnetic interaction in between. To determine the influence of rotor axial motion on the performance of rotary armature, the characteristics of electromagnetic torque (T_{em}) and mechanical power (P_m) versus rotary slip at three different linear speeds (u = 0 m/s, u =3 m/s and u =6 m/s) along with experimental results at zero linear speed (u = 0 m/s) are plotted in Figs. 30 and 31. These effects contribute to diminishing of torque and all other rotary motor performances. One can observe that, the higher axial speed leads to lower rotary torque and mechanical power.

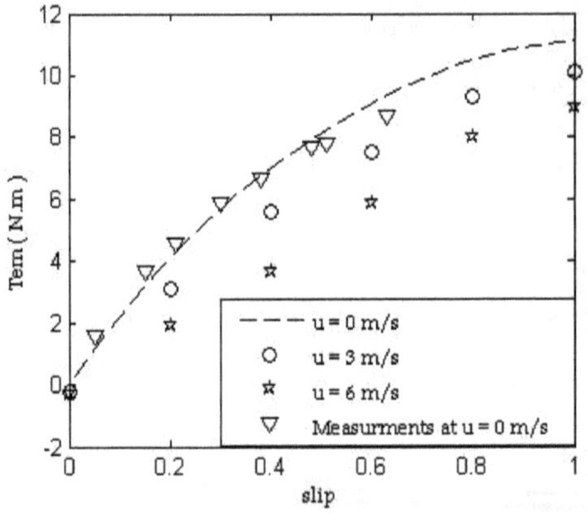

Figure 30. Characteristic of Torque vs rotary slip with and without linear motion.

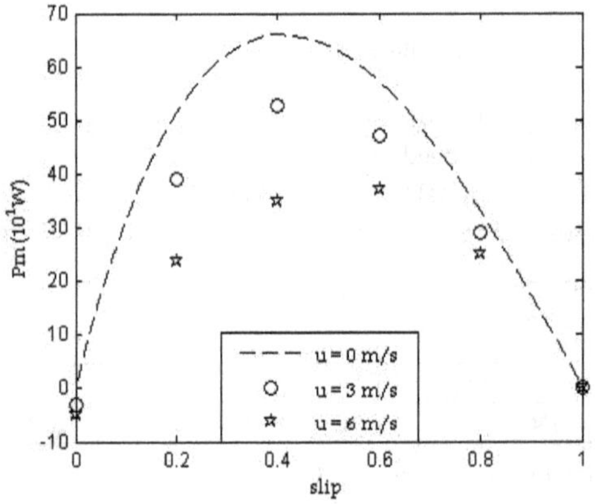

Figure 31. Characteristic of output power vs linear slip with and without linear motion.

Note, that the output quantities are extremely dependent on the property of materials. The conductivity of the materials used in 3D FEM analysis is kept constant, but in reality it might be influenced by the temperature. Therefore, minimal mismatch between experimental measurements, where the temperature changes during the experiment and FEM results is expected. On the other hand, FEM needs as dense mesh as possible to compute quantities accurately, but the execution time of such a complicated model is enormous. Therefore, some trade-off between accuracy and execution time is required to obtain a good solution at reasonable cost. However, the discrepancies between test and simulation results are relatively small which validates the simulation models.

6. Conclusion

Rotary-linear induction motor is one of a few types of motors with two degrees of mechanical freedom. It may find application in robotics and special types of drives like machine tools and drilling machines. One of its representatives, the TARLIM, with two solid layer rotor was modelled in 3-D FEMM and its performance has been determined. The operation of the motor does not differ from the operation of machine set consisting rotary and tubular linear motor of which rotors are firmly coupled. The electromechanical performances of the motor are affected by the end effects which are familiar phenomena in linear machines. Practically, the impact of these phenomena is not significant in low axial speed of the rotor, what is expected for these types of motors.

The results obtained from the test carried out on the experimental model do not differ much from the ones got from simulations. Thus they validate the theoretical modeling of the motor.

Motor with the rotor cage made in form of grid placed on the cylindrical core is another version of TARLIM and is expected to have a better performance with respect to the solid two layer rotor.

Author details

Ebrahim Amiri and Ernest Mendrela
Louisiana State University, USA

7. References

Mendrela, E, Fleszar, J, Gierczak, E. (2003). *Modeling of Induction Motors With One and Two Degrees of Mechanical Freedom*, Norwell, MA: Kluwer Academic Publishers.

Krebs, G, Tounzi, A, Pauwels, B and Willemot, D. (2008). *General overview of integrated linear rotary actuators, in Proc. ICEM Conf.*

Bolognesi P, Bruno O, Landi A, Sani L, Taponecco L. (2004). *Electromagnetic actuators featuring multiple degrees of freedom:* a survey. In: ICEM conference, Krakow (Poland), 5–8 September.

Giancarlo, B and Tellini, B. (2003). *Helicoidal electromagnetic field for coilgun armature stabilization, IEEE Trans. Magn.*, vol. 39, no. 1, pp. 108–111, Jan. 2003.

Anorad Corp., New York, USA. (2001). *Rotary linear actuator*, U.S. Patent 6 215 206.

Yamamura, S. (1972). *Theory of Linear Induction Motors*, John Wiley & Sons.

Creppe, R. C, Ulson, J. A. C, Rodrigues, J. F. (2008). *Influence of Design Parameters on Linear Induction Motor End Effects, IEEE* Trans. Energy Convers., vol. 23, no. 2, pp. 358–362 June. 2008.

Faiz, J Jafari, H. (2000) *Accurate Modelling of Single-Sided Linear Induction Motor Considers End Effect and Equivalent Thickness,* IEEE Transactions on Magnetics, vol. 36, No. 5, September 2000, pp. 3785-3790.

Turowski, J. (1982). *Electromagnetic calculations of machine elements and electro mechanics.* WNT Warsaw, (in Polish).

Gierczak, E, Mendrela, E. (1985). *Magnetic flux, current, force and power loss distribution in twin-armature rotary-linear induction motor,* Scientia Electrica, Vo. 31, pp. 65-74, 1985.

Mosebach, H, Huhns, T, Pierson, E.S, Herman, D. (1977). *Finite Length Effects in Linear Induction Machines with Different Iron Contours,* IEEE Trans. On PAS-96, 1977, pp. 1087-1093.

Poloujadoff, M, Morel, B, Bolopion, A (1980). *Simultaneous Consideration of Finite Length and Finite Width of Linear Induction Motors,* IEEE Trans. On PAS, Vol. PAS-99, No. 3, 1980, pp. 1172-1179.

Pai, R.M., Boldea, I., and Nasar, S.A. (1988). *A complete equivalent circuit of a linear induction motor with sheet secondary,* IEEE. Trans., 1988, MAG-24, (1), pp. 639–654.

Gieras, J. F. , Dawson, G. E. and Eastham, A. R. (1987). *A new longitudinal end effect factor for linear induction motors, IEEE Trans. Energy Convers.,* vol. 2, no. 1, pp. 152–159, Mar. 1987.

Hirasa, T, Ishikawa, S and Yamamuro, T. (1980). *Equivalent circuit of linear induction motors with end effect taken into account, Trans. IEE Jpn.,* vol. 100, no. 2, pp. 65–71, 1980.

Duncan, J., and Eng, C. (1983) *Linear induction motor-equivalent circuit model',* IEE Proc., Electr. Power Appl., 1983, 130, (1), pp. 51–57.

Mirsalim, M, Doroudi, A and Moghani, J. S. (2002). *Obtaining the operating characteristics of linear induction motors: A new approach,* IEEE Trans. Magn., vol. 38, no. 2, pp. 1365–1370, Mar. 2002.

Kwon, B. I, Woo, K. I. and Kim, S (1999). *Finite element analysis of direct thrust controlled linear induction motor,* IEEE Trans. Magn., vol. 35, no. 3, pp. 1306–1309, May 1999.

Fujii, N and Harada, T. (2000). *Basic consideration of end effect compensator of linear induction motor for transit,* in Industry Applications Conf., Oct. 8–12, 2000, pp. 1–6.

Kim, D.-k. and Kwon, B.-I. (2006). *A novel equivalent circuit model of linear induction motor based on finite element analysis and and its coupling with external circuits,* IEEE Trans. Magn., vol. 42, no. 10, pp. 3407–3409, Oct. 2006.

Krebs, G, Tounzi, A, Pauwels, B, Willemot, D and Piriou, F. (2008). *Modeling of a linear and rotary permanent magnet actuator,* IEEE Trans. Magn., vol. 44, no. 10, pp. 4357-4360, Nov. 2008.

Amiri, E, Gottipati, P, Mendrela, E. (2011). *3-D Space Modeling of Rotary-Linear Induction Motor With Twin-Armature* The 1st IEEE International Conference on Electrical Energy Systems, Chennai, Tamil Nadu, India, Jan 2011.

Poloujadoff, M. (1980). *The Theory of Linear Induction Machinery,* Oxford University Press, 1980.

Mendrela, E. (2004) *Electric Machines", Course Pack "Advanced Electric Machine,* Louisiana State University.

Boldea, I, Nasar, S.A.. (2001). *Linear Motion Electromagnetic Devices,* Taylor and Francis, 2001.

Chapter 10

Analysis of Natural Frequency, Radial Force and Vibration of Induction Motors Fed by PWM Inverter

Takeo Ishikawa

Additional information is available at the end of the chapter

1. Introduction

Lately, as engineers have recognized the importance of having a high-quality working place, the effect of the noise and vibration emitted by inverter-fed induction machines has become a subject to study. Economic considerations force to use less active material. Since the encasing is less stiff, the machine becomes more sensitive to vibrations and noise. Less use of iron in the stator not only yields to a weaker structure but also higher field levels, thus causing higher magnetic forces, which yields to increased vibrations. Then the first aim of this work is to reach a wide knowledge how the levels of noise and vibration generated by the induction motor vary under different working conditions.

Electromagnetic noise is generated when the natural frequencies of vibration of induction motors match or are close to the frequencies present in the electromagnetic force spectrum. In order to avoid such noise and vibration, it is necessary to estimate the amplitude of the radial electromagnetic forces as well as the natural frequencies of the structure. For this reason, several papers have been published to analyze the natural frequencies, electromagnetic force, vibration and acoustic noise. For the analysis of the natural frequencies, a lot of papers have analyzed the stator core without winding. However, it is known that it is difficult to estimate the Young's modulus of winding. For the analysis of the radial force, vibration and acoustic noise, several papers have been published (Ishibashi et al., 2003, Shiohata et al., 1998, Munoz et al., 2003). They gave the amplitudes as well as the frequencies of the radial electromagnetic force. However, they mainly treated the case when the slip was 0. Ishibashi et al. did not consider the rotor current (Ishibashi et al., 2003), and Munoz et al. specified stator currents calculated by MATLAB/Simulink as input data not stator voltages (Munoz et al., 2003).

This paper investigates the vibration of induction motors fed by a Pulse Width Modulation (PWM) inverter. First we analyze the natural frequencies of the stator by considering the stator coil, and compare with the measured ones. Next, we analyze the radial electromagnetic force by using two-dimensional (2D) non-linear finite element method (FEM) which is considering the rotor current and is coupled with voltage equations, and discuss the calculated result with the measured vibration velocity. We clarify the influence of slip, the distributed stator winding and the PWM inverter on the radial force. Moreover, it is well known that a random PWM reduces the acoustic noise emitted from an inverter drive motor (Trzynadlowski et al., 1994). Then, we investigate the radial force of the motor fed by two types of random PWM method, namely, a randomized pulse position PWM and a randomized switching frequency PWM.

2. Natural frequencies

2.1. Analysis method of natural frequencies

The mechanical equation for the stator model with the free boundary condition is expressed as

$$[\mathbf{M}]\{\ddot{x}\}+[\mathbf{K}]\{\mathbf{x}\}=\{\mathbf{0}\} \tag{1}$$

where, $\{\mathbf{x}\}$ is the node displacement, $[\mathbf{M}]$ and $[\mathbf{K}]$ are the global mass matrix and stiffness matrix. In the two-dimensional (2D) case, the plate is assumed to have constant mass density ϱ, area A, uniform thickness h, and motion restricted to the {x, y} plane. The element mass matrix is expressed as a 6 × 6 matrix

$$\mathbf{M}^e=\frac{\rho Ah}{12}\begin{bmatrix}2&0&1&0&1&0\\0&2&0&1&0&1\\1&0&2&0&1&0\\0&1&0&2&0&1\\1&0&1&0&2&0\\0&1&0&1&0&2\end{bmatrix} \tag{2}$$

The element stiffness matrix for plane strain is given by

$$\mathbf{K}^e=\frac{E}{(1+\nu)(1-2\nu)}\begin{bmatrix}1-\nu&\nu&0\\\nu&1-\nu&0\\0&0&(1-2\nu)/2\end{bmatrix} \tag{3}$$

where E is Young's modulus and ν is Poisson's ratio.

Equation (1) leads to the eigenvalue problem,

$$([\mathbf{K}]-\omega_i^2[\mathbf{M}])\{\Phi\}_i=\{\mathbf{0}\} \tag{4}$$

where, $\{\Phi\}_i$ is eigenvector representing the mode shape of the i-th natural angular frequency ω_i. We solve (4) by discretizing the stator into a finite element mesh and using an eigenvalue subroutine utilized in International Mathematics and Statistics Library (IMSL).

In the calculation of natural frequencies using FEM, the most important but unknown constant is Young's modulus of winding which is composed of the enameled wires, insulation films and vanish. Itori et al. has given the equivalent Young's modulus of winding in slot by the experimental investigation (Itori et al., 2002)

$$E = (0.0319S - 1.05) \times 10^9 \left[\mathrm{N/m^2}\right] \tag{5}$$

where, S is the space factor of winding.

2.2. Experimental motors

This chapter investigates the vibration characteristics of two motors, hereafter K-model and M-model, whose properties and characteristics are as follows. For M-model, 1.5 kW, 200 volt, 50 Hz, 6.8 A, 4 poles, number of stator slots: 36, number of rotor slots: 44, and one slot-pitch skewing, see Fig.1 (Mori et al., 2005, 2005). For K-model, 100 volt, 50 Hz, 4 poles, number of stator slots: 24, number of rotor slots: 34, stator winding: 66 turns, rotor bar: aluminium, and no skewing, see Fig. 2 (IEEJ, 2002).

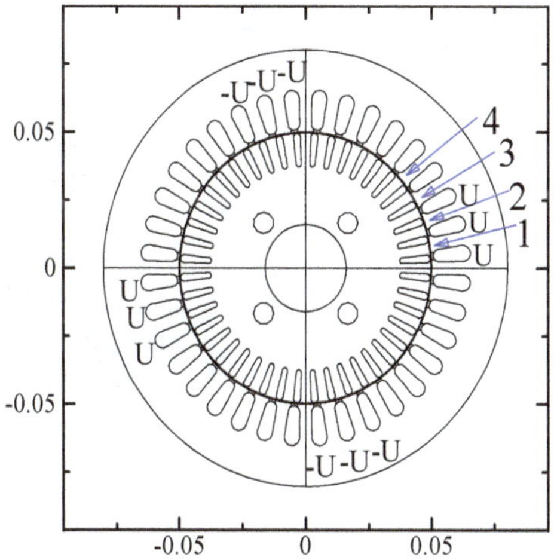

Figure 1. Experimental motor, M-model.

2.3. Measurement of natural frequencies

Natural frequencies are obtained by measuring the transfer function of the stator core. Fig. 3 shows an experimental setup to measure the natural frequencies. A piezoelectric accelerometer PV08A is placed at the top of the stator and is connected to one channel of a charge amplifier UV-06. An impulse hammer PH-51 is connected to the other channel. The charge amplifier is connected to a signal analyzer SA-01A4, and then to a PC where a software for SA-01A4 is installed.

The transfer function is measured by hammering the stator surface. First, the natural frequencies of the stator core only of M-model are measured. We have removed the stator windings from the stator. Table 1 shows the four lowest measured natural frequencies.

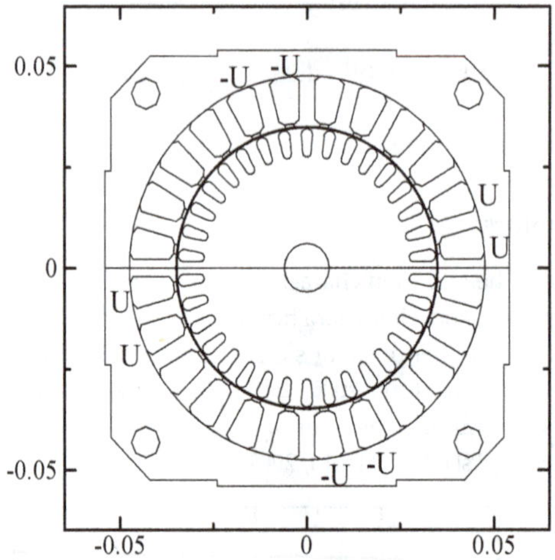

Figure 2. Experimental motor, K-model.

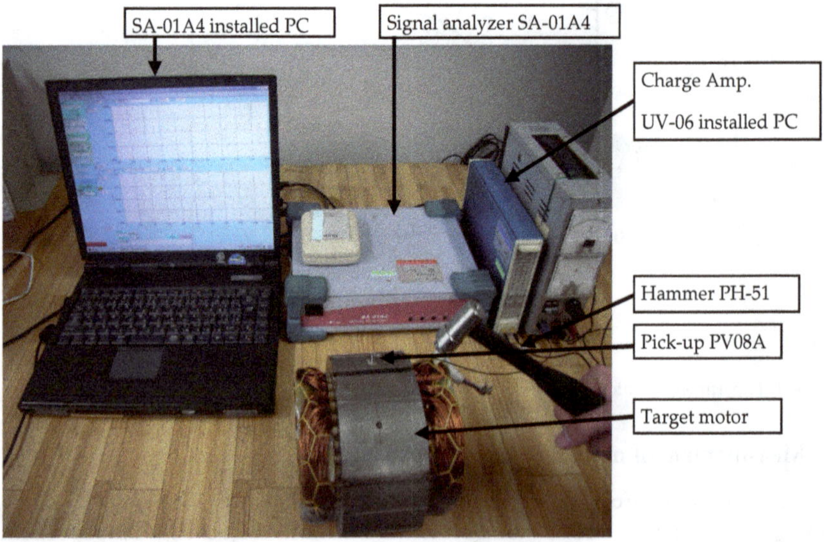

Figure 3. Experimental setup for measurement of natural frequencies.

Mode	Frequency [Hz]
2	1,325
2	1,337
3	3,425
3	3,875

Table 1. Measured natural frequencies of the stator core of M-model motor.

Next, the natural frequencies of the stator with winding of M-model are measured as shown in Fig. 4. The natural frequencies around 1,200Hz are generated from rotor. Three lowest natural frequencies except around 1,200Hz are shown in Table 2.

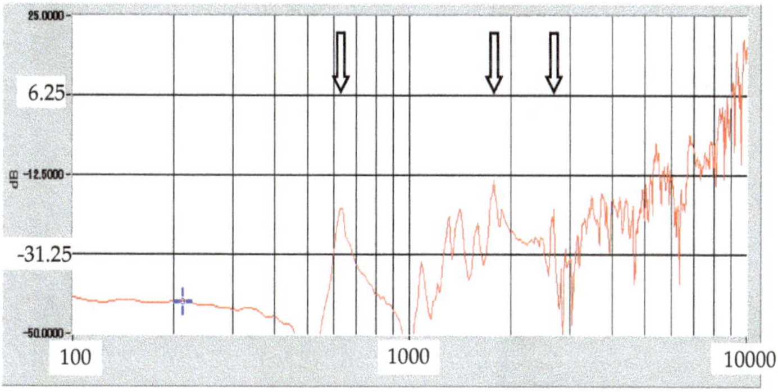

Figure 4. Natural frequencies measured for the whole motor M-model.

Mode	Frequency [Hz]
2	637
3	1,770
4	2,694

Table 2. Three lowest natural frequencies of the M-model motor with stator winding.

2.4. Calculation of natural frequencies

First, we calculate the natural frequencies for the stator core only of the M-model motor, whose mechanical properties include mass density of 7,850kg/m^3, Young's modulus of 2.1x 10^{10}N/m^2 and Poisson's ratio of 0.3. Table 3 shows the comparison of the calculated natural frequencies with the measured ones. It shows a good agreement between the measured values and the calculated ones. In this calculation, we use 18,811 finite element nodes. If we calculate the natural frequencies with a rough mesh, they become higher values. Fig. 5 shows the modes of stator due to each harmonic. The natural frequencies of 1,369 and 1,425Hz have mode 2, and 3,446 and 3,926Hz have mode 3.

Mode	Measured [Hz]	Calculated [Hz]
2	1,325	1,369
2	1,337	1,425
3	3,425	3,446
3	3,875	3,926

Table 3. Comparison of the calculated natural frequencies with the measured ones for the stator core only.

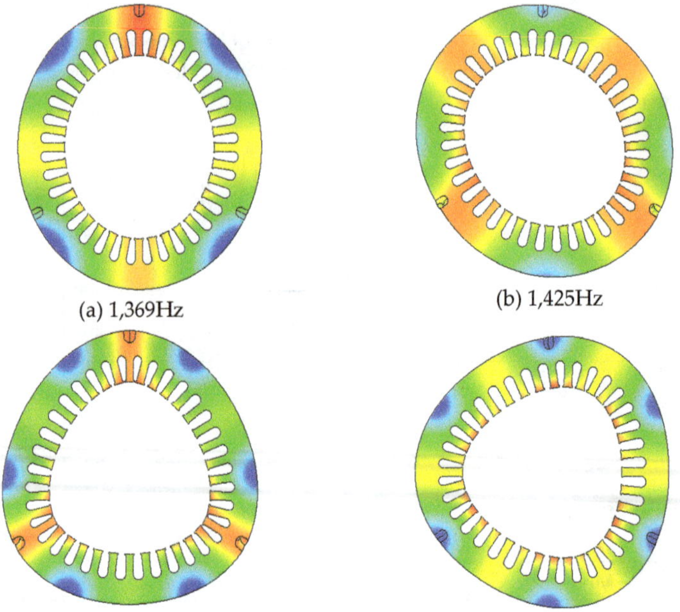

(a) 1,369Hz (b) 1,425Hz

(c) 3,446Hz (d) 3,926Hz

Figure 5. Natural vibration modes for stator core only.

Next, we calculate the natural frequencies of the stator with winding, where the space factor of winding is chosen to be 0.43 by considering the enameled wires. Three lowest natural frequencies and the natural vibration modes are shown in Table 4 and Fig. 6. The natural frequencies of 587, 1,545 and 2,739Hz have mode 2, 3 and 4, respectively.

Mode	Measured [Hz]	Calculated [Hz]
2	637	587.0
3	1,770	1,544.6
4	2,694	2,739.0

Table 4. Comparison of the calculated natural frequencies with the measured ones for the stator with winding.

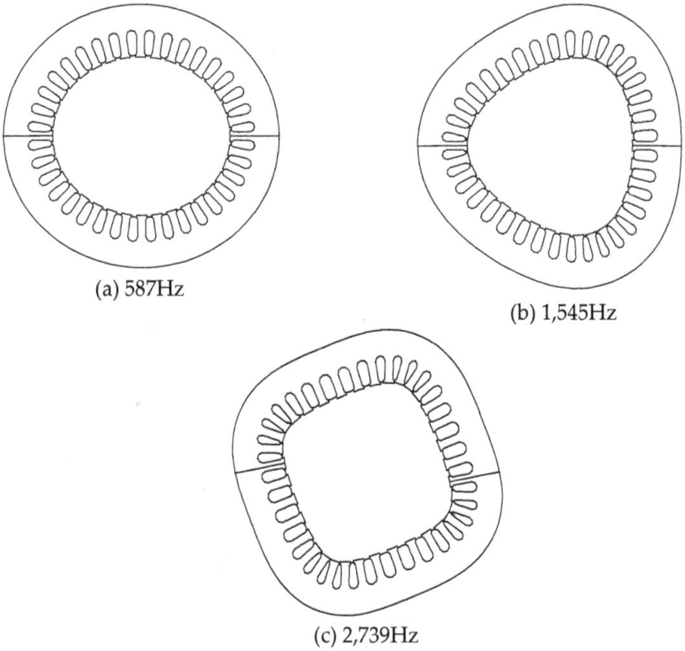

Figure 6. Natural vibration modes for stator with winding.

As the natural frequencies around 1,200Hz are generated from rotor. Three smallest natural frequencies except around 1,200Hz are shown in Table 4 as well as the measured ones. It is shown that the calculated natural frequencies are a little smaller than the measured ones. This is because we calculate the space factor of winding composed of the enameled wires only. If the insulation films and vanish are taken into account, the space factor is larger. Fig. 7 shows the lowest natural frequency by changing the space factor. Therefore, if the insulation films and vanish are taken into account, the smallest natural frequency becomes large, that is, close to the measured one.

3. Radial magnetic force

3.1. Analysis method of radial magnetic force

The simulation of the electromagnetic force is implemented by using a 2D non-linear finite element method considering the rotor current coupled with voltage equations. As we consider the force and vibration at a steady state, the rotating speed is assumed to be constant. Then, the government equations are as follows,

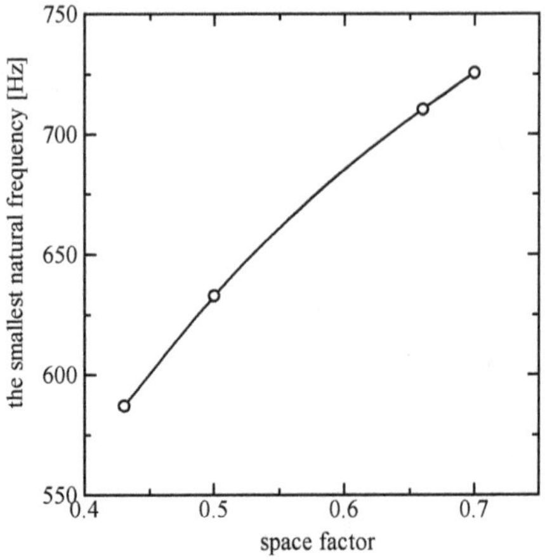

Figure 7. Relationship between the smallest natural frequency and space factor.

$$\frac{\partial}{\partial x}(\nu\frac{\partial A}{\partial x})+\frac{\partial}{\partial y}(\nu\frac{\partial A}{\partial y})=-\sum_{k}J_{0k}+\sigma\frac{\partial A}{\partial t}, \tag{6}$$

$$V_k = r\, i_k + l\frac{d\, i_k}{d\, t}+\frac{d\Phi_k}{d\, t}, \qquad (k=a,b,c) \tag{7}$$

$$\text{where } \Phi_k = \frac{n}{S_k}\iint_{S_k} AL\, dxdy\,, \qquad J_{0k}=\frac{n}{S_k}i_k\,, \tag{8}$$

where, $A,\nu,J_0,\sigma,V,i,r,l,\Phi,n,S,L$ are magnetic vector potential, reluctivity, current density, conductivity, stator phase voltage, stator current, resistance of the stator winding, leakage inductance of the stator end winding, flux linkage, number of turns of stator winding, cross section area of the stator winding, and stack length, respectively. We solve equations (6) and (7) by using the time-stepping FEM. In the case where the motor is driven by the line voltage, the time step Δt is constant so that the step of rotation $\Delta\theta$ is about $2\pi/500$ at slip=0. In the case of PWM inverter, Δt is calculated from the intersection point of a sine wave and a jagged wave with a carrier frequency of 5 kHz. The transient state converged at about five cycles of the input voltage on our simulation. The radial electromagnetic force is calculated by the Maxwell's stress tensor method,

$$F_n=\frac{1}{2\mu_0}(B_n^2-B_t^2) \tag{9}$$

where, μ_0 is the permeability of air, B_n and B_t are the normal and tangential component of the flux density in the air gap. In order to take into account the 3D effects, the resistance of the end ring of the rotor is considered in the 2D model by modifying the conductivity of the rotor bars. Resistances of bar and end ring can be written as

$$R_b = \frac{1}{\sigma}\frac{l_b}{S_b}$$
$$R_r = \frac{2}{\sigma}\frac{l_r}{S_r}\frac{N_2}{(p\pi)} \tag{10}$$

where, σ, l_b, S_b, l_r, S_r, N_2, and p are the conductivity of aluminium, the longitude of the rotor bar, the cross section area of rotor bar, the longitude of the end ring, the cross section area of end ring, the number of rotor slots, and the number of poles, respectively. It is assumed that the rotor resistance is expressed by an equivalent bar with a modified conductivity

$$R_b' = \frac{1}{\sigma'}\frac{l_b}{S_b} = R_b + R_r \tag{11}$$

Therefore, the modified conductivity is obtained by using the next formula (IEE Japan, 2000)

$$\sigma' = \frac{\frac{l_b}{S_b}}{\frac{l_b}{S_b} + 2\times\frac{l_r}{S_r}\times\frac{N_2}{(p\pi)^2}}\sigma \tag{12}$$

Stator end leakage inductance l is also taken into account and given by a traditional method (Horii, 1978)

$$l = \frac{2.3}{2\pi}\mu_0 N^2 l_f\left(\log_{10}\frac{l_f}{d_s} - 0.5\right) \tag{13}$$

where, N, l_f, and d_s are number of stator windings, total length of coil end, and diameter of an equivalent circle whose area equals to the cross section of stator coils. If the motor has skewed slots, we should use one of the multi-slice model, the coupled method of 2D and 3D models (Yamasaki, 1996) and the full 3D model (Kometani et al., 1996). The influence that the skewing has on the radial force and vibration is not taken into account in this paper.

3.2. Steady state characteristics

Electromagnetic force is calculated by the 2D non-linear finite element method coupled with voltage equations. The models are created using a triangular mesh with 13,665 elements and 6,907 nodes for the M-model see Fig. 8. One fourth of the motor is calculated because of symmetry. For the M-model these numbers are 14,498 elements and 7,333 nodes, and half of the motor is calculated, see Fig. 9. The values obtained for the aluminium relative conductivity are $\sigma' = 0.737\sigma$ for the K-model, and $\sigma' = 0.351\sigma$ for the M-model.

To corroborate the validity of the model, the measured and calculated values of the output torque and current are compared, and the results are presented in Figs. 10 and 11. The graphic shows a good agreement between the measured values and the calculated values for both models. This paper does not consider the effect of skewing, then this produces some error around 1400min^{-1} rotating speed in the M-model.

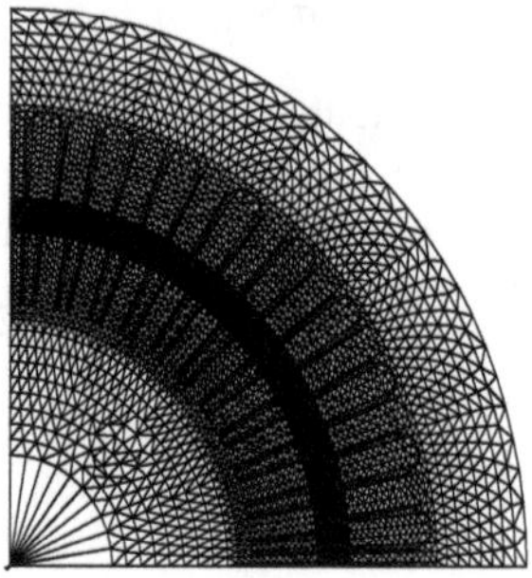

Figure 8. Mesh partition for M-model motor

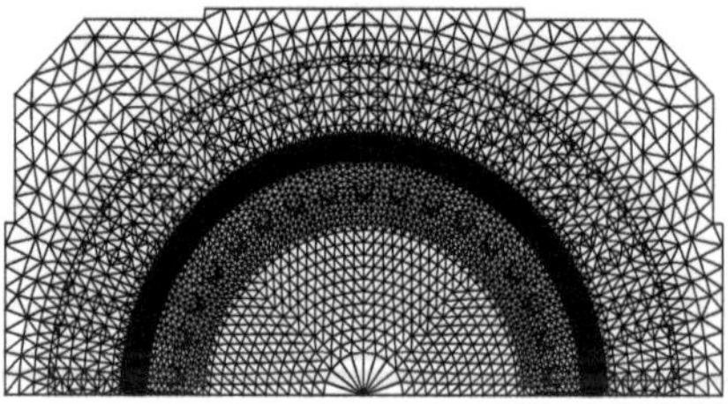

Figure 9. Mesh partition for K-model motor

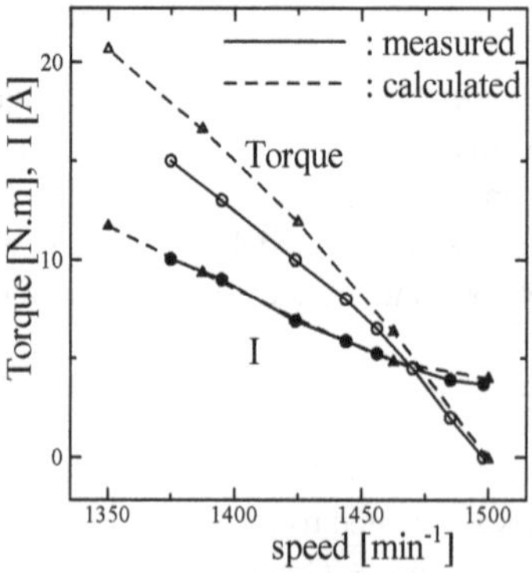

Figure 10. Steady state characteristic for M-model motor

3.3. Radial magnetic force under line source

The space variation of the radial electromagnetic force is presented in Fig. 12. It is shown that the radial force is big at the position where the flux density is big as shown in Fig. 13, and is approximately flat in the teeth and becomes a small value at the positions where the rotor slot exists.

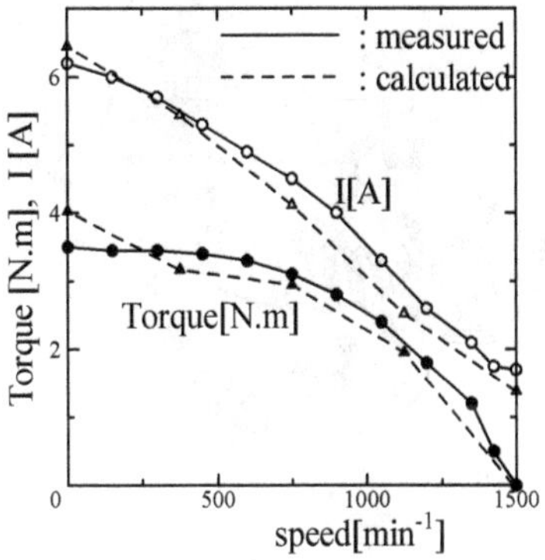

Figure 11. Steady state characteristic for K-model motor

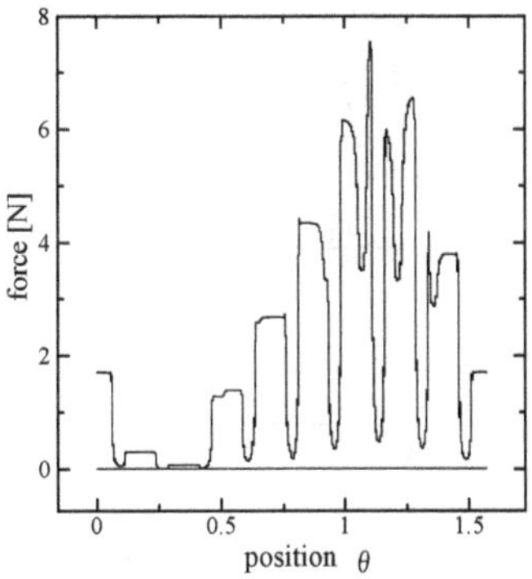

Figure 12. Space variation of radial force.

Fig. 14 shows the time variation of the force at the different teeth. It is shown that the force at tooth 1 is the same as that at tooth 4 and is bigger than those at teeth 2 and 3, because the stator winding is distributed in three slots as shown in Fig. 1.

Figs. 15 and 16 show the radial force and its spectrum at slip=0. It is shown that the force at the teeth is bigger than that at the slots and has a fundamental frequency of 2 times the line frequency of 50Hz. Figs. 17 and 18 show the radial force and its spectrum at slip=0.05. It is shown that the radial force at the slip of 0.05 is very different from that at 0.

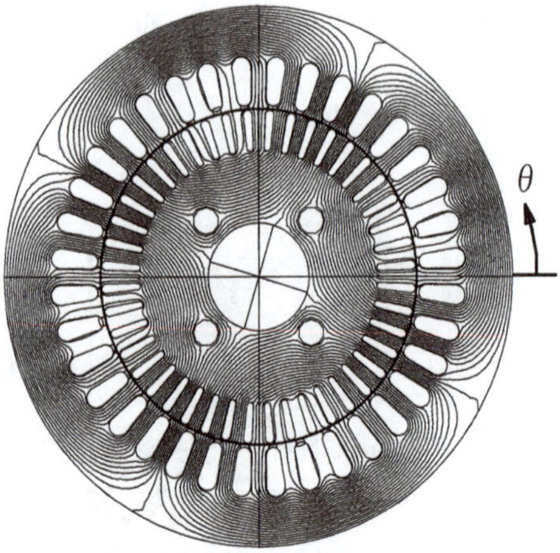

Figure 13. Flux distribution at slip=0.

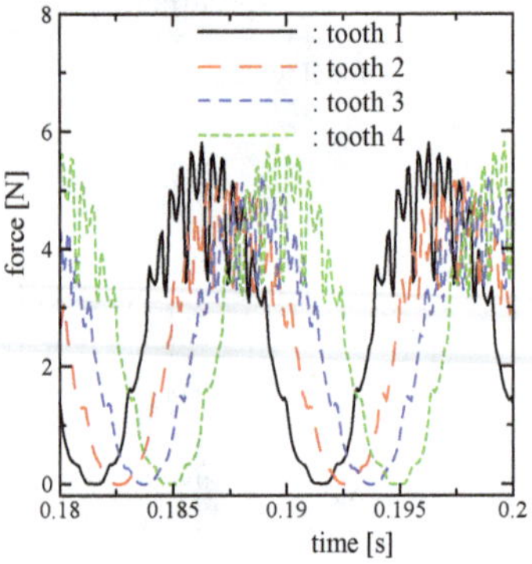

Figure 14. Radial magnetic force at different teeth.

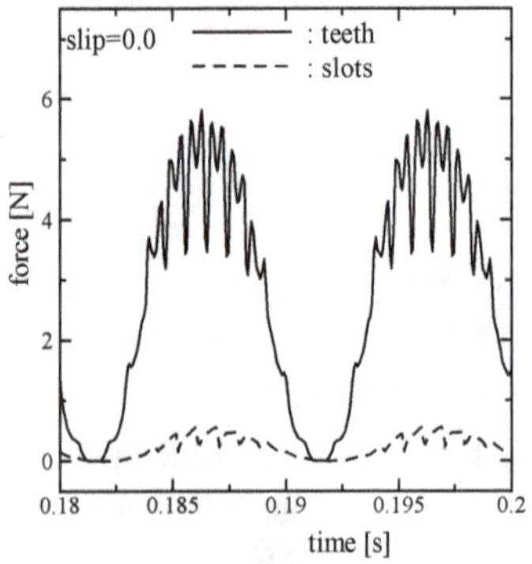

Figure 15. Waveform of radial magnetic force of M model motor at slip=0.

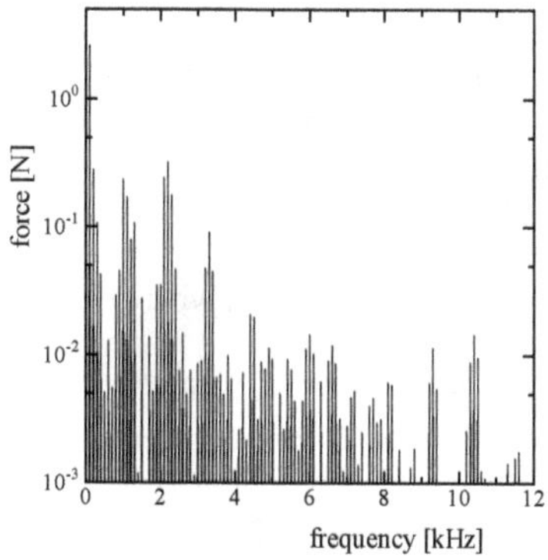

Figure 16. Spectrum of radial magnetic force of M model motor at slip=0.

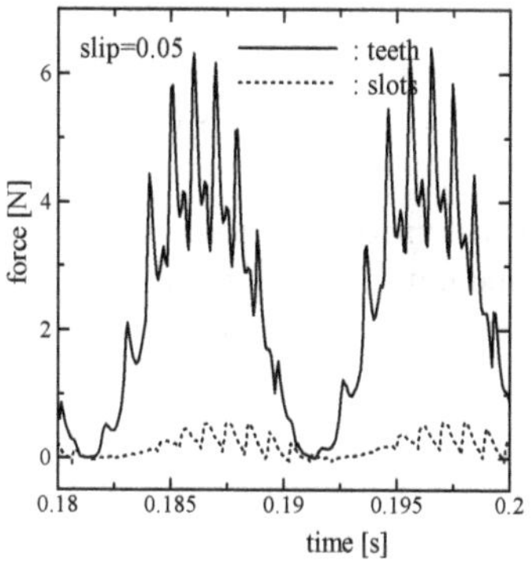

Figure 17. Waveform of radial magnetic force of M model motor at slip=0.05.

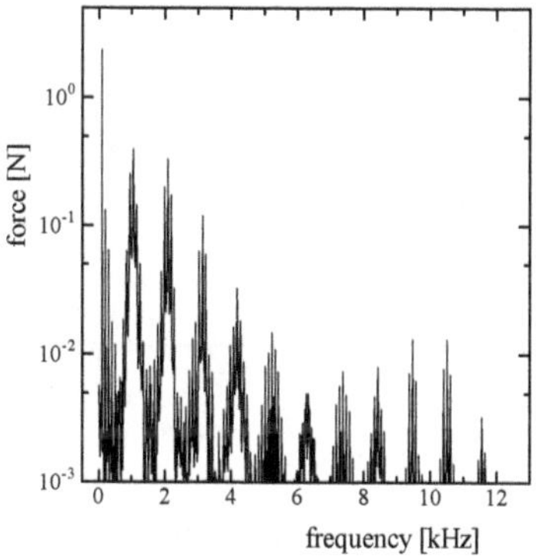

Figure 18. Spectrum of radial magnetic force of M model motor at slip=0.05.

Here we discuss the frequencies of radial force. The electromagnetic flux harmonics are produced due to the relative movement between the rotor and stator. Seeing it from the stator's side where the main flux is generated, the permeance varies periodically due to the presence of the slots in the rotor. Following this reason, the frequency of the harmonics in the electromagnetic flux is obtained by the product of the fundamental stator magnet-motive force (MMF) and the rotor slot permeance. The fundamental stator MMF F is proportional to $\cos(p/2\cdot\theta-2\pi ft)$, where θ is the stator angle. The permeance P is proportional to $1+\Sigma A_k \cos kN_2\left(\theta-\frac{2}{p}(1-s)2\pi ft\right)$, where N_2, s and k are the number of rotor slots, slip and the order of space harmonics, respectively. Considering that the radial electromagnetic force is proportional to $(F\cdot P)^2$, the next three frequencies are obtained,

$$2f\,,\ \frac{2kN_2}{p}(1-s)f\,,\ \frac{2kN_2}{p}(1-s)f\pm 2f \tag{14}$$

Since the rotor has 44 slots, when slip is 0, the combination of the slot permeance and the fundamental stator MMF produces the peaks at 100, {1000, 1100, 1200}, {2100, 2200, 2300}, and so on, see Fig. 16. When the slip is 0.05, the frequencies are 100, {945, 1045, 1145}, {1190, 2090, 2190}, see Fig. 18.

In the vibration problems, small space harmonics, namely, small modes are important. Then, we calculate the space and time spectrum of the radial electromagnetic force in the air gap, and show the time spectrum for several space harmonics in Figs. 19 and 20. It is shown that time harmonics of mode 4 are 100, 200, 400, and so on, and the mode of harmonics of 300, 600 and 900Hz is 12.

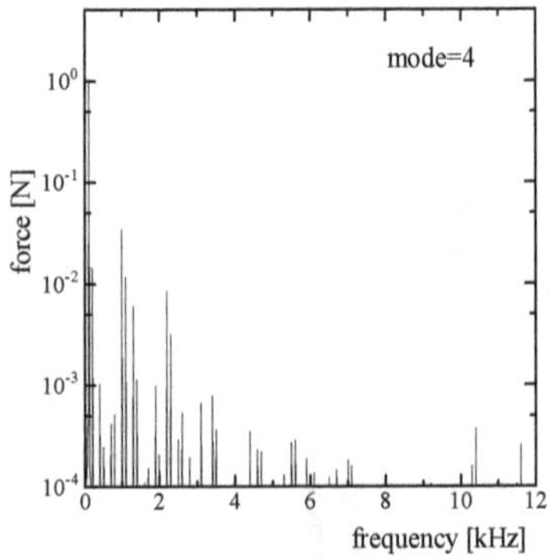

Figure 19. Frequency spectrum for different mode.

For the K-model the rotor has 34 slots, when slip is 0, the combination of the slot permeance and the fundamental stator MMF produces the peaks at 100, {750, 850, 950}, {1600,1700,1800}, {2450,2550,2650}, and so on, see Fig. 21. When the slip is 0.5, the frequencies are 100, {325,425,525}, {750,850,950}, see Fig. 22. When slip=1.0, only the first frequency remains and this is appreciated in Fig. 23.

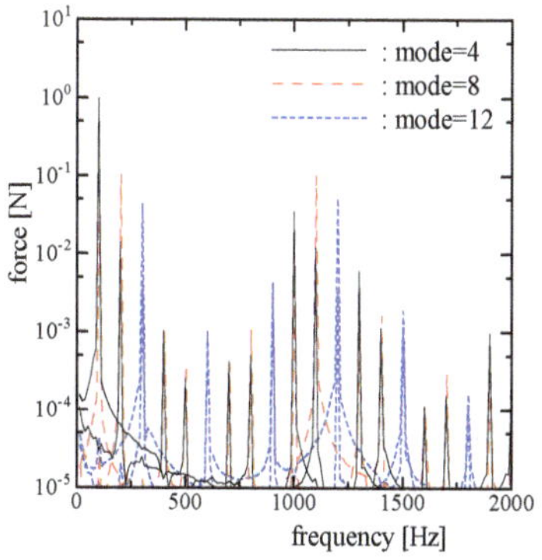

Figure 20. Enlarged one of Fig. 19.

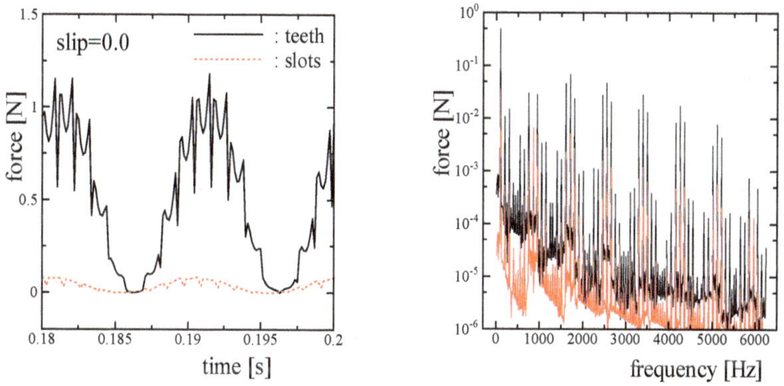

Figure 21. Radial force and its spectrum of K model motor at slip=0.

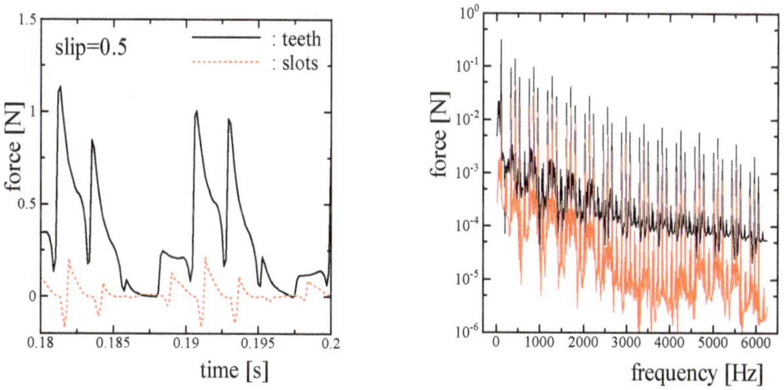

Figure 22. Radial force and its spectrum of K model motor at slip=0.5.

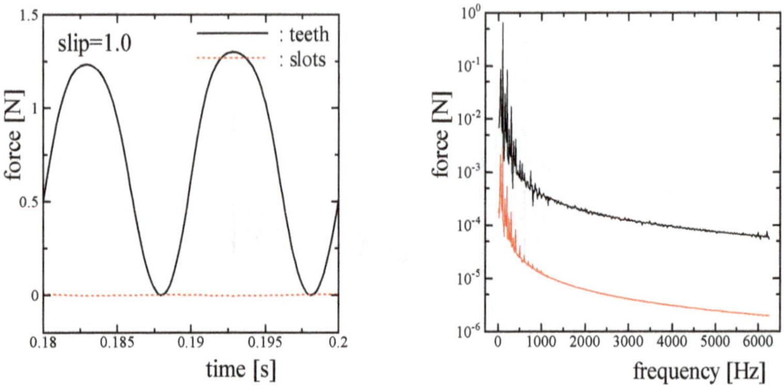

Figure 23. Radial force and its spectrum of K model motor at slip=1.00.

3.4. Radial magnetic force under PWM inverter source

Next, to clarify the difference between the line source and the PWM inverter, Figs. 24 and 25 show the waveforms of torque and stator current at slip=0 and 0.5 for the K-model. The PWM inverter has a currier frequency of 5kHz and the fundamental amplitude is equal to the line source. It is shown that the current and torque contain the component of the carrier frequency.

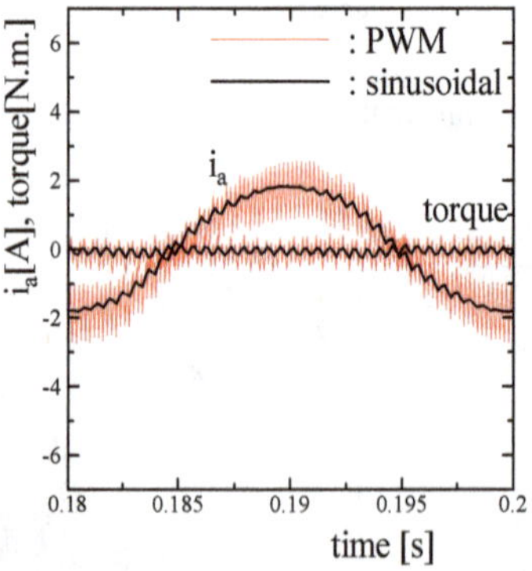

Figure 24. Torque and stator current waveforms of K model driven by PWM inverter, slip=0.

Fig. 26 shows the radial force and its spectrum at slip=0.5 for the K-model. The waveform of radial force driven by the PWM inverter is approximately the same as that driven by the line source. We can find a small noise in the waveform, and find that the amplitude around 5 kHz, that is, carrier frequency is bigger than that of the line source in the spectrum.

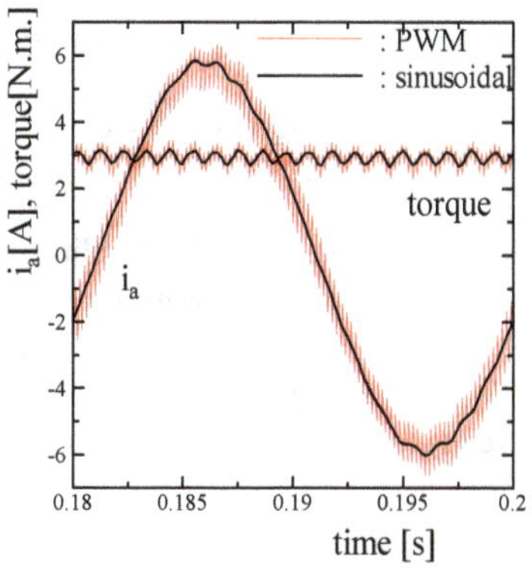

Figure 25. Torque and stator current waveforms of K model driven by PWM inverter, slip=0.5

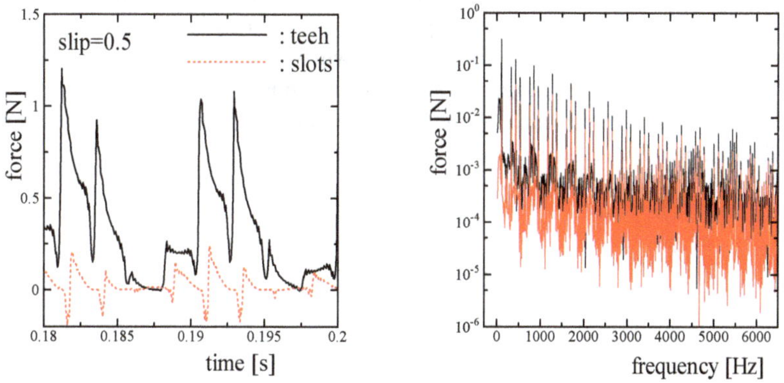

Figure 26. Radial force and frequency spectrum of K-model driven by PWM inverter at slip = 0.5.

3.5. Vibration velocity

Fig. 27 shows the vibration velocity measured at the centre of stator surface, when the motor is running at no-load. The vibration of 600 through 650 Hz is mainly emitted from the natural frequency, and 100, 200, 400, 500, 700, 1000 and 1200 Hz are corresponding to the frequency of the radial force with mode 4. We think that the vibration of 25Hz is emitted by the eccentricity of the rotor. Fig. 28 shows the vibration velocity emitted from the inverter-fed induction motor. We can see the vibration at around nf_c, where n is an integer and f_c is the carrier frequency.

3.6. Radial magnetic force under randomized PWM inverter source

It is well known that a random PWM method reduces the acoustic noise emitted from an inverter drive motor. Then, we analyze the radial force of the motor fed by two types of random PWM method, namely, a randomized pulse position PWM and a randomized switching frequency PWM. The randomized pulse position PWM changes the pulse width as

$$duty = \frac{v + V_{max}}{2V_{max}} + (x - 0.5) * k \tag{15}$$

where, v, V_{max}, x and k are the voltage reference, the amplitude of the jagged wave with 5 kHz carrier frequency, a random number and the maximum variation of pulse position, respectively. This means that the interval of switching signals is changed by DT, where $-0.5k/5000 < DT < 0.5k/5000$ [sec]. The randomized switching frequency PWM changes the switching frequency as

$$f_c = 5000 + (x - 0.5) * (\alpha - 1) * 50 \tag{16}$$

Fig. 29 shows the time spectrum of the radial force of the motor fed by the randomized pulse position PWM, where the end of the interval of switching signals is changed by DT, $-0.2/5000 < DT < 0.2/5000$ [sec]. Fig. 30 shows the time spectrum by the randomized switching frequency PWM, where the switching frequency is change by DF, $-500 < DF < 500$ [Hz]. The time spectrum shown in Figs 29 and 30 are approximately the same as that under line source, except for the reduction of radial forces at around nf_c. The reduction of radial forces is summarized in Tables 5 and 6.

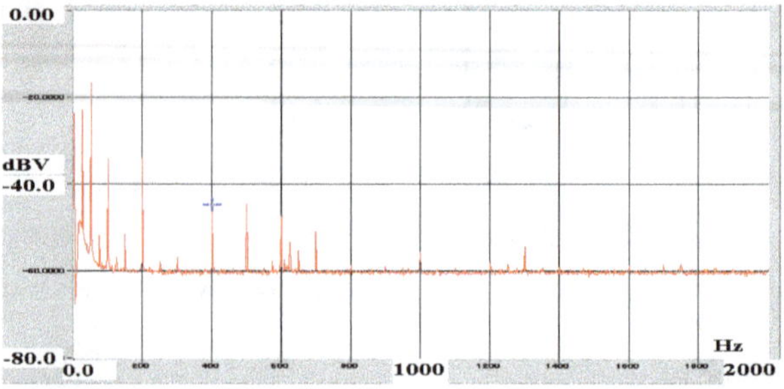

Figure 27. Vibration velocity emitted from M model driven by the line source (measured)

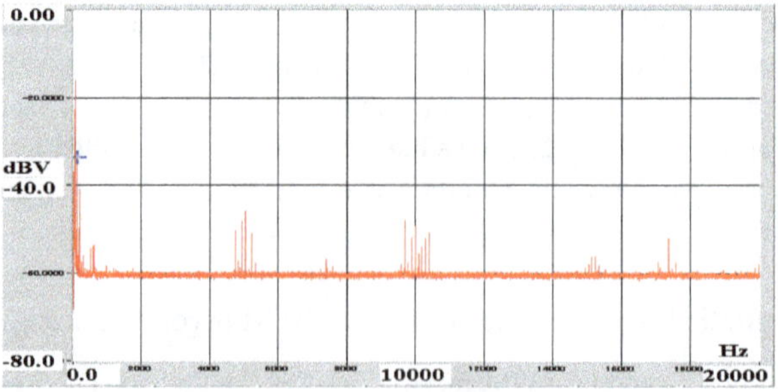

Figure 28. Vibration velocity emitted from M model driven by the PWM inverter source (measured)

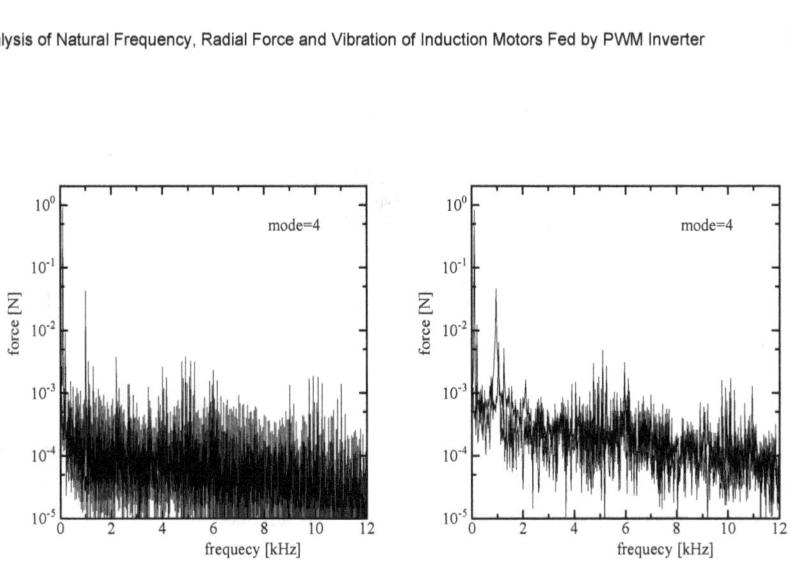

Figure 29. Spectrum of radial force for the randomized pulse position PWM.

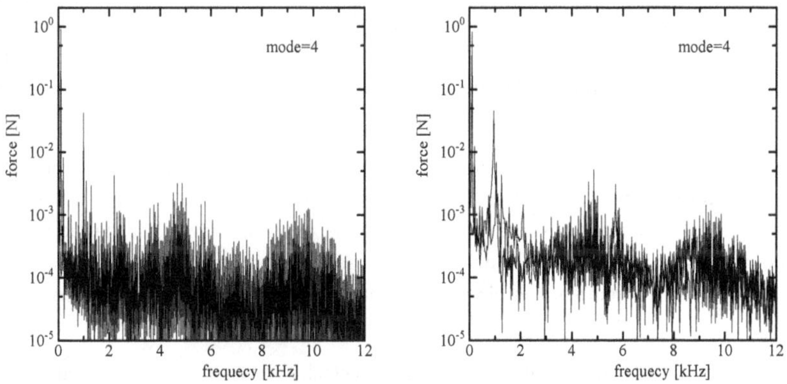

Figure 30. Spectrum of radial force spectrum for the randomized switching frequency PWM.

Frequency	100Hz	400Hz	f_c	$2f_c$
Line	0.897	0.0437	--	--
PWM	0.896	0.0423	0.00366	0.00260
P PWM	0.904	0.0418	0.00375	0.00185
F PWM	0.899	0.0423	0.00321	0.00125

P PWM: Randomized pulse position PWM,
F PWM: Randomized switching frequency PWM

Table 5. Comparison of radial force at slip=0.0.

Frequency	100Hz	400Hz	f_c	$2f_c$
Line	0.815	0.0467	--	--
PWM	0.813	0.0455	0.00560	0.00263
P PWM	0.821	0.0455	0.00480	0.00171
F PWM	0.815	0.0454	0.00516	0.00142

Table 6. Comparison of radial force at slip=0.05.

3.7. Discussions

The steady state characteristics of the induction motor can be calculated by the 2D FEM considering the modified conductivity in the rotor slot by (12) and the leakage inductance of stator coil end by (13). However, for the motor with one-slot skewing there is some error in the low speed range, that is, high torque region.

In the section 3.3, it is shown that the space variation of the radial force is approximately flat in the teeth and becomes a small value at the positions where the rotor slot exists. The radial force at one tooth is bigger than that at the other two teeth, when the stator winding is distributed in three slots as shown in Fig. 14. It is also shown that the radial force has a fundamental frequency of 2 times the line frequency of 50Hz, because this motor is two pole pairs machine. Moreover, some frequencies of the radial force are obtained by considering the stator magnet-motive force and the rotor slot permeance.

In the section 3.4, when the motor is driven by the PWM inverter, the current and torque contain the component of the carrier frequency as well known. As a result, the waveform of the radial force also includes small noise. The Fourier analysis shows that the fundamental component is almost same as that driven by line source, and that the amplitude around 5 kHz, that is, carrier frequency is bigger than that of the line source as shown in Figs. 22 and 26.

In the section 3.5, the measurement of vibration velocity shows that there are vibration at around natural frequency and some frequencies corresponding to the radial force. However, there are the other frequencies corresponding to the harmonics of line frequency and the frequency produce by the eccentricity of the rotor

In the section 3.6, the effect of the randomized PWM inverter on the radial force is calculated. Two types of random PWM method, namely, a randomized pulse position PWM and a randomized switching frequency PWM are taken into account.

However, the time spectrum shown in Figs 29 and 30 are approximately the same as that under line source, except for the reduction of radial forces at two times carrier frequency. The calculation was carried out for about five cycles of the fundamental frequency. If a very long period is calculated, the reduction of radial force at the carrier frequency becomes larger.

4. Conclusion

The natural frequencies of the motor can be estimated by considering the equivalent Young's modulus of the stator windings. For example, the lowest measured and calculated natural frequencies are 1,325 and 1,369 Hz for the stator core only, and are 637 and 587 Hz for the stator core with winding. They agree well with each other.

The steady state characteristics of the induction motor can be calculated by the 2D FEM considering the modified conductivity in the rotor slot and the leakage inductance of stator coil end. Using this simulation model, the radial force of the induction motor fed by the line source has been analyzed. It is shown that the frequencies are explained by the product of the fundamental stator MMF and the rotor slot permeance, and that the radial force is different at each tooth because of the distributed stator winding.

When the motor is driven by the PWM inverter, the fundamental component of radial force is almost same as that driven by line source and the amplitude around the carrier frequency is bigger than that of the line source. Moreover, the effect of the randomized PWM inverter on the radial force is calculated. The radial forces at two times carrier frequency can be reduced by using the randomized pulse position PWM or by the randomized switching frequency PWM.

Author details

Takeo Ishikawa
Gunma University, Japan

5. References

F. Ishibashi, K. Kamimoto, S. Noda, and K. Itomi, Small induction motor noise calculation, IEEE Trans. on Energy Conversion, vol.18, no.3 pp.357-361, 2003.

K. Shiohata, K. Nemoto, Y. Nagawa, S. Sakamoto, T. Kobayashi, M. Itou, and H. Koharagi, A method for analyzing electromagnetic-force-induced vibration and noise analysis, (in Japanese) Trans. IEE Japan, vol.118-D, no.11 pp.1301-1307, 1998.

D. M. Munoz, J. C. S. Lai, Acoustic noise prediction in a vector controlled induction machine, 2003 IEEE international Electric Machines and Drives Conference, pp.104-110, 2003.

Technical Report on Computational method of Electromagnetics for Virtual Engineering of Rotating Machinery, (in Japanese) IEE Japan, vol. 776, March, 2000.

D. Mori and T. Ishikawa, Force and Vibration Analysis of Induction Motors, IEEE Transactions on Magnetics, Vol. 41, No.5, pp. 1948-1951, 2005.

D. Mori and T. Ishikawa, Force and Vibration Analysis of a PWM Inverter-Fed Induction Motor, The 2005 International Power Electronics Conference,pp.644-650, Niigata, 2005

T. Horii, Electrical Machines Outline, Corona, publishing Co., Ltd., 1978

K. Yamasaki, Modification of 2D Nonlinear time-stepping analysis by limited 3D analysis for induction machines, IEEE Trans. on Magnetics, vol.33, no.22, 1997

H. Kometani, S. Sakabe, and N.Nakanishi, 3-D Electromagnetic analysis of a cage induction motor with rotor skew, IEEE Trans. on Energy Conversion, vol.11, no.2, 1996

A. M. Trzynadlowski, F. Blaabjerg, J. K. Pedersen, R. L. Kirlin, and S. Legowski, Random Pulse Width Modulation Techniques for Converter-Fed Drive Systems - A Review, IEEE Trans. on Industry Applications, vol. 30, No.5, pp.1166-1175, 1994.

K. Itori, S. Noda, F. Ishibashi, and H. Yamawaki, Young's modulus of windings on finite element method for natural frequency analysis of stator core in induction motor, (in Japanese) Trans. of the Japan Society of Mechanical Engineers, Series C, vol. 68, no.669, pp.1-6, 2002.

Chapter 11

Electrical Parameter Identification of Single-Phase Induction Motor by RLS Algorithm

Rodrigo Padilha Vieira, Rodrigo Zelir Azzolin, Cristiane Cauduro Gastaldini and Hilton Abílio Gründling

Additional information is available at the end of the chapter

1. Introduction

This chapter addresses the problem of the electrical parameter identification of Single-Phase Induction Motor (SPIM). The knowledge of correct electrical parameters of SPIM allows a better representation of dynamic simulation of this machine. In addition, the identified parameters can improve the performance of the Field Oriented Control (FOC) and sensorless techniques used in these systems.

Controlled induction motor drives have been employed on several appliances in the last decades. Commonly, the control schemes are based on the FOC and sensorless techniques. These methods are mainly applied to three-phase induction machine drives, and a wide number of papers, such as [5, 9, 10, 15, 19, 23, 26] have described such drives. On the other hand, for several years the SPIM has been used in residential appliances, mainly in low power and low cost applications such as in freezers and air conditioning, consuming extensive rate of electrical energy generated in the world. In most of these applications, the SPIM operates at fixed speed and is supplied directly from source grid. However, in the last few years several works have illustrated that the operation with variable speed can enhance the process efficiency achieved by the SPIM ([1, 4, 8, 31]). Furthermore, some other studies have presented high performance drives for SPIM using vector control and sensorless techniques, such as is presented in [7, 12, 18, 24] and [29]. However, these schemes applied on single-phase and three-phase induction motor drives need an accurate knowledge of all electrical parameters machine to have a good performance.

As a consequence of the parameter variation and uncertainties of the machine, literature presents algorithms for computational parameter estimation of induction machines, mainly about three-phase induction machines ([3, 13, 20, 21, 27]). Some authors proposed an on-line parameter estimation, for adaptive systems and self-tuning controllers due to the fact that the parameters of induction machine change with temperature, saturation, and frequency ([22]).

Differently from the three-phase induction motors, the SPIM is an asymmetrical and coupled machine; these features make the electrical parameter estimation by classical methods difficult, and these characteristics complicate the use of high performance techniques, such as vector and sensorless control. Thus, the use of Recursive Least Square (RLS) algorithm can be a solution for the parameter estimation or self-tuning and adaptive controllers, such as presented in [28] and [30]. Other studies have also been reported in literature describing the parameter estimation of SPIM ([2, 11, 17, 25]).

The aim of this chapter is to provide a methodology to identify a set of parameters for an equivalent SPIM model, and to obtain an improved SPIM representation, as consequence it is possible to design a high performance sensorless SPIM controllers. Here, from the machine model, a classical RLS algorithm is applied at q and d axes based on the current measurements and information of fed voltages with a standstill rotor. The automatized test with standstill rotor can be a good alternative in some applications, such as hermetic compressor systems, where the estimation by conventional methods is a hard task due the fact of the machine is sealed.

An equivalent SPIM behavior representation is obtained with this methodology in comparison with the SPIM model obtained by classical tests. However, some types of SPIM drives, for instance a hermetic system, it is impossible to carried out classical tests. In addition, the proposed methodology has a simple implementation.

This chapter is organized as follows: Section 2 presents the SPIM model, Section 3 gives the RLS parameter algorithm, Section 4 presents and discusses the experimental results obtained with the proposed methodology, and Section 5 gives the main conclusions of this study.

2. Single-phase induction motor model

The commercial SPIM commonly used in low power applications is usually a two-phase induction machine with asymmetrical windings, whose equivalent circuit without the permanent split-capacitor can be represented as in Fig. 1.

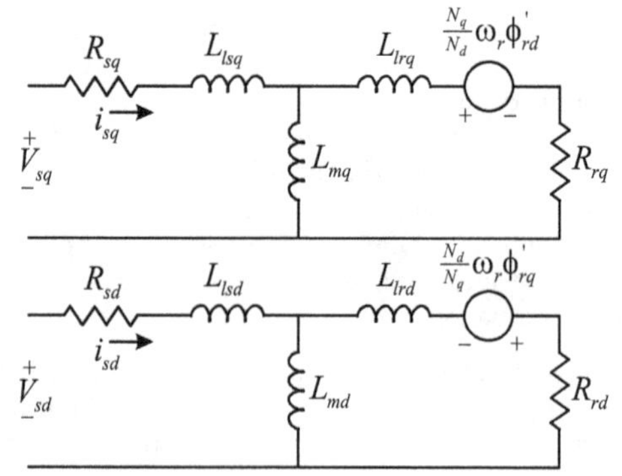

Figure 1. Equivalent circuit of SPIM.

As in [14], in this chapter the squirrel cage SPIM mathematical model is described in a stationary reference-frame by the following equations

$$\begin{bmatrix} V_{sq} \\ V_{sd} \end{bmatrix} = \begin{bmatrix} R_{sq} & 0 \\ 0 & R_{sd} \end{bmatrix} \begin{bmatrix} i_{sq} \\ i_{sd} \end{bmatrix} + \frac{d}{dt} \begin{bmatrix} \phi_{sq} \\ \phi_{sd} \end{bmatrix} \tag{1}$$

$$\begin{aligned} \begin{bmatrix} V_{rq} \\ V_{rd} \end{bmatrix} = \begin{bmatrix} R_{rq} & 0 \\ 0 & R_{rd} \end{bmatrix} \begin{bmatrix} i_{rq} \\ i_{rd} \end{bmatrix} + \frac{d}{dt} \begin{bmatrix} \phi_{rq} \\ \phi_{rd} \end{bmatrix} \\ +\omega_r \begin{bmatrix} 0 & -1/n \\ n & 0 \end{bmatrix} \begin{bmatrix} \phi_{rq} \\ \phi_{rd} \end{bmatrix} = \begin{bmatrix} 0 \\ 0 \end{bmatrix} \end{aligned} \tag{2}$$

$$\begin{bmatrix} \phi_{sq} \\ \phi_{sd} \end{bmatrix} = \begin{bmatrix} L_{sq} & 0 \\ 0 & L_{sd} \end{bmatrix} \begin{bmatrix} i_{sq} \\ i_{sd} \end{bmatrix} + \begin{bmatrix} L_{mq} & 0 \\ 0 & L_{md} \end{bmatrix} \begin{bmatrix} i_{rq} \\ i_{rd} \end{bmatrix} \tag{3}$$

$$\begin{bmatrix} \phi_{rq} \\ \phi_{rd} \end{bmatrix} = \begin{bmatrix} L_{mq} & 0 \\ 0 & L_{md} \end{bmatrix} \begin{bmatrix} i_{sq} \\ i_{sd} \end{bmatrix} + \begin{bmatrix} L_{rq} & 0 \\ 0 & L_{rd} \end{bmatrix} \begin{bmatrix} i_{rq} \\ i_{rd} \end{bmatrix} \tag{4}$$

$$T_e = p(L_{mq} i_{sq} i_{rd} - L_{md} i_{sd} i_{rq}) \tag{5}$$

$$p(T_e - T_L) = J\frac{d\omega_r}{dt} + B_n \omega_r \tag{6}$$

where, the indexes q and d represent the main winding and auxiliary winding, respectively, the indexes sq and sd represent the stator variables, and the indexes rq and rd are used for the rotor variables. V_{sq}, V_{sd}, V_{rq}, V_{rd}, i_{sq}, i_{sd}, i_{rq}, i_{rd}, ϕ_{sq}, ϕ_{sd}, ϕ_{rq}, and ϕ_{rd} are the stator and rotor voltages, currents, and flux; R_{sq}, R_{sd}, R_{rq}, and R_{rd} are the stator and rotor resistances; L_{lsq}, L_{lsd}, L_{lrq}, and L_{lrd} are the leakage inductances; L_{mq} and L_{md} are the mutual inductances; L_{sq}, L_{sd}, L_{rq}, and L_{rd} are the stator and rotor inductances, and are given by: $L_{sq} = L_{lsq} + L_{mq}$, $L_{sd} = L_{lsd} + L_{md}$, $L_{rq} = L_{lrq} + L_{mq}$, and $L_{rd} = L_{lrd} + L_{md}$; N_q and N_d represent the number of turns for the main and auxiliary windings, respectively; p is the pole pair number and ω_r is the rotor speed, and n is the relationship between the number of turns for auxiliary and for main winding N_d/N_q. T_e is the electromagnetic torque, T_L is the load torque, B_n is the viscous friction coefficient, and J is the inertia coefficient.

From (1) - (4) it is possible to obtain the differential equations that express the dynamical behavior of the SPIM, as follows,

$$\frac{d}{dt} i_{sq} = -\frac{R_{sq} L_{rq}}{\overline{\sigma}_q} i_{sq} - \omega_r \frac{1}{n} \frac{L_{mq} L_{md}}{\overline{\sigma}_q} i_{sd} + \frac{R_{rq} L_{mq}}{\overline{\sigma}_q} i_{rq} - \omega_r \frac{1}{n} \frac{L_{rd} L_{mq}}{\overline{\sigma}_q} i_{rd} + \frac{L_{rq}}{\overline{\sigma}_q} V_{sq} \tag{7}$$

$$\frac{d}{dt} i_{sd} = \omega_r n \frac{L_{md} L_{mq}}{\overline{\sigma}_d} i_{sq} - \frac{L_{rd} R_{sd}}{\overline{\sigma}_d} i_{sd} + \omega_r n \frac{L_{rq} L_{md}}{\overline{\sigma}_d} i_{rq} + \frac{R_{rd} L_{md}}{\overline{\sigma}_d} i_{rd} + \frac{L_{rd}}{\overline{\sigma}_d} V_{sd} \tag{8}$$

$$\frac{d}{dt} i_{rq} = \frac{L_{mq} R_{sq}}{\overline{\sigma}_q} i_{sq} + \omega_r \frac{1}{n} \frac{L_{sq} L_{md}}{\overline{\sigma}_q} i_{sd} - \frac{L_{sq} R_{rq}}{\overline{\sigma}_q} i_{rq} + \omega_r \frac{1}{n} \frac{L_{sq} L_{rd}}{\overline{\sigma}_q} i_{rd} - \frac{L_{mq}}{\overline{\sigma}_q} V_{sq} \tag{9}$$

$$\frac{d}{dt} i_{rd} = -\omega_r n \frac{L_{sd} L_{mq}}{\overline{\sigma}_d} i_{sq} + \frac{L_{md} R_{sd}}{\overline{\sigma}_d} i_{sd} - \omega_r n \frac{L_{sd} L_{rq}}{\overline{\sigma}_d} i_{rq} - \frac{L_{sd} R_{rd}}{\overline{\sigma}_d} i_{rd} - \frac{L_{md}}{\overline{\sigma}_d} V_{sd} \tag{10}$$

where $\overline{\sigma}_q = L_{sq}L_{rq} - L_{mq}^2$, $\overline{\sigma}_d = L_{sd}L_{rd} - L_{md}^2$.

The transfer functions in the axes q and d at a standstill rotor ($\omega_r = 0$) are obtained from (7)-(10), where these functions are decoupled and presented in (11) and (12).

$$H_q(s) = \frac{i_{sq}(s)}{V_{sq}(s)} = \frac{s\overline{\sigma}_q^{-1}L_{rq} + \overline{\sigma}_q^{-1}\tau_{rq}^{-1}L_{rq}}{s^2 + sp_q + \overline{\sigma}_q^{-1}R_{rq}R_{sq}} \tag{11}$$

$$H_d(s) = \frac{i_{sd}(s)}{V_{sd}(s)} = \frac{s\overline{\sigma}_d^{-1}L_{rd} + \overline{\sigma}_d^{-1}\tau_{rd}^{-1}L_{rd}}{s^2 + sp_d + \overline{\sigma}_d^{-1}R_{rd}R_{sd}} \tag{12}$$

where $p_q = \left(R_{sq}L_{rq} + R_{rq}L_{sq}\right)/\overline{\sigma}_q$ and $p_d = \left(R_{sd}L_{rd} + R_{rd}L_{sd}\right)/\overline{\sigma}_d$.

3. Parameter identification of single-phase induction machine

In section 2, the decoupled transfer functions of the SPIM were obtained assuming a standstill rotor ($\omega_r = 0$). Thus, in this section the parameter identification is achieved with SPIM at a standstill rotor by a RLS algorithm. The identification with a standstill rotor is appropriated in some cases such as hermetic refrigeration compressors ([28]). The RLS identification algorithm requires the plant model in a discrete time linear regression form. Assuming the actual sampling index k, the regression model is given by

$$\hat{\mathbf{Y}}(k) = \phi^T(k)\theta(k) \tag{13}$$

The recursive algorithm is achieved with the equations (14)-(17).

$$e(k) = \mathbf{Y}(k) - \hat{\mathbf{Y}}(k) \tag{14}$$

$$\mathbf{K}(k) = \frac{\mathbf{P}(k-1)\phi(k)}{1 + \phi^T(k)\mathbf{P}(k-1)\phi(k)} \tag{15}$$

$$\theta(k) = \theta(k-1) + \mathbf{K}(k)e(k) \tag{16}$$

$$\mathbf{P}(k) = \left(I - \mathbf{K}(k)\phi^T(k)\right)\mathbf{P}(k-1) \tag{17}$$

where $dim\ \mathbf{Y} = \bar{M} \times \bar{N},\ dim\ \phi^T(k) = \bar{M} \times \bar{r}$

$dim\ \theta(k) = \bar{r} \times \bar{N},\ dim\ e(k) = \bar{M} \times \bar{N}$

$dim\ \mathbf{K}(k) = \bar{r} \times \bar{M}\ ,\ dim\ I = dim\ \mathbf{P}(k) = \bar{r} \times \bar{r}$

From the equations (11) and (12) it is possible to reformulate the estimation parameter problem based on a linear regression model. Here, the parameter estimation method is divided into two steps:

***First step**: estimation of (18) and (19) vide equations (20) and (21) :*

This step consists into obtaining a linear-time-invariant model of the SPIM. The identification of b_1, b_0, a_1 and a_0 is done by performing a standstill test. The coefficients presented in (11) and (12) are functions of the machine parameters. For simplicity, the transfer functions given in (11) and (12) are rewritten in two transfer functions given by (18) and (19).

$$H_q(s) = \frac{i_{sq}^s(s)}{V_{sq}^s(s)} = \frac{sb_{1q} + b_{0q}}{s^2 + sa_{1q} + a_{0q}} \tag{18}$$

and

$$H_d(s) = \frac{i_{sd}^s(s)}{V_{sd}^s(s)} = \frac{sb_{1d} + b_{0d}}{s^2 + sa_{1d} + a_{0d}} \tag{19}$$

where

$$b_{1q} = \frac{L_{rq}}{\overline{\sigma}_q}, \, b_{0q} = \frac{L_{rq}}{\overline{\sigma}_q \tau_{rq}}, \quad a_{1q} = \frac{R_{sq}L_{rq} + R_{rq}L_{sq}}{\overline{\sigma}_q}, \, a_{0q} = \frac{R_{sq}R_{rq}}{\overline{\sigma}_q} \tag{20}$$

and

$$b_{1d} = \frac{L_{rd}}{\overline{\sigma}_d}, \, b_{0d} = \frac{L_{rd}}{\overline{\sigma}_d \tau_{rd}}, \quad a_{1d} = \frac{R_{sd}L_{rd} + R_{rd}L_{sd}}{\overline{\sigma}_d}, \, a_{0d} = \frac{R_{sd}R_{rd}}{\overline{\sigma}_d} \tag{21}$$

In order to obtain the regression linear model the transfer functions of (18) and (19) can be generalized and rewritten as,

$$\frac{d^2 i_{sq}}{dt^2} + a_{1q}\frac{di_{sq}}{dt} + a_{0q}i_{sq} = b_{1q}\frac{dV_{sq}}{dt} + b_{0q}V_{sq} \tag{22}$$

and

$$\frac{d^2 i_{sd}}{dt^2} + a_{1d}\frac{di_{sd}}{dt} + a_{0d}i_{sd} = b_{1d}\frac{dV_{sd}}{dt} + b_{0d}V_{sd} \tag{23}$$

Solving for the second derivative of the stator current,

$$\frac{d^2 i_{sq}}{dt^2} = \begin{bmatrix} -\frac{di_{sq}}{dt} & -i_{sq} & \frac{dV_{sq}}{dt} & V_{sq} \end{bmatrix} \begin{bmatrix} a_{1q} \\ a_{0q} \\ b_{1q} \\ b_{0q} \end{bmatrix} \tag{24}$$

and

$$\frac{d^2 i_{sd}}{dt^2} = \begin{bmatrix} -\frac{di_{sd}}{dt} & -i_{sd} & \frac{dV_{sd}}{dt} & V_{sd} \end{bmatrix} \begin{bmatrix} a_{1d} \\ a_{0d} \\ b_{1d} \\ b_{0d} \end{bmatrix} \tag{25}$$

The estimation of coefficients b_{1q}, b_{0q}, a_{1q}, a_{0q}, b_{1d}, b_{0d}, a_{1d} and a_{0d} is done by using RLS estimation algorithm described in the equations (13)-(17). The linear regression model form (13) is given for the q axis by the following equations,

$$\mathbf{Y}_q(k) = \frac{d^2 i_{sq}}{dt^2} \tag{26}$$

$$\phi_q^T(k) = \begin{bmatrix} -\frac{di_{sq}}{dt} & -i_{sq} & \frac{dV_{sq}}{dt} & V_{sq} \end{bmatrix} \tag{27}$$

$$\theta_q^T(k) = \begin{bmatrix} a_{1q} \; a_{0q} \; b_{1q} \; b_{0q} \end{bmatrix} \tag{28}$$

And, for the d axis,

$$\mathbf{Y}_d(k) = \frac{d^2 i_{sd}}{dt^2} \tag{29}$$

$$\phi_d^T(k) = \begin{bmatrix} -\frac{di_{sd}}{dt} & -i_{sd} & \frac{dV_{sd}}{dt} & V_{sd} \end{bmatrix} \tag{30}$$

$$\theta_d^T(k) = \begin{bmatrix} a_{1d} \; a_{0d} \; b_{1d} \; b_{0d} \end{bmatrix} \tag{31}$$

where, let us assume that the derivatives presented in (26)-(27) and (29)-(30) are measurable quantities. In the implementation, these quantities are obtained by State Variable Filters (SVF) such as in [6]. Four SVF filters were developed by discretization of the continuous-time transfer function given by,

$$\frac{V_{sqf}}{V_{sq}} = \frac{V_{sdf}}{V_{sd}} = \frac{i_{sqf}}{i_{sq}} = \frac{i_{sdf}}{i_{sd}} = G_{svf}(s) = \frac{\omega_{svf}^3}{(s+\omega_{svf})^3} \tag{32}$$

where, ω_{svf} is the filter bandwidth defined at around 5 to 10 times the input frequency signal. Here, $\omega_{svf} = \omega$, and ω is signal frequency, and the signals V_{sq}, V_{sd}, i_{sq}, and i_{sd} are used to obtain the filtered signals V_{sqf}, V_{sdf}, i_{sqf} and i_{sdf}.

The discretized transfer function, using the Euler method and sampling time of T_s, can be performed in state-space as

$$\mathbf{x}_{svf(k+1)} = \left(1 + \mathbf{A}_{svf}T_s\right)\mathbf{x}_{svf(k)} + T_s\mathbf{B}_{svf}u_{svf(k)} \tag{33}$$

where, $\mathbf{A}_{svf} = \begin{bmatrix} 0 & 1 & 0 \\ 0 & 0 & 1 \\ -\omega_{svf}^3 & -3\omega_{svf}^2 & 3\omega_{svf} \end{bmatrix}$, $\mathbf{B}_{svf} = \begin{bmatrix} 0 \\ 0 \\ \omega_{svf}^3 \end{bmatrix}$, $x_{svf} = \begin{bmatrix} x_1 \\ x_2 \\ x_3 \end{bmatrix}$ The variable u_{svf} represents the input signal, while the state variables x_1, x_2 and x_3 represent the input filtered signal, first derivative signal and second derivative signal, respectively.

With the SVF it is possible to avoid the use o low-pass filters in signals of the currents and the voltages, due the fact of the SVF has behavior of a low-pass filter. In addition, due this characteristic the SVF atenues the impact of pulsed and noise signals on the measurements of stator currents. For instance, to demonstrate the response of SVF Figure 2 presents the Module and the Phase Bode diagram with bandwidth of $\omega_{svf} = 5 * 2\pi * 30Hz\ [rad/s]$. It is possible to observer that the high frequency signals are attenuated.

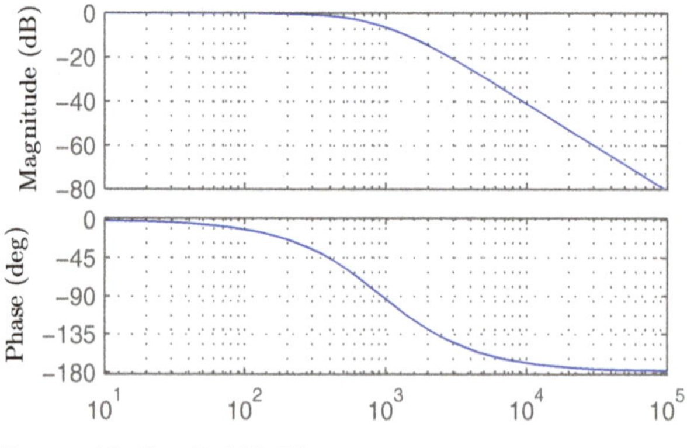

Figure 2. Bode diagram of the State Variable Filter.

Second step*: identification of machine parameters,* R_{sq}, R_{sd}, R_{rq}, R_{rd}, L_{mq}, L_{md}, L_{sq}, L_{sd}, L_{rq} *and,* L_{rd}:

The electrical parameters of the SPIM are obtained combining the identified coefficients of $\theta_q(k)$ and $\theta_d(k)$ in (28) and (31) with the coefficients of the transfer functions (20) and (21), respectively, after the convergence of the RLS algorithm by the equations (34) and (35). In the numerical solution, the stator and rotor inductances are considered to have the same values in each winding.

$$\begin{aligned}
\hat{R}_{sq} &= \frac{a_{0q}}{b_{0q}} \\
\hat{R}_{rq} &= \frac{a_{1q}}{b_{1q}} - \hat{R}_{sq} \\
\hat{L}_{mq} &= \frac{\sqrt{\hat{R}_{rq}\left(b_{1q}^2\hat{R}_{rq} - b_{0q}\right)}}{b_{0q}} \\
\hat{L}_{sq} &= \hat{L}_{rq} = \hat{R}_{rq}\frac{b_{1q}}{b_{0q}}
\end{aligned} \tag{34}$$

$$\begin{aligned}
\hat{R}_{sd} &= \frac{a_{0d}}{b_{0d}} \\
\hat{R}_{rd} &= \frac{a_{1d}}{b_{1d}} - \hat{R}_{sd} \\
\hat{L}_{md} &= \frac{\sqrt{\hat{R}_{rd}\left(b_{1d}^2\hat{R}_{rd} - b_{0d}\right)}}{b_{0d}} \\
\hat{L}_{sd} &= \hat{L}_{rd} = \hat{R}_{rd}\frac{b_{1d}}{b_{0d}}
\end{aligned} \tag{35}$$

4. Results and discussions

The RLS parameter identification algorithm presented in this chapter is implemented in a DSP based platform using TMS320F2812 DSP and a three-leg voltage source inverter. Figure 3 shows the diagram of the system to obtain the experimental results. The machine used for the validation of this methodology is a SPIM of a commercial hermetic refrigeration compressor of an air conditioning. The SPIM was removed from the hermetic compressor to achieve classical tests. The machine is two-pole, 220 V type. In the implementation of RLS identification parameter algorithm, the DC bus was limited in 177V/5A, and the sampling time of $400\mu s$ was used.

As described in Sections 2 and 3, the experiment for the parameter identification is achieved with a standstill rotor. In the first step, a square wave with variable frequency and reduced voltage supplies the main winding, while the auxiliary winding is opened. The stator current in the main winding is measured using hall effect sensor. The voltage used in this algorithm is estimated by the product of modulation and DC bus indexes. After the identification for the main winding, the same procedure is repeated for the auxiliary winding. The square wave is used for better excitation of the plant.

The frequency of the supply voltage is 5 Hz in the estimation of resistances to minimize the skin effect, and it is 30 Hz for the estimation of inductances.

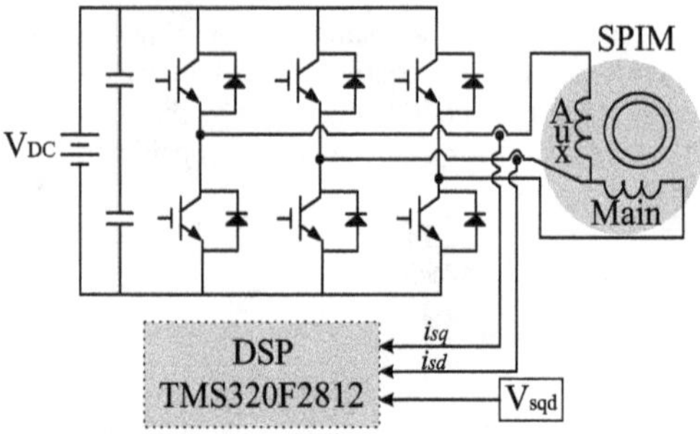

Figure 3. System diagram for the experimental results.

The first test for the identification of stator and rotor resistances is achieved in the main winding. The convergence of coefficients θ_q for this winding is presented in Fig. 4. Fig. 4 (a) presents the coefficients a_{1q} and a_{0q}, while Fig. 4 (b) gives the coefficients b_{1q} and b_{0q}. In this experiment the frequency of the supply voltage is selected on 5 Hz. As presented in figures, the coefficient convergence is fast and it is excited by the reset of the covariance matrix (**P**). Here, the reset of the covariance matrix (**P**) is used to avoid that this matrix reach zero and consequently loses the ability of to update the parameter matrix ($\theta_q(k)$). Some oscillations are introduced every time that the covariance matrix (**P**) is reset, but the convergence of coefficients $\theta_q(k)$ is stable around a region.

The aim of the algorithm (13)-(17) is to identify a set of parameters that produce a good dynamic response such as real dynamic response of the induction machine. If it is necessary to ensure the parameter convergence to the true values the Lemma 12.5.2 (Persistent Excitation) of [16] must be satisfied. However, in practical implementations the designer normally unknown the exactly parameter values due the assumptions and approximations made to develop the mathematical model of the plant, and due the presence of unmodulated dynamics in the system, such as the measurement and drive systems.

The coefficients used in the electrical parameter calculation are the final values of Fig 4. The coefficient convergence for d axis has behavior similar to coefficient convergence for q axis in Fig. 4.

Table 1 shows the final obtained coefficients a_{1q}, a_{0q}, b_{1q}, and b_{0q} in a q axis when the voltage frequency is 5 Hz and 30 Hz. Table 2 shows the final obtained coefficients a_{1d}, a_{0d}, b_{1d}, and b_{0d} in a d axis when the voltage frequency is 5 Hz and 30 Hz. From Fig. 4 it can be observed that the coefficient convergence is fast.

	5 Hz	30 Hz
a_{1q}	244.27	707.64
a_{0q}	447.96	8084.14
b_{1q}	24.82	27.73
b_{0q}	128.58	2913.7

Table 1. Final convergence for coefficients of the main winding.

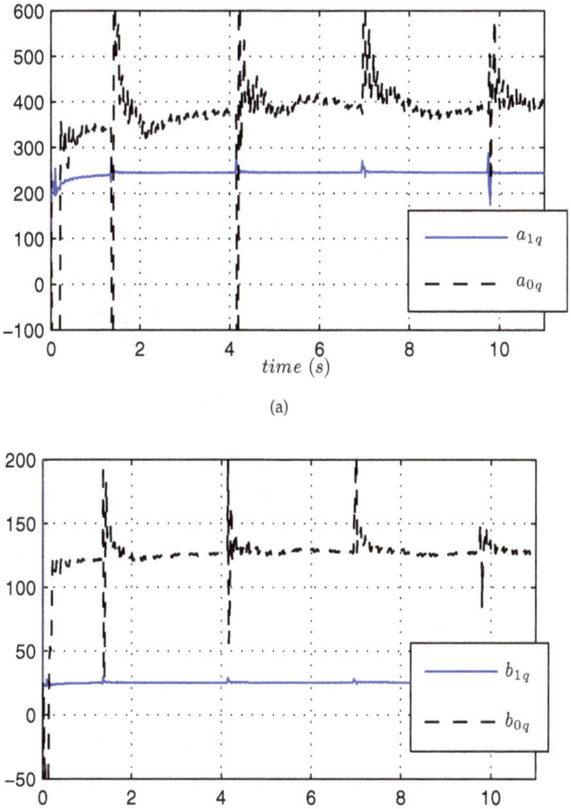

Figure 4. Coefficients convergence.

	5 Hz	30 Hz
a_{1d}	246.15	779.65
a_{0d}	559.4	3304.5
b_{1d}	9.83	17.62
b_{0d}	43.67	2448.17

Table 2. Final convergence for coefficients of the auxiliary winding.

The electrical parameters of the induction machine are obtained combining the final value of coefficients in Table 1 and Table 2 with the equations (34) and (35), respectively. Table 3 presents the identified electrical parameters of SPIM. In this study we also make a comparison with the results obtained by classical methods, thus, the SPIM identified by RLS algorithm is also tested using no-load and standstill classical methods. The electrical parameters estimated when the SPIM is tested by classical methods are shown in Table 4.

	R_{sq}	R_{rq}	L_{mq}	$L_{sq} = L_{rq}$
Identified	3.62Ω	6.27Ω	0.2186H	0.2376H
	R_{sd}	R_{rd}	L_{md}	$L_{sd} = L_{rd}$
Identified	12.79Ω	12.23Ω	0.2849H	0.3142H

Table 3. Experimental identified electrical parameters by RLS algorithm.

	R_{sq}	R_{rq}	L_{mq}	$L_{sq} = L_{rq}$
Estimated	3.95Ω	5.1506Ω	0.2149H	0.2292H
	R_{sd}	R_{rd}	L_{md}	$L_{sd} = L_{rd}$
Estimated	11.95Ω	8.6463Ω	0.382H	0.401H

Table 4. Electrical parameters estimated by classical tests.

Aiming the model validation, two experiments are carried out. In the first experiment, the SPIM is driven by a *v/f* strategy at no-load operation, thus, the main and the auxiliary windings are supplied by controlled voltages varying the frequency from zero until a steady-state condition. The rotor speed and the stator currents are also measured. These measurements are recorded for posterior comparison with simulated values.

The model of single-phase induction machine presented in equations (7)-(10) is simulated using the estimated parameters of the machine by RLS algorithm given in Table 3. The model is also simulated using the parameters estimated by the classical methods presented in Table 4. The simulated stator currents are compared with the measured stator currents. The induction motor model equations (7)-(10) are discretized using the Euler Method in the same frequency of the experimental commutation for this test at 5 kHz. The recorded rotor speed is used in the simulated model to make it independent of the mechanical parameters.

In the first experimental result, the frequency of stator voltages varies from zero until 25 Hz and it is fixed in 25 Hz, by the *v/f* method. The recorded voltage values are used to supply the SPIM model in the simulation. Thus, from the same input voltages, Fig. 5 shows currents simulated and measured for the main winding in the first experiment. Fig. 5 (a) presents the comparison between the simulation of i_{sq} current with parameters estimated by RLS algorithm, and the experimental measurement of i_{sq} current, and Fig 5 (b) presents the comparison of the i_{sq} current when the simulations is carried out with parameters estimated by classical tests.

Fig. 6 presents a detail of the comparison between the measured and the simulated currents in steady-state condition for the main winding. In Fig. 6 (a) the comparison between experimental current and simulated current is presented when the SPIM parameters are estimated by RLS algorithm, while Fig. 6 (b) presents the comparison between experimental current and simulated current when the SPIM parameters are estimated by classical tests.

Fig. 7 presents a comparison among i_{sd} currents on the first experiment. Fig. 7 (a) shows the simulated current using parameters estimated by RLS algorithm in the SPIM model and the

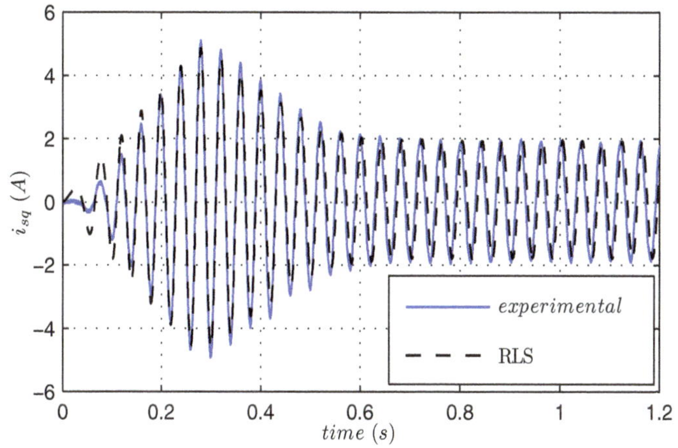

(a)

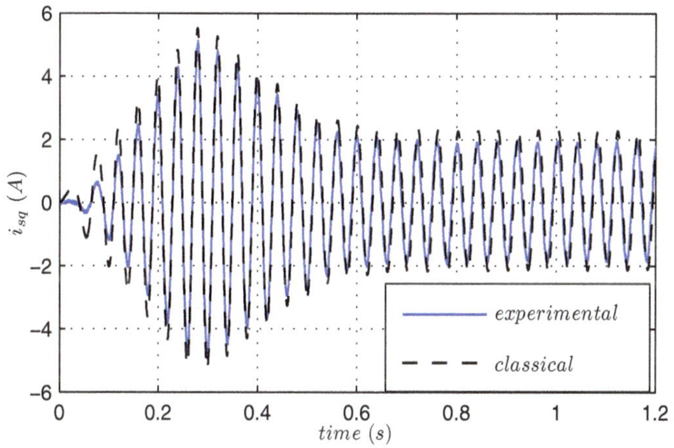

(b)

Figure 5. Comparison among measured and simulated currents for the first test. (a) Simulated i_{sq} using parameters estimated with RLS algorithm and measured i_{sq} current. (b) Simulated i_{sq} current using parameters estimated by classical tests and measured i_{sq} current.

measured i_{sd} current. Fig. 7 (b) gives the simulated current using parameters estimated by classical tests and the measured current.

Fig. 8 shows the comparison between i_{sd} currents in steady-state condition, Fig. 8 (a) gives the comparison between the measured i_{sd} current and the simulated i_{sd} current using SPIM parameters estimated by the RLS algorithm, whereas, Fig. 8 (b) presents the comparison

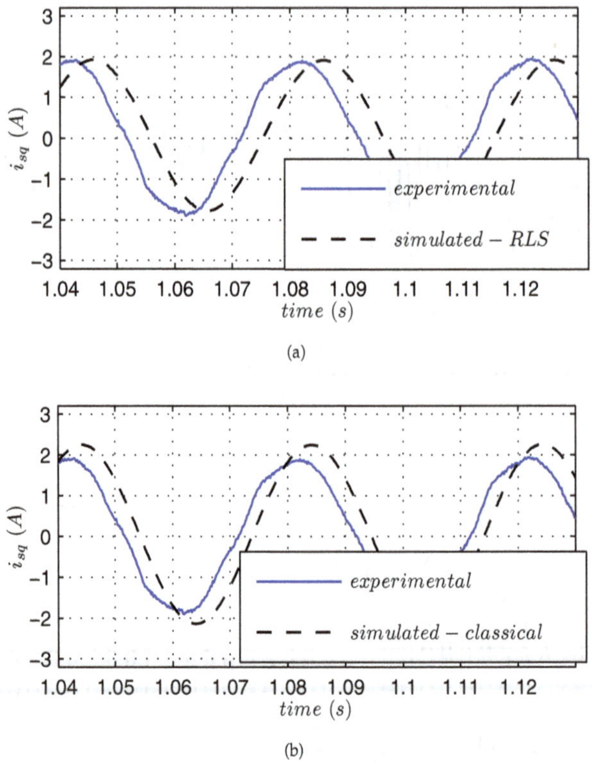

Figure 6. Comparison between measured and simulated currents at 25 Hz. (a) Simulated (RLS - Parameters) and measured i_{sq} currents. (b) Simulated (Classical - Parameters) and measured i_{sq} currents.

between the measured i_{sd} current and the simulated i_{sd} current using classical method for parameter estimation.

From Figures 5-8 it is possible to observer the good matching between the simulated and experimental currents for the q and d axes. In addition, some small discrepancies are found in these figures due the parameter inaccuracies and unmodulated effects (for instance the measurement and drive systems).

The advantage of the methodology presented in this chapter employing the RLS algorithm is that some types of applications, such as hermetic compressor, it is impossible carried out classical tests for estimation of the electrical parameters of the SPIM. In addition, the methodology has simplicity in implementation.

In the second experiment, the frequency of the supplied voltages of SPIM varies from zero until 30 Hz, and it is fixed at 30 Hz. As in the previous test, this is a no-load test and the SPIM is driven by a *v/f* strategy. The stator currents and rotor speed are measured and recorded.

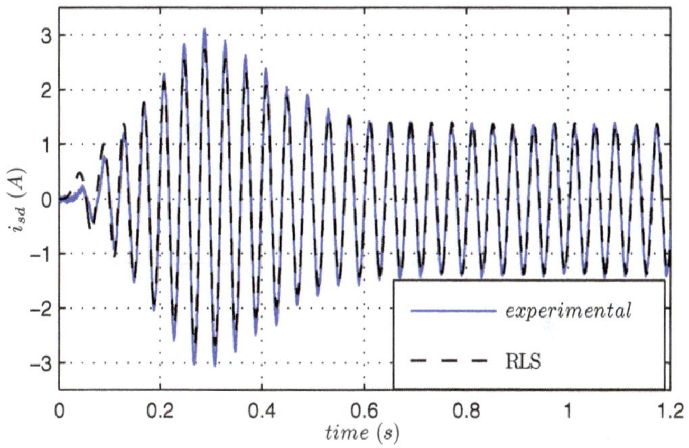

(a)

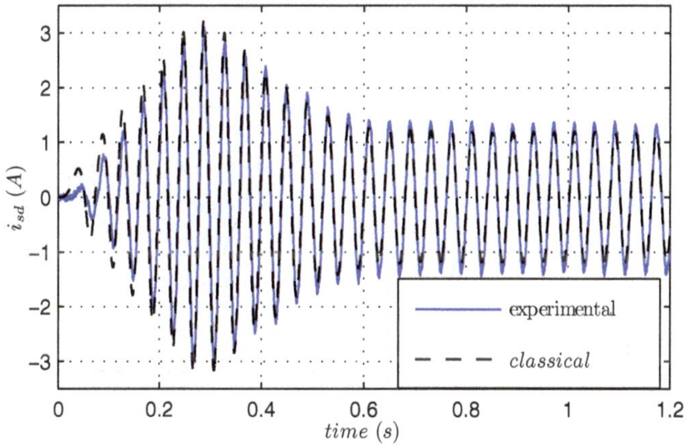

(b)

Figure 7. Comparison among i_{sd} currents in first test. (a) Simulated i_{sd} current by RLS parameter estimation and measured i_{sd} current. (b) Simulated i_{sd} current by classical tests and measured i_{sd} current.

Fig. 9 presents a comparison between measured and simulated i_{sq} currents in steady-stead condition. In Fig. 9 (a) shows the measured i_{sq} current and the simulated i_{sq} current using parameters estimated by RLS algorithm on the SPIM model, while Fig. 9 (b) gives the measured i_{sq} and the simulated i_{sq} current using parameters estimated by classical tests.

Fig. 10 presents a comparison between i_{sd} currents in steady-state condition for the second test.

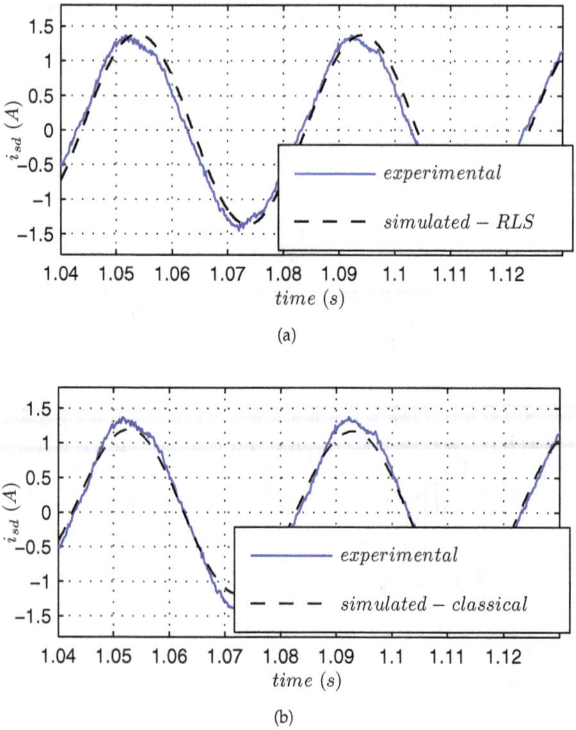

Figure 8. Comparison between i_{sd} currents at 25 Hz. (a) Simulated (RLS - parameter) and measured i_{sd} currents. (b) Simulated (classical tests) and measured i_{sd} currents.

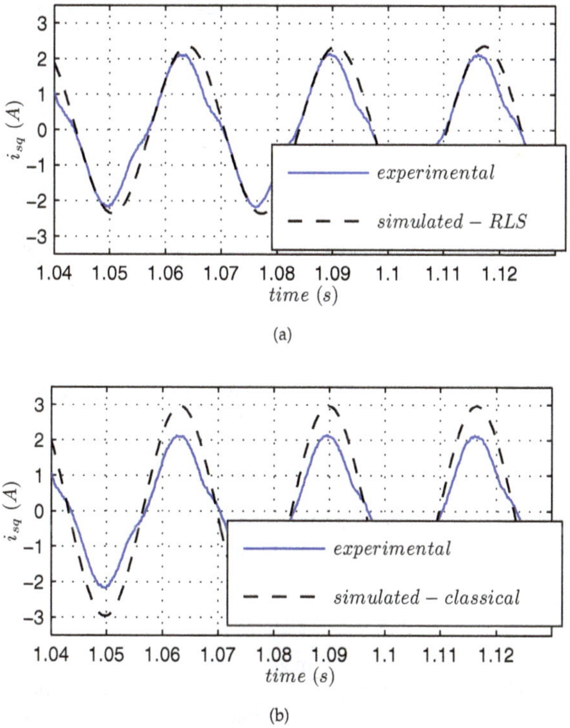

Figure 9. Comparison between i_{sq} currents at 30 Hz in steady-state condition. (a) Simulated (RLS - parameter) and measured i_{sq} currents. (b) Simulated (classical tests) and measured i_{sq} currents.

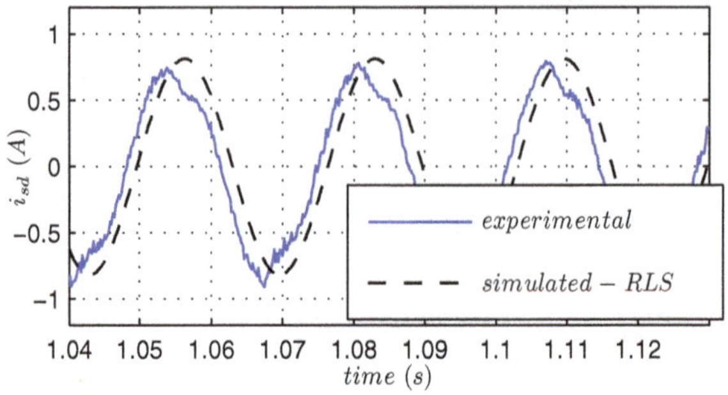

(a)

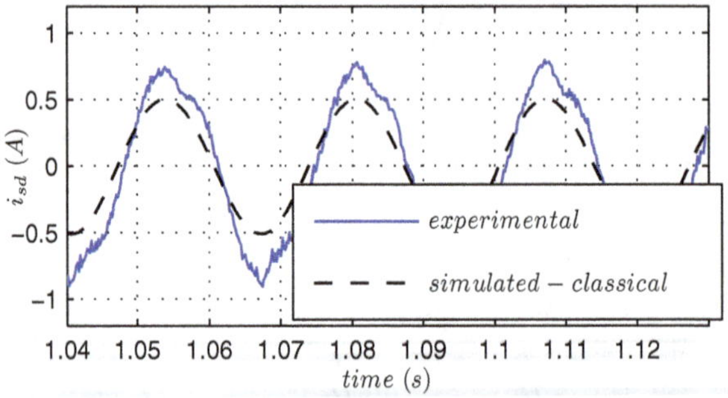

(b)

Figure 10. Comparison between i_{sd} currents at 30 Hz. (a) Simulated (RLS - parameter) and measured i_{sd} currents. (b) Simulated (classical tests) and measured i_{sd} currents.

5. Conclusion

A methodology for single-phase induction machine parameter identification was presented and discussed in this chapter. The machine tested was a SPIM used in a hermetic compressor of air conditioning. Using the proposed methodology it is possible to obtain all electrical parameters of SPIM for simulation and design of high performance vector control and sensorless SPIM drives. The main contribution of this study is the development of an automatized procedure for the identification of all electrical parameters of SPIM, such as the SPIM used in hermetic conditioning compressor. Experimental results demonstrate the effectiveness of the method. Some experimental comparisons among measurements and simulations using parameters estimated by classical tests and simulations using parameters obtained by RLS algorithm are presented. From Table 3 and Table 4 it is possible to observe that the parameters obtained with RLS algorithm converge to different values compared to classical tests. However, the results in Fig. 5 - 10 show that the parameters estimated with RLS algorithm present equivalent dynamical behavior compared with parameters estimated by classical methods. The methodology proposed in this chapter can be extended to be applied in other SPIM drives and three-phase induction motor drives.

Author details

Rodrigo Padilha Vieira
Federal University of Pampa - UNIPAMPA, Federal University of Santa Maria - UFSM, Power Electronics and Control Research Group - GEPOC, Brazil

Rodrigo Zelir Azzolin
Federal University of Rio Grande - FURG, Federal University of Santa Maria - UFSM, Power Electronics and Control Research Group - GEPOC, Brazil

Cristiane Cauduro Gastaldini
Federal University of Pampa - UNIPAMPA, Federal University of Santa Maria - UFSM, Power Electronics and Control Research Group - GEPOC, Brazil

Hilton Abílio Gründling
Federal University of Santa Maria - UFSM, Power Electronics and Control Research Group - GEPOC, Brazil

6. References

[1] Amin, A., El Korfally, M., Sayed, A. & Hegazy, O. [2009]. Efficiency optimization of two-asymmetrical-winding induction motor based on swarm intelligence, *IEEE Transactions on Energy Conversion* 24(1): 12 –20.

[2] Azzolin, R. Z., Gastaldini, C. C., Vieira, R. P. & Gründling, H. A. [2011]. *A RMRAC Parameter Identification Algorithm Applied to Induction Machines, Electric Machines and Drives, Miroslav Chomat (Ed.)*, InTech.

[3] Azzolin, R. Z. & Gründling, H. A. [2009]. A MRAC parameter identification algorithm for three-phase induction motors, *IEEE International Electric Machines and Drives Conference, IEMDC '09.*, pp. 273 –278.

[4] Blaabjerg, F., Lungeanu, F., Skaug, K. & Tonnes, M. [2004]. Two-phase induction motor drives, *IEEE Industry Applications Magazine* 10(4): 24–32.

[5] Bose, B. [2009]. Power electronics and motor drives recent progress and perspective, *IEEE Transactions on Industrial Electronics* 56(2): 581 –588.

[6] Câmara, H. T., Cardoso, R. C., Azzolin, R. Z., Pinheiro, H. & Gründling, H. A. [2006]. Low-cost sensorless induction motor speed control, *IEEE 32nd Annual Conference on Industrial Electronics, IECON 2006*, pp. 1200 –1205.

[7] de Rossiter Corrêa, M., Jacobina, C., Lima, A. & da Silva, E. [2000]. Rotor-flux-oriented control of a single-phase induction motor drive, *IEEE Transactions on Industrial Electronics* 47(4): 832–841.

[8] Donlon, J., Achhammer, J., Iwamoto, H. & Iwasaki, M. [2002]. Power modules for appliance motor control, *IEEE Industry Applications Magazine* 8(4): 26–34.

[9] Finch, J. & Giaouris, D. [2008]. Controlled AC electrical drives, *IEEE Transactions on Industrial Electronics* 55(2): 481 –491.

[10] Holtz, J. [2005]. Sensorless control of induction machines: With or without signal injection?, *IEEE Transactions on Industrial Electronics* 53(1): 7 – 30.

[11] Hrabovcova, V., Kalamen, L., Sekerak, P. & Rafajdus, P. [2010]. Determination of single phase induction motor parameters, *International Symposium on Power Electronics Electrical Drives Automation and Motion (SPEEDAM)*, pp. 287 –292.

[12] Jemli, M., Ben Azza, H., Boussak, M. & Gossa, M. [2009]. Sensorless indirect stator field orientation speed control for single-phase induction motor drive, *IEEE Transactions on Power Electronics* 24(6): 1618–1627

[13] Koubaa, Y. [2004]. Recursive identification of induction motor parameter, *Simulation Modelling Practice and Theory* 12(5): 368–381.

[14] Krause, P. C., Wasynczuk, O. & Sudhoff, S. D. [2002]. *Analysis of Electric Machinery and Drive Systems*, 2 edn, Wiley-IEEE Press.

[15] Lascu, C., Boldea, I. & Blaabjerg, F. [2005]. Comparative study of adaptive and inherently sensorless observers for variable-speed induction-motor drives, *IEEE Transactions on Industrial Electronics* 53(1): 57 – 65.

[16] Middleton, R. H. & Goodwin, G. C. [1990]. *Digital Control and Estimation - A Unified Approach*, 1 edn, Prentice Hall.

[17] Myers, M., Bodson, M. & Khan, F. [2011]. Determination of the parameters of non-symmetric induction machines, *Annual IEEE Applied Power Electronics Conference and Exposition (APEC)*, pp. 1028 –1033.

[18] Nied, A., de Oliveira, J., de Farias Campos, R., Jr., S. I. S. & de Souza Marques, L. C. [2011]. *Space Vector PWM-DTC Strategy for Single-Phase Induction Motor Control, Electric Machines and Drives, Miroslav Chomat (Ed.)*, InTech.

[19] Orlowska-Kowalska, T. & Dybkowski, M. [2010]. Stator-current-based mras estimator for a wide range speed-sensorless induction-motor drive, *IEEE Transactions on Industrial Electronics* 57(4): 1296 –1308.

[20] Rao, S., Buss, M. & Utkin, V. [2009]. Simultaneous state and parameter estimation in induction motors using first- and second-order sliding modes, *IEEE Transactions on Industrial Electronics* 56(9): 3369 –3376.

[21] Ribeiro, L. A. S., Jacobina, C. B. & Lima, A. M. N. [1995]. Dynamic estimation of the induction machine parameters and speed, *26th Annual IEEE Power Electronics Specialists Conference, PESC '95.* 2: 1281–1287 vol.2.

[22] Toliyat, H., Levi, E. & Raina, M. [2003]. A review of RFO induction motor parameter estimation techniques, *IEEE transactions on Energy conversion* 18(2): 271–283.

[23] Utkin, V. [1993]. Sliding mode control design principles and applications to electric drives, *IEEE Transactions on Industrial Electronics* 40(1): 23 –36.

[24] Vaez-Zadeh, S. & Reicy, S. [2005]. Sensorless vector control of single-phase induction motor drives, *Proceedings of the Eighth International Conference on Electrical Machines and Systems, 2005. ICEMS 2005* 3: 1838–1842 Vol. 3.

[25] van der Merwe, C. & van der Merwe, F. [1995]. A study of methods to measure the parameters of single-phase induction motors, *IEEE Transactions on Energy Conversion* 10(2): 248 –253.

[26] Vas, P. [1998]. *Sensorless Vector and Direct Torque Control*, Oxford Univ. Press.

[27] Velez-Reyes, M., Minami, K. & Verghese, G. [1989]. Recursive speed and parameter estimation for induction machines, *Conference Record of the Industry Applications Society Annual Meeting, 1989.* pp. 607–611 vol.1.

[28] Vieira, R. P., Azzolin, R. Z., Gastaldini, C. C. & Gründling, H. [2010]. Electrical parameters identification of hermetic refrigeration compressors with single-phase induction machines using RLS algorithm, *International Conference on Electrical Machines, 2010. ICEM 2010.*

[29] Vieira, R. P., Azzolin, R. Z. & Gründling, H. A. [2009]. A sensorless single-phase induction motor drive with a MRAC controller, *35st Annual Conference of IEEE Industrial Electronics Society, 2009.*, pp. 1003 –1008.

[30] Vieira, R. P., Azzolin, R. Z. & Gründling, H. A. [2009]. Parameter identification of a single-phase induction motor using RLS algorithm, *Brazilian Power Electronics Conference, 2009. COBEP '09.*, pp. 517 –523.

[31] Zahedi, B. & Vaez-Zadeh, S. [2009]. Efficiency optimization control of single-phase induction motor drives, *IEEE Transactions on Power Electronics* 24(4): 1062 –1070.

Permissions

The contributors of this book come from diverse backgrounds, making this book a truly international effort. This book will bring forth new frontiers with its revolutionizing research information and detailed analysis of the nascent developments around the world.

We would like to thank Rui Esteves Araújo, for lending his expertise to make the book truly unique. He has played a crucial role in the development of this book. Without his invaluable contribution this book wouldn't have been possible. He has made vital efforts to compile up to date information on the varied aspects of this subject to make this book a valuable addition to the collection of many professionals and students.

This book was conceptualized with the vision of imparting up-to-date information and advanced data in this field. To ensure the same, a matchless editorial board was set up. Every individual on the board went through rigorous rounds of assessment to prove their worth. After which they invested a large part of their time researching and compiling the most relevant data for our readers. Conferences and sessions were held from time to time between the editorial board and the contributing authors to present the data in the most comprehensible form. The editorial team has worked tirelessly to provide valuable and valid information to help people across the globe.

Every chapter published in this book has been scrutinized by our experts. Their significance has been extensively debated. The topics covered herein carry significant findings which will fuel the growth of the discipline. They may even be implemented as practical applications or may be referred to as a beginning point for another development. Chapters in this book were first published by InTech; hereby published with permission under the Creative Commons Attribution License or equivalent.

The editorial board has been involved in producing this book since its inception. They have spent rigorous hours researching and exploring the diverse topics which have resulted in the successful publishing of this book. They have passed on their knowledge of decades through this book. To expedite this challenging task, the publisher supported the team at every step. A small team of assistant editors was also appointed to further simplify the editing procedure and attain best results for the readers.

Our editorial team has been hand-picked from every corner of the world. Their multi-ethnicity adds dynamic inputs to the discussions which result in innovative outcomes. These outcomes are then further discussed with the researchers and contributors who give their valuable feedback and opinion regarding the same. The feedback is then collaborated with the researches and they are edited in a comprehensive manner to aid the understanding of the subject.

Apart from the editorial board, the designing team has also invested a significant amount of their time in understanding the subject and creating the most relevant covers. They scrutinized every image to scout for the most suitable representation of the subject and create an appropriate cover for the book.

The publishing team has been involved in this book since its early stages. They were actively engaged in every process, be it collecting the data, connecting with the contributors or procuring relevant information. The team has been an ardent support to the editorial, designing and production team. Their endless efforts to recruit the best for this project, has resulted in the accomplishment of this book. They are a veteran in the field of academics and their pool of knowledge is as vast as their experience in printing. Their expertise and guidance has proved useful at every step. Their uncompromising quality standards have made this book an exceptional effort. Their encouragement from time to time has been an inspiration for everyone.

The publisher and the editorial board hope that this book will prove to be a valuable piece of knowledge for researchers, students, practitioners and scholars across the globe.

List of Contributors

Alecsandru Simion, Leonard Livadaru and Adrian Munteanu
"Gh. Asachi" Technical University of Iaşi, Electrical Engineering Faculty, Romania

Muşuroi Sorin
Politehnica University of Timişoara, Romania

Adisa A. Jimoh, Pierre-Jac Venter and Edward K. Appiah
Tshwane University of Technology, Pretoria, South Africa

Marija Mirošević
University of Dubrovnik, Department of Electrical Engineering and Computing, Croatia

Miloje Kostic
Electrical Engineering Institute "Nikola Tesla", Belgrade University, Belgrade, Serbia

Houssem Rafik El-Hana Bouchekara
Department of Electrical Engineering, College of Engineering and Islamic Architecture, Umm Al-Qura University, Makkah, Saudi Arabia
Electrical Laboratory of Constantine "LEC", Department of Electrical Engineering, Mentouri University - Constantine, Constantine, Algeria

Mohammed Simsim and Makbul Anwari
Department of Electrical Engineering, College of Engineering and Islamic Architecture, Umm Al-Qura University, Makkah, Saudi Arabia

Venkat Krishnan and James D. McCalley
Iowa State University, USA

Marcel Janda, Ondrej Vitek and Vitezslav Hajek
Brno University of Technology, Czech Republic

Ebrahim Amiri and Ernest Mendrela
Louisiana State University, USA

Takeo Ishikawa
Gunma University, Japan

Rodrigo Padilha Vieira
Federal University of Pampa - UNIPAMPA, Federal University of Santa Maria - UFSM, Power Electronics and Control Research Group - GEPOC, Brazil

Rodrigo Zelir Azzolin
Federal University of Rio Grande - FURG, Federal University of Santa Maria - UFSM, Power Electronics and Control Research Group - GEPOC, Brazil

Cristiane Cauduro Gastaldini
Federal University of Pampa - UNIPAMPA, Federal University of Santa Maria - UFSM, Power Electronics and Control Research Group - GEPOC, Brazil

Hilton Abílio Gründling
Federal University of SantaMaria - UFSM, Power Electronics and Control Research Group - GEPOC, Brazil

www.ingramcontent.com/pod-product-compliance
Lightning Source LLC
Chambersburg PA
CBHW070737190326
41458CB00004B/1202